The CASSELL
DICTIONARY of
BIOLOGY

GILLIAN WAITES

CASSELL

A CASSELL BOOK

This edition first published in the UK in 1998

by
Cassell
Wellington House
125 Strand
London WC2R 0BB

British Library Cataloguing-in-Publication Data
A catalogue record for this book is available from the British Library

ISBN 0-304-35036-2

Designed, edited and typeset by Book Creation Services, London

Printed and bound in Great Britain by
Mackays of Chatham PLC, Chatham, Kent

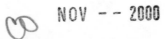

Contents

How to use *The Cassell Dictionary of Biology*

Arrangement of the dictionary
Entries are arranged alphabetically on a letter by letter basis, ignoring hyphens and spaces between words. Headwords – or main entries – are shown in **bold** type; **bold italics** are used to indicate an alternative form of the main headword.

Cross-references
Words that appear in SMALL CAPITALS in articles have their own entries elsewhere in the dictionary. Certain very common scientific words, such as 'element' or 'atom', are not automatically cross-referenced each time they are mentioned in the text.

See denotes a direct cross-reference to another article. *See also* indicates related articles or entries that contain more information about a particular subject.

Units
SI and metric units are used throughout the dictionary.

Abbreviations
In those cases where the part of speech of a headword is specified, the abbreviations used are as follows:

adj.   adjective
n.   noun
vb.   verb

A

abdomen In vertebrates, the part of the body below the THORAX containing the intestines, liver, kidneys and other organs except the heart and lungs. In mammals it is separated from the thorax by a muscular DIAPHRAGM. The lower region of the abdomen, called the PELVIS, is bounded by a set of bones called the PELVIC GIRDLE, to which the lower limbs are attached.

In invertebrates the abdomen is the hind part of the body. In insects and spiders the abdomen is characterized by the absence of limbs.

abiogenesis *See* SPONTANEOUS GENERATION THEORY.

abiotic (*adj.*) Describing the non-living or non-organic elements in an ECOSYSTEM, such as light, temperature, soil factors and rainfall.

ABO system A classification of human blood types based on the presence or absence of two ANTIGENS, A and B. This gives rise to four possible blood groups, A, B, AB and O, that determines their compatibility in transfusion. *See also* BLOOD GROUP SYSTEM.

abscissic acid A PLANT GROWTH SUBSTANCE that inhibits growth, in contrast to the other plant growth substances AUXIN, GIBBERELLIN and CYTOKININ. Abscissic acid inhibits NUCLEIC ACID and PROTEIN SYNTHESIS. Its main effect is on leaf or fruit drop (ABSCISSION).

abscission The shedding of a leaf, fruit, or other part from a plant. Abscission is controlled by PLANT GROWTH SUBSTANCES, particularly ABSCISSIC ACID, which causes the formation of an abscission layer between the main body of the plant and the part that is to fall. Another plant growth substance, AUXIN, inhibits this process, so it is a balance of auxin and abscissic acid that eventually causes abscission. Auxin can be applied to a crop to prevent premature fruit drop. Abscissic acid can be applied to a crop to regulate the fruit drop.

abscission layer *See* ABSCISSIC ACID.

accommodation In the EYE, the ability of the LENS to focus on objects at different distances. The lens changes its shape under the control of CILIARY MUSCLES surrounding it, so that it becomes more spherical for close objects and thinner for distant objects. The flexibility of the lens decreases with age, making accommodation less possible and spectacles may be then required.

acellular (*adj.*) Describing organisms or their parts that have no real cellular structure, for example tissues forming a SYNCYTIUM and many fungi. *Compare* MULTICELLULAR, UNICELLULAR.

acetyl CoA *See* ACETYL COENZYME A.

acetyl coenzyme A, acetyl CoA COENZYME A that is carrying an ACYL GROUP.

acetylcholine A NEUROTRANSMITTER that is mostly concerned with the transmission of NERVE IMPULSES across NEUROMUSCULAR JUNCTIONS, resulting in MUSCULAR CONTRACTION.

achene A dry, one-seeded fruit, for example the fruit of the buttercup, dandelion, grasses and daisies. Achenes do not split open to release their seeds (they are INDEHISCENT), and develop from a single CARPEL. The fruit wall can be winged, as in ash and sycamore, or can have hairs attached, as in the dandelion. Both features help DISPERSAL.

acid Any compound that contains hydrogen and liberates hydrogen ions when dissolved in water. Acids release hydrogen ions when they react with metals and they react with bases to form a SALT plus water.

acid rain Rain with a high acidity, caused mainly by sulphur dioxide (from the burning of FOSSIL FUELS) dissolving in water to form sulphuric and sulphurous acids. Nitrogen oxides (from industry and car exhausts) also contribute to acid rain. Acid rain causes damage in particular to coniferous forest species and some aquatic species, either directly due to the acidity, or indirectly when the rain leaches toxic aluminium from soils (*see* LEACHING).

Levels of sulphur dioxide in the air can be measured by the use of INDICATOR SPECIES, such as LICHENS and MOSSES. These have variable tolerances to sulphur dioxide levels, so their survival is indicative of the sulphur dioxide concentration.

See also POLLUTION.

acinus In certain GLANDS, such as the liver or pancreas, the functional part of any of the lobules that make up the gland.

acquired immune deficiency syndrome *See* AIDS.

acquired immunity *See* IMMUNITY.

acrosome A specialized structure at the tip of a SPERM that is formed from the GOLGI APPARATUS. The acrosome contains enzymes that are involved in the ACROSOME REACTION, which enables the sperm to penetrate the OVUM. *See also* FERTILIZATION.

acrosome reaction The process by which a SPERM is able to penetrate an OVUM. Enzymes are released from the ACROSOME at the tip of the head of the sperm, and soften the outer membrane of the ovum. The acrosome then inverts and a fine filament develops at the tip of the sperm that pierces the ovum, allowing the sperm to enter the ovum. *See also* FERTILIZATION.

ACTH *See* ADRENOCORTICOTROPHIC HORMONE.

actin A EUKARYOTIC protein that is a major constituent of muscle fibres and plays a vital role in a cell's CYTOSKELETON. Two actin filaments twist around one another to form the thin filaments characteristic of muscle MYOFIBRILS. Filaments of actin and MYOSIN can contract together as ACTOMYOSIN, which is vital for MUSCULAR CONTRACTION.

Actinobacteria, *Actinomycetes, Actinomycota* A phylum of BACTERIA. They are Gram positive (*see* GRAM'S STAIN), mostly anaerobic and non-motile. Like fungi, they have filamentous cells that produce reproductive spores on aerial branches. The phylum includes bacteria of the genus *Actinomyces,* some of which cause disease in animals and humans, and *Streptomyces,* which is a source of many ANTIBIOTICS, including streptomycin.

actinoid (*adj.*) Describing a radial or tentacled structure, such as a starfish.

actinomorphic (*adj.*) A term describing flowers that exhibit RADIAL SYMMETRY and can be cut vertically through two or more planes into similar halves. An example is buttercup, whose petals and sepals are of similar size. *Compare* ZYGOMORPHIC.

Actinomycetes *See* ACTINOBACTERIA.

Actinomycota *See* ACTINOBACTERIA.

Actinozoa, *Anthozoa* A class of the phylum CNIDARIA, including corals and sea anemones. Actinozoans are mostly marine animals.

action potential A change in the potential difference across the membrane of a NEURONE or muscle fibre that occurs when a NERVE IMPULSE travels along it. This is accompanied by the passage of sodium and potassium ions across the membrane. *See also* MUSCULAR CONTRACTION.

activator A substance that enables an ENZYME to bind to a SUBSTRATE. *See also* COFACTOR.

active immunity *See* IMMUNITY.

active site The part of an ENZYME molecule to which the SUBSTRATE binds. The active site is formed by the three-dimensional structure of the enzyme and the distribution of electric charge in the molecule. The substrate that binds to an enzyme is specific to that enzyme and is determined by the active site. An enzyme can have more than one active site, and therefore bind more than one substrate.

Some active sites require metal ions as PROSTHETIC GROUPS to provide the charge needed in the active site. Some other active sites change their conformation only after the enzyme has combined with a molecule at another site. Enzyme inhibitors can also reversibly bind to an active site, thereby blocking the action of the enzyme. *See also* ENZYME INHIBITION.

active transport An energy-requiring process (usually involving ATP) where substances, usually molecules or ions, are moved across a membrane against a concentration gradient – that is, from a region of low concentration to one of higher concentration. The process involves 'pumps' of protein molecules in the membrane that carry specific ions across. An example is the SODIUM PUMP. *Compare* DIFFUSION.

actomyosin A complex biochemical compound formed by the interaction of the proteins ACTIN and MYOSIN. This interaction occurs in muscle in the presence of calcium ions and is the basis of MUSCULAR CONTRACTION. Actomyosin dissociates in the presence of ATP.

acyl group The organic group that remains when the –OH group is removed from a CARBOXYLIC ACID. Acyl groups are named after the carboxylic acid from which they are derived, for example ethanoyl, CH_3CO-, from ethanoic acid CH_3COOH.

Adam's apple *See* LARYNX.

adaptation Any change occurring in the structure or function of an organism which benefits its survival and reproduction in its environment. In EVOLUTION, adaptations occur as a result of NATURAL SELECTION.

adaptive radiation A variation on the theory of EVOLUTION, that suggests that new species can evolve from a single ancestral type as a result of MIGRATION to new unoccupied ECOLOGICAL NICHES. New adaptations develop to accommodate the different ways of life, resulting eventually in new species. For example, DARWIN'S FINCHES on the Galapagos Islands probably descended from a single species on the South American mainland that adapted to suit different niches on the islands. *See also* SPECIATION.

Addison's disease A rare disorder in which the outer layer (the cortex) of the ADRENAL GLAND fails to function. This results in a deficiency of the hormone GLUCOCORTICOID, which regulates glucose metabolism. The symptoms of this deficiency are low blood pressure, low blood sugar, extreme fatigue and a darkening of the skin. Addison's disease may be due to infection, but it is more commonly a result of autoimmune destruction of the adrenal cortex (*see* AUTOIMMUNITY). Treatment to replace lost glucocorticoids is generally successful.

additive In food, any substance added to improve the colour, flavour or nutrient value, or to prolong the shelf-life of the food. Additives can be natural or artificial. Their use is regulated, as some can cause side-effects, including asthma, hyperactivity and cancer, in some people. Additives approved by the European Union are called 'E NUMBERS'.

Most additives used are synthetic flavourings or flavour enhancers that do not have to be listed in detail on a product label. Texture enhancers, such as thickeners, emulsifiers (which bind fat and water) and stabilizers (which prevent separation of fat and water), are useful for producing low fat spreads. Added minerals and vitamins also classify as additives. *See also* FOOD PRESERVATION.

adenine An organic base called a PURINE that occurs in NUCLEOTIDES. *See also* BASE PAIR, DNA, RNA.

adenoids A mass of LYMPHOID TISSUE at the back of the throat. The adenoids form part of a child's natural defence against infection, but can block the breathing passages when swollen and may be removed if recurrent infections occur. Adenoids shrink during childhood and normally disappear by the age of ten. *See also* LYMPH NODE, TONSILS.

adenosine A PURINE NUCLEOSIDE, consisting of the organic base ADENINE and the sugar RIBOSE. In its phosphorylated forms (*see* PHOSPHORYLATION) adenosine is AMP, ADP and ATP.

adenosine diphosphate *See* ADP.

adenosine monophosphate *See* AMP.

adenosine triphosphate *See* ATP.

adenylate cyclase An enzyme that catalyses the formation of CYCLIC AMP from ATP.

ADH *See* ANTI-DIURETIC HORMONE.

adhesion 1. The attractive force between molecules of one substance and those of another. Considerable adhesive forces exist between the walls of XYLEM vessels and the water within them, contributing to the movement of water up a plant. *See also* CAPILLARY EFFECT, SURFACE TENSION.

2. An abnormal joining together of two normally unconnected surfaces within the body. This can be due to inflammation of an area or to abnormal healing of a wound.

3. *See* CELL ADHESION.

adipose tissue, *fatty tissue* CONNECTIVE TISSUE consisting of cells containing large globules of fat that provide insulation, particularly around the kidneys, heart and in the inner layer of skin. Adipose tissue serves as an energy reserve.

ADP (adenosine diphosphate) The product formed by the PHOSPHORYLATION of AMP during energy-yielding biochemical reactions, and produced from the HYDROLYSIS of ATP. ADP can be phosphorylated to form ATP.

adrenal gland, *suprarenal gland* In vertebrates, either one of a pair of ENDOCRINE GLANDS found on top of the KIDNEY that produces hormones in response to stress situations.

Adrenal glands consist of two independent regions, the outer adrenal cortex and the inner adrenal medulla. The adrenal cortex consists of the majority of the glands and is itself divided into three zones, secreting a number of STEROID hormones collectively called CORTICOSTEROIDS (or corticoids). The outermost layer of the cortex secretes ALDOSTERONE, which regulates water retention by the kidney. The middle zone of the adrenal cortex secretes GLUCOCORTICOIDS, including CORTISOL, in response to internal stress, such as low blood temperature or volume. The adrenal medulla produces the hormones ADRENALINE and NORA-DRENALINE, which prepare the body for action in response to the external stress, such as fear, anger and pain.

adrenaline, *epinephrine* A hormone, derived from the amino acid tyrosine, that is secreted by the ADRENAL GLAND in response to external stress, such as fear, anger and pain. It prepares the body for action by increasing blood flow to the heart and muscles, causing the heart rate to quicken, and dilating airways in the lungs to enable more oxygen to be delivered to cells of the body, while constricting blood vessels in the skin and gut. Adrenaline also increases the amount of sweat produced, causes hair to stand up, pupils to dilate and increases breakdown of GLYCOGEN to GLUCOSE in the liver. Both adrenaline and the similar NORADRENALINE are important links between the ENDOCRINE SYSTEM and the NERVOUS SYSTEM because they are also NEUROTRANSMITTERS. Their production by the adrenal glands is regulated by the HYPOTHALAMUS via nervous connections. They mimic effects of the SYMPATHETIC NERVOUS SYSTEM.

adrenocorticotrophic hormone (ACTH), *corticotrophin* A POLYPEPTIDE secreted by the anterior PITUITARY GLAND that regulates growth of the cortex of the ADRENAL GLAND and stimulates production of its hormones. ACTH is commonly produced as a result of stress, which causes a substance, known as corticotrophin-releasing factor (CRF), to be released from the HYPOTHALAMUS. CRF initiates production of ACTH.

adventitious root A root of a plant that grows from a stem instead of other roots, or that grows in an unusual position. For example, adventitious roots of ivy grow sideways out of the stem to find support. *See also* FIBROUS ROOT, PROP ROOT.

aerobe Any organism that requires oxygen for RESPIRATION. Most organisms are aerobic, with the exception of some bacteria. Some cells can function for short periods without oxygen but most die. Those organisms that can survive without oxygen are called ANAEROBES.

aerobic respiration *See* AEROBE, RESPIRATION.

affinity chromatography A CHROMATOGRAPHY technique that depends on the affinity between specific molecules. The stationary phase, usually packed into a column, contains a substance that the molecule under investigation will attach to, thus separating it from the

mixture. The bound molecule can then be eluted from the stationary phase using a different solvent. This technique is useful, for example, in ANTIBODY purification, where a specific ANTIGEN is attached to the matrix. It can also be used to separate groups of substances, for example sugars, from a mixture. Sometimes it is easier to bind unwanted molecules to the stationary phase.

afterbirth In mammals, the material shed from the UTERUS following the birth of their young. It includes the PLACENTA and EXTRAEMBRYONIC MEMBRANES.

agar An extract of red seaweeds (RHODOPHYTA) that forms a gel at room temperature. It is widely used as a gelling agent in foodstuffs, medicine and cosmetics and as a culture media in MICROBIOLOGY.

agarose A POLYSACCHARIDE used in CHROMATOGRAPHY and ELECTROPHORESIS. It is one of the constituents of AGAR.

agglutination The clumping or sticking together of cells such as bacteria and RED BLOOD CELLS. Agglutination is caused by ANTIBODIES reacting with their specific ANTIGEN. It can occur to help the body remove foreign cells, such as bacteria, or it can occur due to a mismatch of antigens of the BLOOD GROUP SYSTEM, for example during blood transfusion.

agglutinin A cell-surface or PLASMA PROTEIN that acts as an ANTIBODY, causing clumping (AGGLUTINATION) of ANTIGENS on foreign cells such as bacteria. Agglutinins are often LECTINS.

Agnatha The class of vertebrates that consists of the jawless fish. *See* FISH.

agranulocyte A type of WHITE BLOOD CELL that has a non-granular CYTOPLASM and compact nucleus and can be phagocytic (*see* PHAGOCYTE). The main agranulocyte cell type in humans is the LYMPHOCYTE, which is involved in the body's defence mechanisms. *See also* GRANULOCYTE.

AIDS *(acquired immune deficiency syndrome)* A disease caused by the human immunodeficiency virus (HIV) in which the IMMUNE SYSTEM becomes deficient. Victims of AIDS usually die within 3 years of developing the disease, often from secondary infections such as pneumonia. At present only about 50 per cent of HIV-positive individuals develop fully blown AIDS within 10 years, but there is no evidence to suggest that HIV infection does not lead eventually to AIDS in 100 per cent of cases.

There is no cure for AIDS at present, but the drug AZT (zidovudine) interferes with the virus's ability to divide and can delay the onset and severity of AIDS. VACCINE development is difficult because there are many strains of HIV.

See also IMMUNOCOMPROMISATION.

alanine An AMINO ACID that occurs in most proteins.

albumin A group of PROTEINS produced by the LIVER that constitutes up to half of the content of human PLASMA PROTEIN. They are also found in egg white (albumen), milk and various animal and plant tissues. Albumins are coagulated by heat.

alcohol Any one of a group of organic compounds in which there is one or more hydroxyl groups (–OH) attached directly to carbon atoms. An alcohol therefore has the general formula $C_nH_{2n+1}OH$. An alcohol with one hydroxyl group is called monohydric, one with two is dihydric, one with three is trihydric and so on.

Alcohols are classified by the position of their alkyl groups (C_nH_{2n+1}); primary alcohols contain CH_2OH, secondary alcohols contain CHOH and tertiary alcohols contain COH.

$$CH_3-\overset{\overset{\displaystyle H}{|}}{\underset{\underset{\displaystyle H}{|}}{C}}-OH \qquad CH_3-\overset{\overset{\displaystyle H}{|}}{\underset{\underset{\displaystyle CH_3}{|}}{C}}-OH$$

Primary   Secondary

$$CH_3-\overset{\overset{\displaystyle CH_3}{|}}{\underset{\underset{\displaystyle CH_3}{|}}{C}}-OH$$

Tertiary

Alcohols are named after the HYDROCARBON from which they are derived, with the ending -anol. For example, ethanol is derived from ethane (ethyl alcohol).

Oxidation of primary alcohols yields ALDEHYDES, which can in turn be oxidized to CARBOXYLIC ACIDS. Oxidation of secondary alcohols yields KETONES, which are not easily oxidized further.

Alcohols react with ACIDS to form ESTERS. The lower alcohols – methanol, ethanol, propanol, butanol – are liquids that mix with water. The higher members are oily liquids and the highest are waxy solids. Ethanol is produced naturally during FERMENTATION and is used to manufacture alcoholic beverages.

alcoholic fermentation *See* FERMENTATION.

aldehyde, *alkanal* Any of a group of ORGANIC compounds containing the group

$$-C\overset{\displaystyle O}{\underset{\displaystyle H}{}}$$

The carbon of the carbonyl group (C=O) can be attached to another hydrogen atom (as in methanal HCHO), or to a carbon atom (as in ethanal CH_3CHO). (*Compare* KETONE.) Aldehydes are named after the HYDROCARBON from which they are derived, with the ending -anal, for example, ethanal. Most aldehydes are colourless liquids but the higher members are solids.

Aldehydes are formed by the oxidation of primary ALCOHOLS, hence their name since the alcohol loses a hydrogen atom (alcohol dehydrogenation). Aldehydes can be reduced back to primary alcohols and are themselves readily oxidized to the corresponding CARBOXYLIC ACID.

FEHLING'S TEST is used to test for aldehydes.

aldose, *aldo-sugar* A sugar containing an ALDEHYDE group (CHO). *See* MONOSACCHARIDE.

aldosterone A MINERALOCORTICOID hormone of the ADRENAL GLAND that regulates water retention in the kidney by controlling the distribution of sodium in the body tissue. It also affects centres in the brain, creating a sensation of thirst to stimulate the animal to seek water. *See also* ANGIOTENSIN.

aldo-sugar *See* ALDOSE.

aleuroplast A type of LEUCOPLAST that stores proteins in plants.

algae (*sing.* **alga**) A collective term for a number of varied photosynthetic organisms (*see* PHOTOSYNTHESIS) normally living in aquatic or damp terrestrial conditions. They used to be classified with bacteria and fungi, but are now classified in the kingdom PROTOCTISTA. All algae are EUKARYOTES. There are several subdivisions according to their pigmentation: green algae (CHLOROPHYTA), which contain the same CHLOROPHYLL as higher plants; brown algae (PHAEOPHYTA), including the seaweeds; red algae (RHODOPHYTA); yellow-green algae

(Xanthophyta); stoneworts (CHAROPHYTA); golden-brown algae (Chrysophyta); and the diatoms (BACILLARIOPHYTA). The CHLOROPLASTS of algae are varied, and most algae can reproduce sexually.

Algae are of considerable importance to humans because they cover the surface of the oceans and in total use more carbon dioxide from the atmosphere for photosynthesis than all land plants combined. This is important for carbon fixation at the first level of aquatic FOOD CHAINS, and to counteract GLOBAL WARMING (*see* GREENHOUSE EFFECT). Algae also produce more than half the oxygen released by plants and algae into the atmosphere.

Algae can be used as fertilizers, to assist in sewage breakdown, and as a direct food source, e.g. SINGLE CELL PROTEIN. Brown algae have a non-toxic acid in their cell walls that readily forms gels, and is used to thicken, for example, ice-cream, hand-cream, paint and confectionery.

CYANOBACTERIA were formerly known as blue-green algae and were grouped with algae under Prokaryotae. They are now classified as bacteria.

algal bloom An increased growth of ALGAE that may form in waters where there is an excess of NITRATES and PHOSPHATES from fertilizers and detergents. Algal blooms cause the water to smell and taste unpleasant, deplete oxygen and so cause the death of fish. *See also* EUTROPHICATION, SEWAGE DISPOSAL.

alimentary canal *See* DIGESTIVE SYSTEM.

alkali A BASE that dissolves in, or reacts with, water to produce hydroxide ions, OH⁻. An example is sodium hydroxide, NaOH, which dissolves in water to form Na⁺ and OH⁻ ions.

alkaloid One of a group of organic substances found in plants, especially flowering plants, and which are usually poisonous. Many drugs used in medicine owe their properties to the presence of alkaloids. Examples include morphine, caffeine and nicotine. Alkaloids vary in their constitution but all are basic and combine with acids to form salts that are usually water-soluble.

allantois In birds and reptiles, an EXTRAEMBRYONIC MEMBRANE that acts as a bladder to store waste products. In humans it is less important and combines with the CHORION.

allele Either one of two (or sometimes more) variants of a GENE at a given position (LOCUS)

on a CHROMOSOME. Variants are caused by a difference in the DNA of the gene. Each allele specifies a particular form of the characteristic coded for by the gene, for example, blue or brown eyes. DIPLOID organisms, such as humans, have two sets of chromosomes in the nucleus of each cell and two copies of each gene. If the two alleles occurring at a particular locus are the same they are called HOMOZYGOUS and if the two alleles are different they are called HETEROZYGOUS.

Some alleles are DOMINANT, which means that they hide the effects of other, RECESSIVE, alleles. For example, the allele for blue eyes is recessive and the allele for brown eyes is dominant. Therefore a heterozygous individual with one blue and one brown allele will have brown eyes. A recessive allele will only be expressed by an individual if both alleles are of the recessive type; that is, the individual is homozygous recessive.

See also CODOMINANCE, HARDY–WEINBERG PRINCIPLE, LINKAGE, MENDEL'S LAWS, SEX LINKAGE.

allergen A substance that causes an ALLERGY. An allergen is a particular type of ANTIGEN.

allergic reaction An apparently pointless IMMUNE RESPONSE initiated by a non-threatening foreign protein. Such an ALLERGEN activates IgE-bearing (*see* IGE) B CELLS to secrete IgE antibody, which binds to MAST CELLS and BASOPHILS causing the release of allergic mediators, such as HISTAMINE. These mediators cause the typical symptoms of ALLERGY. Symptoms vary according to where the allergen has entered the body. Release of histamine into the skin causes irritation, itching, reddening and often a rash – this is URTICARIA. The eyes and nose can be affected, as in HAYFEVER, and if the allergen enters the lungs an ASTHMA attack is triggered. Food allergies where the allergen has entered the intestines cause muscle cramps and diarrhoea and often a skin rash elsewhere as the food is absorbed into the blood and carried to the skin. A severe systemic reaction termed ANAPHYLAXIS can occur if the allergen is injected directly into the blood, as in a bee sting. This reaction is immediate and dramatic.

The best treatment for allergies is avoidance of the cause. Antihistamine drugs are used as a treatment, although they are not effective against asthma. Respiratory allergies are usually treated with CORTICOSTEROIDS given as sprays or inhalers, which reduce inflammation caused by allergies. Another drug, sodium cromoglycate, prevents the mast cells from releasing histamine and can be used as drops for eyes or as an inhaler for hayfever or asthma.

allergy Excessive sensitivity to one or more substances encountered in everyday life. Such substances include pollen, foods (such as dairy or wheat products, shellfish), food additives, house dust mites or animal hair. In a person with an allergy, such substances trigger an adverse physical ALLERGIC REACTION. Examples of allergies include HAYFEVER, ASTHMA, ECZEMA and URTICARIA.

allometric growth Where the growth of a given feature progresses at a different rate to the growth of the entire organism. For example, the head of a human baby is relatively larger than the head of an adult. Some organs in an individual show allometric growth; for example, the organs of the reproductive system grow very little early in life but then develop rapidly at the onset of puberty. The ratio of body surface area to total body volume is another example of allometric growth, as this ratio decreases as body size gets larger. *Compare* ISOMETRIC GROWTH. *See also* GROWTH.

alpha actinin, α-*actinin* A protein that cross-links ACTIN filaments. Actinin is important in MUSCLE, where it anchors and cross-links actin filaments, and in the CYTOSKELETON, where it cross-links actin in many cells or at junctions between cells and the EXTRACELLULAR MATRIX. *See also* CELL ADHESION.

alpha helix, α-*helix* A common type of structure of PROTEIN in which the POLYPEPTIDE chain is coiled into a corkscrew shape (helix). HYDROGEN BONDS form between successive turns of the helix, stabilizing the structure. *See also* BETA-PLEATED SHEET.

alpha particles Positively charged, high energy particles that are emitted from the nucleus of a radioactive atom during its decay. Alpha particles have a short range of a few centimetres in air and can be stopped by a sheet of paper. However, they are strongly ionizing (*see* IONIZING RADIATION) and so can severely damage living cells.

alternation of generations The occurrence of two distinct forms in the life cycle of an organism. The forms (or generations) differ from each other in appearance, habit and method of reproduction. Alternation of generations occurs in certain protozoans, some lower animals and plants. In plants, the first form, the DIPLOID generation, produces HAPLOID SPORES by MEIOSIS and is called the SPOROPHYTE form. The second form, the haploid generation, produces GAMETES and is called the GAMETOPHYTE form. The gametes of the gametophyte fuse to form a diploid ZYGOTE, which develops into a new sporophyte. The sporophyte is dependent on the gametophyte for water and nutrients, so the generations alternate.

alveolus (*pl. alveoli*) **1.** Any one of millions of air sacs in the lungs in which exchange of oxygen and carbon dioxide takes place between the air and the blood. *See also* RESPIRATORY SYSTEM.
   2. A sac of secretory EPITHELIUM, for example in MAMMARY GLANDS.

amino acid Any one of a group of water-soluble molecules, mainly composed of carbon, oxygen, hydrogen and nitrogen, containing a basic amino group (NH_2) and an acidic carboxyl group (COOH). There are over a hundred amino acids of which 20 make up all the different proteins known. Nine of these 20 amino acids are termed the ESSENTIAL AMINO ACIDS, which cannot be synthesized by humans and must be obtained from the diet.

All amino acids have the same core structure (two carbon atoms, two oxygen atoms, a nitrogen and four hydrogen atoms) with a variable group attached to this. This can be as simple as another hydrogen atom, as in GLYCINE (NH_2CH_2COOH), or more complex, as in TYROSINE $C_6H_4OH.CH_2CH.(NH_2).COOH$. Some amino acids, such as CYSTEINE, METHIONINE and CYSTINE, contain sulphur groups.

Many amino acids are neutral since they have one acidic and one basic group, for example VALINE and ALANINE. Some have more basic (NH_2) groups, for example ARGININE and LYSINE, and some have more acidic (COOH) groups, for example ASPARTIC ACID and GLUTAMIC ACID. All amino acids except glycerine form ISOMERS; all naturally occurring amino acids are of the L(−) form (*see* LAEVOROTATORY). Amino acids can join together to form a PEPTIDE or POLYPEPTIDE.

Other important amino acids include ISOLEUCINE, LEUCINE, PHENYLALANINE, THREONINE, TRYPTOPHAN, SERINE, ASPARAGINE, GLUTAMINE and HISTIDINE.

amino group The NH$_2$ group. *See* AMINO ACIDS.

ammonia (NH$_3$) A colourless, irritant gas, with a characteristic smell, highly soluble in water. In nature, the formation of ammonia is vitally important in the NITROGEN CYCLE. Nitrogen-fixing bacteria (*see* NITROGEN FIXATION) convert atmospheric nitrogen into ammonia, which is used by nitrifying bacteria (*see* NITRIFICATION) to produce NITRITES and NITRATES. Ammonia is a waste product derived from the breakdown of proteins and can be excreted by many aquatic animals and insects. However, it is toxic and in most vertebrates ammonia is converted into UREA for excretion. Ammonia is manufactured commercially for use in fertilizers.

ammonification The breakdown of proteins and amino acids by bacteria to produce ammonia. *See* NITROGEN CYCLE.

ammoniotelic (*adj.*) Describing an animal that excretes ammonia as its main nitrogenous waste product. *See* UREA.

ammonite An extinct aquatic mollusc, characterized by a coiled shell divided into chambers. Ammonites evolved quickly during the Mesozoic era (225 to 65 million years ago), and their fossils are used to date the rock strata in which they are found.

amniocentesis A medical procedure in which a fine needle is inserted through the AMNION surrounding the foetus during pregnancy, and a sample of the AMNIOTIC FLUID removed. This fluid contains some foetal cells, therefore allowing the foetal chromosomes to be examined for abnormalities, such as DOWN'S SYNDROME. This is performed at about 16 weeks of pregnancy and there is some risk of miscarriage. *See also* CHORIONIC VILLUS SAMPLING.

amnion In most mammals, the innermost EXTRA-EMBRYONIC MEMBRANE that usually expands to reach the CHORION and encloses the embryo. It contains AMNIOTIC FLUID.

amniotic fluid In mammals, the fluid surrounding the foetus during pregnancy, contained by the AMNION. The fluid provides a cushioning pad to protect the foetus from physical impact and also maintains a constant internal environment. The fluid is swallowed by the foetus and so is circulated, allowing some waste to be removed by the PLACENTA. *See also* AMNIOCENTESIS.

Amoeba (*pl. amoebae*) One of the simplest organisms, a protozoan (*see* PROTOZOA) belonging to the kingdom PROTOCTISTA. It consists of a colourless PROTOPLASM from which extensions called pseudopodia (*see* PSEUDOPODIUM) form and engulf food. Amoebae reproduce by BINARY FISSION and possess a contractile VACUOLE for OSMOREGULATION. The pseudopodia are also used for movement. Movement similar to that of an amoeba by pseudopodia is called 'amoeboid'. An example of an amoeba is *Entamoeba histolytica*, which causes dysentery in humans.

amoebocyte Any animal cell with no fixed location that is free to move through body tissue. *See also* PORIFERA.

AMP (adenosine monophosphate) A NUCLEOTIDE component of DNA and RNA, and the product formed from the HYDROLYSIS of ATP and ADP. PHOSPHORYLATION of AMP yields ADP. AMP can be converted to CYCLIC AMP by the enzyme adenylate cyclase in response to the appropriate extracellular signals. Cyclic AMP is important in many biochemical pathways.

Amphibia A class of vertebrates, including toads, frogs, newts and salamanders. Amphibians live partly in water and partly on land. They have four legs (each with five digits), a moist smooth skin with scales and lay eggs that are not protected by a shell. Amphibians begin their life in water in the larval stages (as tadpoles). After METAMORPHOSIS they live on land as adults, with lungs, and return to the water to breed. They are poikilothermic (cold-blooded; *see* POIKILOTHERMY) animals, so cannot maintain their own body temperature. They continue to grow throughout their life.

amphibian A member of the vertebrate class AMPHIBIA.

ampulla A swelling in the SEMI-CIRCULAR CANALS of the inner ear. It is concerned with balance.

amylase One of a group of enzymes that breaks down STARCH into its constituent sugars. It is found in humans in SALIVA and PANCREATIC JUICES. Plants also contain amylases such as diastase, which is a component of malt and important in the brewing industry.

amylopectin A POLYSACCHARIDE made up of GLUCOSE molecules in a branched structure. Amylopectin is a component of STARCH. *See also* AMYLOSE.

amyloplast A type of LEUCOPLAST that stores STARCH in plants.

amylose A straight-chained POLYSACCHARIDE made up of hundreds of GLUCOSE molecules. Amylose is a component of STARCH. *See also* AMYLOPECTIN.

anabolic steroid A STEROID HORMONE that has anabolic effects (*see* ANABOLISM); that is, to speed up tissue growth, particularly muscle. The male sex hormone TESTOSTERONE is a natural anabolic steroid and many synthetic variants exist. Anabolic steroids are given as replacement therapy where natural hormone production is deficient or to treat certain types of cancer of the breast or uterus.

There has been a widespread abuse of anabolic steroids by athletes, as they help build up body muscles. There are considerable risks to the health of the user. Side-effects include liver damage, acne, baldness, reduced fertility for both men and women, hardening of the arteries and an increased risk of certain cancers. In young boys these drugs stunt growth and in women they have a masculinizing effect. The use of anabolic steroids has now been banned by athletic organizations.

anabolism The building up of body tissue. *See* METABOLISM.

anaemia A deficiency in the number of RED BLOOD CELLS in the blood or in their HAEMOGLOBIN content. Anaemia results in pallor, shortness of breath, lack of energy, dizziness and digestive disorders. *See also* PERNICIOUS ANAEMIA, SICKLE-CELL DISEASE.

anaerobe An organism that does not require oxygen for RESPIRATION. *Compare* AEROBE. *See also* GLYCOLYSIS.

anaerobic respiration *See* ANAEROBE, RESPIRATION.

anaesthetic A drug that is used to render a person insensitive to pain. Anaesthetics can be applied locally by freezing or by injection of a drug at the site to be treated, or generally to cause loss of consciousness during operations. General anaesthesia has been induced by a number of different agents in the past, including ether, chloroform and dinitrogen oxide. Halothane is now commonly used since it has less side-effects.

anal fin *See* FIN.

analgesic A pain-relieving agent. OPIATE analgesics are the strongest of these drugs, whilst non-opiates, such as ASPIRIN and PARACETAMOL, are useful for less severe pain. Analgesics act by both preventing NERVE stimuli being sent to the brain and also by removing awareness of the pain.

anaphase A stage of MITOSIS and MEIOSIS.

anaphylaxis Hypersensitivity to a foreign substance, causing an immediate and dramatic ALLERGIC REACTION. This type of response can occur when the foreign substance or ALLERGEN is injected directly into the blood, as in a bee sting. The allergen quickly spreads all over the body causing the release of HISTAMINE. Within a few minutes, weals can develop all over the skin, soft facial tissues become swollen so the eyes may close, the lips and throat may swell, blood pressure drops and the person may collapse. Breathing may be difficult or even stop. Anaphylactic shock can be the result of a severe drug allergy.

anatomy The study of the structure of the body and its component parts.

androecium The collective name for the STAMENS, the male part of a flower.

anemophily POLLINATION of flowers by the wind. Flowers pollinated in this way are usually unscented and lack petals. The male and female flowers are often separate and are formed before the leaves to allow the pollen to be easily transported.

angiosperm Any flowering plant. They are classified in the phylum ANGIOSPERMOPHYTA.

Angiospermophyta A phylum of the plant kingdom comprising all flowering plants. Flowering plants produce SEEDS protected within an ovary (*see* CARPEL). ANGIOSPERMS form most of the terrestrial vegetation found today.

There are two groups of angiosperms, MONOCOTYLEDONS and DICOTYLEDONS, which have one and two seed leaves respectively in the embryo. Most flowering plants, for example, the buttercup, daisy and wallflower, are dicotyledons. They are broad-leaved, with flower parts arranged in fours or fives, and usually pollinated by insects.

Flowering plants are unique in undergoing DOUBLE FERTILIZATION, after which the OVULE develops into the seed and the ovary into a fruit. Angiosperms are found in a variety of habitats and more than 250,000 species exist.

angiotensin A PLASMA PROTEIN produced by the action of the enzyme RENIN. It is made by the kidney in response to low levels of sodium in the blood or reduced blood volume. Angiotensin stimulates production of ALDOSTERONE, which is involved in the regulation of water retention by the kidney.

angstrom (Å) A non-SI UNIT of length, equal to 10^{-10} m. It is still sometimes used to specify the wavelengths and intermolecular distances, but has largely been superseded by the nanometre (nm), which is 10^{-9} m.

animal A multicellular, EUKARYOTIC organism of the kingdom Animalia, lacking the rigid cell walls of plants and usually capable of movement for at least part of its life cycle. All animals are HETEROTROPHS. The oldest animal fossil on land was found in 1990 in Ludlow (Shropshire, UK); it was 440 million years old. There are 18 phyla of animals, from the most primitive PORIFERA (SPONGES), to the largest phylum CHORDATA, which includes the VERTEBRATA. Vertebrates are the dominant animals of land, sea and air, not in numbers but in BIOMASS and other ecological terms. *See also* ANNELIDA, ARTHROPODA, CNIDARIA, ECHINODERMATA, MOLLUSCA, NEMATODA, PLATYHELMINTHES.

Animalia The kingdom consisting of ANIMALS.

anion A negatively charged ION.

annelid A member of the phylum ANNELIDA.

Annelida A phylum consisting of invertebrate animals that have a soft, segmented body with an outer CUTICLE and possess bristles (CHAETAE). Examples include earthworms, lugworms and leeches. Some annelids have a distinct head, for example the lugworm. There are about 9,000 species living in water and soil.

The body consists of a series of similar segments separated from one another by internal membranes. This is called 'metameric' segmentation. There is repetition of nerves, blood vessels and muscles in each segment, but some features, for example reproductive organs, are only repeated in a few segments. Annelidae possess longitudinal and circular muscles by which some move, aided by the bristles, and they feed on a variety of organisms and organic debris. Reproduction can be sexual or asexual (*see* BUDDING, FRAGMENTATION). Some are parasitic (*see* PARASITE) but they do not cause major problems for humans. *See also* HIRUDINEA, OLIGOCHAETA, POLYCHAETA.

annual plant Any plant that completes its life cycle in one year and then dies. Annual plants, for example the common poppy, germinate from seeds, grow to maturity and produce seeds within one year or season. *Compare* BIENNIAL PLANT, PERENNIAL PLANT.

anorexia nervosa An eating disorder, usually due to psychological problems or stress. It is characterized by an abnormal fear of obesity, leading to an excessive reduction in food intake and consequent wasting of the muscles.

antagonistic (*adj.*) Describing opposing actions or forces. Two muscles operating together to enable movement in opposite directions, for example at a joint, are antagonistic. In this case, one muscle contracts while the other relaxes. Substances such as drugs or hormones are termed antagonistic if the action of one inhibits the action of the other. *Compare* SYNERGISTIC.

antenna (*pl.* **antennae**) Either one of a pair of appendages on the heads of insects, crustaceans, etc., that respond to touch and taste. Some antennae are specialized for swimming or attachment.

anterior (*adj.*) Referring to the front of an organism. In lower animals this is the head end and in higher animals it is the front of the body. The term is also used to refer to the front of an organ or gland, for example, the anterior PITUITARY GLAND, or to a chamber, for example, in the eye. In plants, anterior refers to leaves or flowers that are in front of and face away from the main stem. *Compare* POSTERIOR.

anther A structure in flowers that is responsible for the production of pollen grains. It is part of the STAMEN (the male reproductive structure) and consists of two lobes, containing pollen sacs, supported by a long stalk called the filament.

antheridium (*pl.* **antheridia**) The male sex organ of fungi and plants without seeds, such as members of BRYOPHYTA. The male GAMETES are released from the antheridium, which is a sac, and swim down into the ARCHEGONIUM to fuse with the female gamete. This is part of the GAMETOPHYTE generation in plants showing ALTERNATION OF GENERATIONS.

Anthocerotae A class of the phylum BRYOPHYTA, consisting of the hornworts. Hornworts are usually found in warm climates in damp conditions. Like the liverworts and mosses, to which they are related, they show ALTERNATION OF GENERATIONS.

Anthozoa *See* ACTINOZOA.

antibiotic A chemical substance produced by a micro-organism that prevents the growth of other micro-organisms (but not viruses) and is used to combat many animal and human illnesses. The first antibiotic, PENICILLIN, was discovered in 1929 by Alexander Fleming (1881–1955). Penicillin now consists of a family of antibiotics obtained from moulds of the genus *Penicillium*. Some antibiotics, including penicillin, are called narrow-spectrum because they are only effective against a few PATHOGENS.

Other antibiotics are broad-spectrum and can inhibit the growth of a wide variety of pathogens.

Many antibiotics have been discovered but only a few are medically useful and commercially viable. Their use may be restricted because of side-effects, for example toxicity and allergy. A pathogen may become resistant to a particular antibiotic (see ANTIBIOTIC RESISTANCE). Antibiotics are secondary METABOLITES, chemicals that are not essential for growth of the organism but often have a secondary role in, for example, the defence of the organism. Their production is complex.

The action of antibiotics varies: streptomycin affects DNA, RNA and PROTEIN SYNTHESIS; penicillin prevents formation of bacterial cell walls.

antibiotic resistance The inability of an ANTIBIOTIC to slow the growth of a PATHOGEN previously affected by it. This occurs when a micro-organism is repeatedly exposed to an antibiotic or is exposed to insufficient doses. Resistance can be inherited, so the micro-organism may not need to be exposed to the antibiotic to exhibit resistance. This has been a problem with PENICILLIN. There is therefore a continual need to find new types of antibiotics to overcome the problem of resistance. As well as searching for new natural antibiotics, more use is being made of GENETIC ENGINEERING to develop new strains or mutated strains of micro-organisms to be antibiotic producers.

antibody A protein secreted by a subclass of LYMPHOCYTES called B CELLS in response to the presence of a foreign substance (called an ANTIGEN), such as a viral or bacterial infection. This is only one of the ways the body can fight an infection. Antibodies are not restricted to the blood, but occur throughout the body. Different B cells produce antibodies with specificities for different antigens. This is called antibody diversity.

On its surface, a B cell possesses a specific antibody; when an appropriate antigen is presented, the B cell is stimulated to divide and secrete its antibody. Once produced, the antibody binds non-covalently to its specific antigen, recognizing the overall three-dimensional shape of the antigen as well as its chemical make-up, and an antigen-antibody complex forms. Some cross-reactivity can occur as antigens can have similarities.

Antibodies can act in a number of ways to remove or destroy the foreign substance, for example by PRECIPITATION of the antigen-antibody complex, AGGLUTINATION of antigens, or neutralization of toxins produced by micro-organisms. Antibodies can remain in the blood after an infection and protect the body against future infection by the same organism.

See also IMMUNE RESPONSE, ANTITOXIN, VACCINATION.

anticoagulant A substance that prevents the clotting of blood. See also BLOOD CLOTTING CASCADE.

anticodon A specific sequence of three NUCLEOTIDES carried by TRANSFER RNA that is complimentary to, and can therefore form BASE PAIRS with, a CODON sequence carried on MESSENGER RNA. See also PROTEIN SYNTHESIS.

anti-diuretic hormone (ADH) A hormone produced by the HYPOTHALAMUS that is responsible for maintaining the correct salt/water balance in vertebrates. ADH is passed to the PITUITARY GLAND, where it is stored and secreted under the control of the hypothalamus. ADH reduces the amount of water lost from the kidney as urine and also raises blood pressure by constricting ARTERIOLES. When water is in short supply, ADH secretion is increased, allowing water to be conserved in the kidney. When water is plentiful, ADH secretion is reduced and the urine is more dilute so that more water can leave the body.

antigen Any substance that induces the production of an ANTIBODY by the body's IMMUNE SYSTEM. Antigens are usually proteins or GLYCOPROTEINS, such as proteins on the surface of bacteria, viruses and pollen grains. An antigen that triggers the IMMUNE RESPONSE is said to be immunogenic. Not all antigens cause the initial induction of the immune response. Some initiate further production of antibodies in a response that has already been induced.

Antigens do not usually bind directly to antibodies but are instead presented to LYMPHOCYTES on the surface of ANTIGEN-PRESENTING CELLS. Body tissues and blood cells can also act as antigens and these have to be matched between donor and recipient for successful blood transfusions or organ transplants.

See also ALLERGIC REACTION, ANTIGENIC VARIATION, HISTAMINE, MAJOR HISTOCOMPATIBILITY COMPLEX.

antigen D See RHESUS FACTOR.

antigen-presenting cell (APC) A cell that presents, on its surface, fragments of ANTIGEN to T CELLS. Antigen-presenting cells are found in the SPLEEN and LYMPH NODES and trap antigens in the blood or LYMPH. The antigen degrades and fragments are presented in combination with molecules of the MAJOR HISTO-COMPATIBILITY COMPLEX, which may then activate T cells to divide. This may in turn activate B CELLS to produce a specific antibody. *See also* MACROPHAGE.

antigenic variation The ability of some PATHOGENS, for example the influenza virus, to change their surface ANTIGENS during an infection. This makes the fight against the disease and the search for a VACCINE more difficult.

antipodal cell In plants, one of three cells within the EMBRYO SAC of a developing OVULE. The cells are at the opposite end to the MICROPYLE.

antiseptic A substance that prevents or inhibits the growth of micro-organisms. Antiseptics were introduced by the English surgeon Joseph Lister (1827–92), who pioneered the use of carbolic acid (phenol).

antiserum A blood SERUM that contains ANTI-BODIES to a specific ANTIGEN.

antitoxin An antibody that works by neutralizing TOXINS produced by micro-organisms so that they become inactive.

anus The opening of the RECTUM to the outside, at the end of the DIGESTIVE SYSTEM. Undigested food is removed (egested) from the body through the anus in the form of FAECES. This is controlled by a muscular SPHINCTER that in adults can be regulated voluntarily. Some simpler organisms have no anus and have only one opening to the digestive system.

anvil *See* INCUS.

aorta The main ARTERY of vertebrates that carries oxygenated blood away from the heart. It branches to form smaller arteries that supply blood to the rest of the body except the lungs. Unlike most other arteries, the aorta has non-return valves to ensure the one-way flow of blood. *See also* CIRCULATORY SYSTEM.

APC *See* ANTIGEN-PRESENTING CELL.

Apicomplexa A phylum from the kingdom PROTOCTISTA. The members (sporozoans) are mostly parasitic (*see* PARASITE) and have little movement. Reproduction is by multiple fission (*see* BINARY FISSION). One member of this phylum is *Plasmodium*, the parasite that causes MALARIA.

apocrine gland An EXOCRINE GLAND in which the apical part (the tip) of a cell breaks down during secretion. An example is the MAMMARY GLAND. *See also* HOLOCRINE GLAND, MEROCRINE GLAND.

apoenzyme An inactive enzyme that needs to associate with a COFACTOR in order to function.

apoplast pathway The most important of the three pathways by which water and minerals move upwards through a plant (*see also* SYM-PLAST and VACUOLAR PATHWAY). In the apoplast pathway, substances are carried from cell to cell via the cell walls. Water enters the air-spaces between the CELLULOSE fibres making up the cell wall and is drawn to the adjacent cell wall when water evaporates through the stomata (*see* STOMA). The pull of water is transmitted through the plant by the strong cohesive forces between water molecules. In the root, water cannot enter the cell wall of endodermal cells (*see* ENDODERMIS) because of a band of impermeable SUBERIN constituting a CASPARIAN STRIP. It is thus forced to go through the cytoplasm of these cells. *See also* TRANSLOCATION, TRANSPIRATION.

appendix In mammals, a small closed sac within the gut leading on from the CAECUM. In humans, the appendix has no real use and frequently becomes inflamed and needs to be removed. However, it is important in HERBI-VORES because it contains micro-organisms needed for digestion of CELLULOSE.

aqueous (*adj.*) Relating to water, particularly a solution in water.

aqueous humour In vertebrates, a watery fluid found in the eye in the space between the CORNEA and the lens. It is similar to CERE-BROSPINAL FLUID in composition and is continuously renewed. It provides a link between the CIRCULATORY SYSTEM and the lens and cornea.

arachidonic acid An ESSENTIAL FATTY ACID which is an important precursor of many biological compounds, particularly PROSTAGLANDINS. Arachidonic acid is a polyunsaturated fatty acid and plays a role in fat metabolism and membrane production.

Arachnida A class of the phylum ARTHROPODA, including spiders, scorpions and ticks.

arachnoid membrane The middle of the three membranes that cover the brain and spinal cord. *See* MENINGES.

archegonium (*pl.* **archegonia**) The female sex organ of mosses, liverworts, ferns and most

GYMNOSPERMS. It is a flask-shaped structure with the OVUM or egg cell at the base. The male GAMETE (formed in the ANTHERIDIUM) swims down the neck of the archegonium and fuses with the ovum. The archegonium and antheridium form part of the GAMETOPHYTE in plants showing ALTERNATION OF GENERATIONS. Once the male and female gametes have fused the SPOROPHYTE generation begins.

archenteron, *gastrocoel* A cavity within an animal embryo developing at GASTRULATION. The archenteron later forms the cavity of the GUT.

area of outstanding natural beauty (AONB) An area of land worthy of CONSERVATION and protection due to its natural beauty.

areolar tissue See CONNECTIVE TISSUE.

arginine An amino acid with a basic side chain that is a component of the ORNITHINE CYCLE by which UREA is produced.

arithmetic mean The sum of a set of numbers divided by the number of elements in the set. *See also* AVERAGE.

arteriogram A RADIOGRAPH of the arteries, made with an injection of a CONTRAST ENHANCING MEDIUM.

arteriole A small branch of an artery.

artery In animals with a CIRCULATORY SYSTEM, a vessel that carries blood away from the heart to the rest of the body ('a' for artery, 'a' for away).

Arteries have thick muscular walls to withstand blood at high pressure, but also contain elastic fibres so they can expand to allow for the increase in blood pressure following contraction of the heart muscles. Arteries (except the PULMONARY arteries supplying the lungs) carry highly oxygenated blood to all the main organs of the body. Unlike veins, which all have valves to ensure the one-way flow of blood, the only arteries to have valves are the AORTA and the pulmonary arteries. *See also* ATHEROSCLEROSIS, PULSE.

arthritis Inflammation of a joint producing pain and stiffness. *See* OSTEOARTHRITIS, RHEUMATOID ARTHRITIS.

arthropod A member of the phylum ARTHROPODA.

Arthropoda A large phylum of invertebrate animals with a hard EXOSKELETON, jointed legs and a segmented body. Among the classes of the Arthropoda phylum are: CRUSTACEA, including Daphnia (the water flea), crabs, prawns and crayfish; INSECTA, including the locust, cockroach, housefly, butterfly and bee; ARACH-NIDA, including the spider, scorpion and tick;

CHILOPODA, including the centipede; and DIPLOPODA, including the millipede.

Arthropods make up three-quarters of all living animals and are well adapted to living in water or on land. They can be free-living or parasitic. The exoskeleton is made mainly from CHITIN. Growth of arthropods occurs in stages after the moulting of their exoskeleton, called ECDYSIS.

articular (*adj.*) Relating to JOINTS or the structural components within joints.

articular cartilage See CARTILAGE.

artificial selection The selected breeding by humans of plants or animals in order to develop particular characteristics, such as disease-resistance in plants, improved milk production in cows, racehorse breeding, cat and dog breeding. Inbreeding (breeding of an animal with its close relatives) can result in harmful genes being expressed, which in an outbreeding population would be RECESSIVE. To avoid this, regular inbreeding is interspersed with occasional new genes by outbreeding.

Eugenics is the study of ways in which the human race can be improved by the selection or elimination of specific characters, for example to control the spread of genetic disorders.

ascomycete A member of the ASCOMYCOTA phylum of FUNGI.

Ascomycota A phylum of the kingdom FUNGI. Ascomycetes are characterized by having septa (partitions) in their HYPHAE. SEXUAL REPRODUCTION is by the formation of ascospores in a structure called an ascus. Asexual reproduction is by non-motile SPORES called conidia (*see* CONIDIUM). The phylum includes yeast (*Saccharomyces*) and the moulds *Aspergillus* and *Penicillium*. Many members of the Ascomycota are a cause of food or crop spoilage.

ascorbic acid See VITAMIN C.

ascospore Any SPORE of an ASCOMYCETE, from the ASCOMYCOTA phylum of FUNGI.

ascus In ASCOMYCETES, a cell within which ASCOSPORES are formed during SEXUAL REPRODUCTION.

asexual reproduction The production of offspring from a single parent, involving no fusion of GAMETES. The offspring are usually genetically identical to each other and to the parent. Since only a single parent is involved, it can lead to a rapid increase in the population, which is a considerable advantage. The disadvantage is that there is no genetic VARIATION and therefore no opportunity to adapt to a changing environment. Many asexual organisms can also reproduce sexually, which increases the chance for genetic variation (*see* FERTILIZATION, SEXUAL REPRODUCTION). Asexual reproduction can be by BINARY FISSION, BUDDING, FRAGMENTATION, SPORULATION or VEGETATIVE REPRODUCTION. *See also* PARTHENOGENESIS.

asparagine A non-essential amino acid which is an uncharged derivative of ASPARTIC ACID. It is found in asparagus, hence its name, and in potatoes and beetroot.

aspartame An artificial sweetener (trade name Nutrasweet). It is 200 times sweeter than sugar and does not have the aftertaste that SACCHARIN has. It cannot be used for cooking, as it breaks down on heating.

aspartic acid A non-essential acidic amino acid, commonly called aspartate to indicate its negative charge at physiological pH. ASPARAGINE is the uncharged derivative. Aspartic acid is involved in the production of UREA through the ORNITHINE CYCLE and also acts as a NEUROTRANSMITTER. It is found in young sugar cane and sugar beet.

aspirin A widely used pain-killing drug that acts by inhibiting PROSTAGLANDINS. It also reduces fever and inflammation, for example in arthritis. Side-effects of long-term usage are stomach bleeding and kidney damage, although recently it has been suggested that an aspirin a day can reduce the risk of heart attacks and thrombosis.

asthma A respiratory disorder characterized by wheezing and shortness of breath. Asthma occurs when the small bronchial tubes in the lungs become swollen with inflammation, blocked by phlegm or narrowed by a tightening of the muscles surrounding them. This is often due to an ALLERGIC REACTION to some common substance, such as house dust mites, animal fur, pollen, feathers or pollutants in the air. Asthma can be caused by an infection or brought on by anxiety. There is a tendency for asthma to run in families and it is often associated with ECZEMA. The treatment of asthma depends on the cause. Antihistamines will help asthma caused by allergy, although often the asthma is not merely a result of HISTAMINE release as in many other allergies. Infections can be treated with antibiotics. Drugs known as bronchodilators may be administered to widen the airways or relax the surrounding muscles. These drugs include the CORTICOSTEROIDS usually given which reduce inflammation. Bronchodilators are given as inhalers delivered directly to the lungs. Some are taken regularly to reduce the likelihood of an attack and others are used at the time of an attack to give immediate relief.

astigmatism A defect in the CORNEA of the eye that arises when the surface is more strongly curved in one direction than another. Parallel rays of light are not brought to a focus at the same point, leading to distortions in the image. Astigmatism can be corrected with spectacles that have a cylindrically curved surface.

atherosclerosis A disease in which FATTY ACIDS build up in the walls of arteries as a person ages. This reduces the flexibility and internal diameter of the artery, so increasing blood pressure and leading to heart disease. Diet plays an important role in this process. The disease is particularly prevalent in the Western world. *See also* CHOLESTEROL.

ATP (adenosine triphosphate) The short-term energy storage and carrier molecule found in all living cells. It transfers energy from where there is plenty to where it is needed for cellular reactions. Energy is released when one of the three phosphate groups of ATP is removed (catalysed by a number of enzymes) by a process called HYDROLYSIS. This yields ADP (adenosine diphosphate). Hydrolysis of ADP then yields AMP (adenosine monophosphate). The phosphate molecules can be added back to AMP to reconvert to ATP by a process called PHOSPHORYLATION.

atrium (*pl. atria*), *auricle* Either one of the two upper chambers in the heart. The walls of the atria are thin so that they can stretch to receive blood that returns from the body. The atria contract to force blood into the VENTRICLES. *See also* HEART.

auditory canal, *ear canal, external auditory meatus* In mammals and birds, a tube leading from the opening of the outer ear to the EARDRUM.

auditory nerve The nerve that transmits impulses from the ear to the brain.

auricle *See* ATRIUM.

autoimmunity A disorder in which an IMMUNE RESPONSE is mounted to an organism's own (self) ANTIGENS. It seems to arise as a result of defects in the B-cell ANTIBODY response rather than the T-cell response. An example is RHEUMATOID ARTHRITIS.

autolysis The self-destruction of a cell or tissue, brought about by the action of enzymes released by the cell itself. *See also* LYSOSOME.

autonomic nervous system Part of the NERVOUS SYSTEM that is self-governing and controls the involuntary responses of the heart, glands and smooth muscle (such as in the digestive tract and blood vessels). It forms part of the EFFECTOR SYSTEM, which receives information from the CENTRAL NERVOUS SYSTEM and transmits it to EFFECTORS (muscles or glands) that stimulate the appropriate action.

The autonomic nervous system can be divided into the sympathetic system and the parasympathetic system. The sympathetic system responds to stress, for example by increasing the heart rate, increasing blood pressure, preparing the body for action. The parasympathetic system is important when the body is resting, for example by slowing the heart rate, decreasing blood pressure and stimulating the digestive tract. These two systems normally oppose one another.

Some control over the activities of the autonomic nervous system, for example, control of bladder and anal SPHINCTERS, can be learned through training.

autophagosome *See* PHAGOSOME.

autoradiography A technique used in the laboratory for visualizing a substance that has been radioactively labelled (i.e. attached to a radioactive ISOTOPE). The substance is placed in contact with a photographic film in a light-tight cassette for a period of time. The radioactivity causes the film to darken and thus the substance to be identified.

Autoradiography can be used, for example to analyse polyacrylamide gels (*see* POLYACRYLAMIDE GEL ELECTROPHORESIS) containing radiolabelled proteins. It can be also used to examine the incorporation of a radiolabelled substance into living tissues and cells, in TISSUE CULTURE or as sections in MICROSCOPY.

autosome Any CHROMOSOME that is the same in males and females. All chromosomes except the sex chromosomes are autosomes. The sex chromosomes are called HETEROSOMES.

autotroph Any organism that can manufacture organic compounds from inorganic molecules using light or chemical energy. Autotrophs can exist independently of any external source of organic compounds, unlike HETEROTROPHS.

All green plants and some bacteria are photoautotrophs. Photoautotrophs obtain their energy from PHOTOSYNTHESIS, which uses light to convert carbon dioxide and water into sugars. Some bacteria use chemical energy, for example from sulphur-containing compounds, to synthesize organic compounds. These are chemoautotrophs.

Autotrophs are the primary producers in the FOOD CHAIN; they provide nourishment for all the other animals in the food chain, which are heterotrophic. *See also* CHEMOSYNTHESIS.

autotrophic nutrition The synthesis of organic compounds from inorganic molecules using light or chemical energy. The organisms that are capable of this are called AUTOTROPHS and do not need any external source of organic compounds. Autotrophic nutrition is self-feeding. *See also* CHEMOSYNTHESIS, FOOD CHAIN, HETEROTROPH, PHOTOSYNTHESIS.

auxin Any one of a group of PLANT GROWTH SUBSTANCES. Auxins influence many aspects of plant growth, including TROPISMS, cell enlargement and growth of roots. They are the most common type of plant growth substance.

The most common naturally occurring auxin is indoleacetic acid, which is made in the shoot and root tips and transported to other parts of the plant. The short-distance cell-to-cell transport of auxin is by DIFFUSION, but long-distance transport is via the PHLOEM. The transport of auxin is polar (in one direction only), away from the tips.

Auxins act by increasing the elasticity of the cell wall, so the cell expands when TURGOR pressure increases, and continues to enlarge until enough resistance is provided by the cell wall. Auxin also affects GENE EXPRESSION (of at least 10 genes). At higher concentrations, auxins can inhibit growth and cause death of a plant. PHOTOTROPISM can be explained in terms of auxin distribution.

Many synthetic auxins have been developed, for example, to help root development in cuttings, to prevent fruit drop in orchards, to achieve synchronous flowering (and therefore fruiting) in pineapple and as weedkillers, where they can cause such rapid growth that the plant dies. Some synthetic auxins have different effects on different plants, which is useful in producing selective weedkillers to kill only the unwanted plants.

See also ABSCISSIC ACID, CYTOKININ.

average The typical member of a set of data. It usually refers to the ARITHMETIC MEAN, which is obtained by adding all of a group of numbers together and dividing the total by the number of samples. A mean value is often expressed plus or minus the STANDARD DEVIATION. The term average is also used to refer to the MEDIAN, which is the middle number in a set of numbers arranged in increasing or decreasing order, or the MODE, which is the most frequently occurring number in a group.

Aves A class of vertebrates, consisting of the birds. Birds possess feathers on their skin, scales on their legs, a beak instead of teeth, wings, lungs and eggs with a large yolk and hard shell from which the young hatch. They form the largest group of land vertebrates and there are 8,500 species. They have two legs, each with three digits; the front legs are modified to form a wing. Most birds can fly but some cannot, including the ostrich. Birds are HOMEOTHERMS, maintaining a body temperature of 41°C. Their hearing and eyesight are good but their sense of smell is poor. Males are often brightly coloured to attract females, and communication is by vision and sound. The eggs are hatched by the female in a nest and cared for over a period of time. The study of birds is called ornithology.

axil In a plant, the angle between the stalk of a leaf or branch and the stem from which it grows.

axon A NERVE FIBRE that is a long extension of a NEURONE and conducts impulses away from the cell body to the SYNAPSE with another neurone or EFFECTOR organ, such as a muscle. Most neurones only have one axon (monopolar), some have two (bipolar) and others have several (multipolar). Some axons are insulated by a MYELIN SHEATH.

AZT, *zidovudine* A drug used in the treatment of AIDS.

B

Bacillariophyta A phylum of the kingdom PRO-
TOCTISTA that consists of the diatoms. They are
unicellular organisms that live in moist soil,
freshwater or on the surface layers of the oceans,
where they form a major component of PLANK-
TON. Their cell walls are composed chiefly of silica and are divided into two halves that fit one
inside the other, like a box with its lid. The fossil
remains of diatom shells are mined and used in
abrasives, filters and fillers for paint and rubber.

bacillus (*pl.* **bacilli**) Any rod-shaped bacterium
(*see* BACTERIA). Bacilli are widespread in soil and
air and cause diseases such as anthrax (*Bacillus
anthracis*).

backbone *See* VERTEBRAL COLUMN.

backcross A mating between a parent and
its offspring, used in genetics to determine
CHROMOSOME MAPS.

background radiation The collective name
for the many sources of IONIZING RADIATION.
The most important of these are naturally
occurring radioactive materials in rocks, soil
and atmosphere. The other main source of
background radiation is that entering the
atmosphere from space.

Compared to these two sources of radiation,
the radiation present from nuclear weapons and
nuclear reactors represents only 1 or 2 per cent
of the total exposure to ionizing radiation for
the average human. In addition, individuals
often experience significant doses of ionizing
radiation, mainly X-rays, from medical sources.
The level of medical exposure can vary widely
from one individual to another, though in the
West it typically accounts for about 13 per cent
of the lifetime dose. *See also* RADIOACTIVITY.

bacteria (*sing.* **bacterium**) Microscopic organisms with a single cell and lacking an organized
nucleus. Bacteria are classified as in the kingdom PROKARYOTAE and are neither plant (most
lack CHLOROPHYLL) nor animal.

Bacteria can be spherical (COCCUS) in
shape, for example *Staphylococcus aureus*; rod-
shaped (BACILLUS), for example *Escherichia coli*;
spiral (spirilla), for example *Spirillum rubrum*;

or comma-shaped (vibrios), for example *Vibrio
cholerae*. Cells can be either Gram-positive or
Gram-negative according to their ability to
stain in Christian Gram's (1855–1938) stain of
1884. This is important in their classification
(*see* GRAM'S STAIN). Their size varies between 0.5
and 2.0 μm, so they are visible with a light
microscope. Reproduction is by BINARY FISSION.
Non-spherical bacteria can possess one or more
FLAGELLA by which they move. Some have
smaller filaments called pili, which make the
cells sticky, and under certain conditions some
secrete a capsule of non-living viscous material
around the cell wall.

Bacteria are found in large numbers every-
where and their activities are very important.
Most are SAPROTROPHS or PARASITES. In the soil,
they break down plant and animal tissues and
their role is crucial to soil fertility and thus to all
life. This ability is also important in SEWAGE DIS-
POSAL. Some bacteria are used in industry to
bring about desirable chemical reactions, while
others cause serious food and drink spoilage.
They cause many serious diseases in animals
and humans, but some are a source of ANTI-
BIOTICS. Some live harmlessly in sites within animals and humans, such as in the skin and gut.

bacterial growth curve The curve obtained when
the changes in the size of a bacterial culture are
plotted against time on a graph. The growth of
bacterial cultures follows the same pattern as
that of a population (*see* POPULATION GROWTH).
Cultures are grown in a sterile medium with the
nutrients required by the bacteria and at the
optimum temperature for growth.

Four phases of growth are observed when a
logarithmic growth curve is plotted. The lag
(latent) phase is the first phase during which
growth is increasing slowly as the bacteria are
adapting to the culture and making enzymes
necessary to utilize the medium. Then follows a
period of rapid growth called the exponential
(logarithmic) phase, where nutrients are plentiful and waste products have not built up to a
limiting level. During this phase of balanced

growth, cell size and protein content remain constant, as does the doubling time. This is seen as a straight line on a logarithmic scale.

As nutrients are used up and toxic waste products accumulate, the growth enters a stationary phase where the number of new cells produced equals the numbers dying. This is a phase of unbalanced growth where the cells are of different sizes and chemical make-up due to the differing composition of the medium. The culture then enters the death phase where the number of living cells is decreasing, although the total number of cells (living and dead) remains the same. Cells die due to lack of nutrients, build up of toxic waste products or lack of space. *See also* EXPONENTIAL GROWTH, GROWTH.

bacteriochlorophyll A type of CHLOROPHYLL found in some photosynthetic bacteria.

bacteriophage A virus that infects a bacterium (*see* BACTERIA). Bacteriophages can contain DNA or RNA as their genetic material in the centre of the phage surrounded by a protein coat. Phages can be very simple (RNA phages) or complex, consisting of head, collar and tail regions. Some phages are spherical in shape, while others are filamentous. The complex

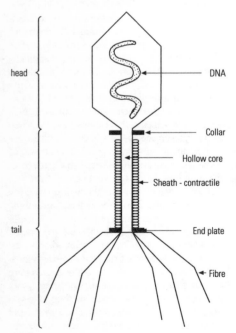

head

tail

DNA

Collar

Hollow core

Sheath - contractile

End plate

Fibre

Structure of a complex bacteriophage.

phages attach to bacterial cell walls by their tail, and inject their DNA into the bacteria cell leaving their protein 'ghost' outside. Virulent phages kill their host cell, whereas temperate phages do not and instead integrate their DNA with the hosts and replicate with it for several generations. *See also* TRANSFECTION.

bacterium *See* BACTERIA.

bar chart A simple visual means of comparing numerical quantities, each quantity being represented by a bar the length of which is proportional to its value. In particular, bar charts are often used in statistics to show the number of times a particular event occurs, such as the number of times a certain total value is obtained when two dice are thrown. Such statistical bar charts are often referred to as HISTOGRAMS.

barium meal A drink containing barium in the form of barium sulphate, given to a patient as a CONTRAST ENHANCING MEDIUM before taking a RADIOGRAPH of the stomach or intestines.

bark The outer protective layer on the stems and roots of WOODY PLANTS. Bark is composed mainly of dead cells and varies in thickness. It is added to from the inside of the plant, and so the outer layer often cracks and is shed to allow for expansion of the stem. A number of chemicals are deposited by trees in their bark, many of which are used in medicine.

basal metabolic rate (BMR) The METABOLIC RATE of a resting animal. It represents the energy needed to maintain vital body functions such as heartbeat, breathing, nervous activity and maintenance of body temperature. Different tissues have different metabolic rates and the BMR therefore depends on the tissue composition of an animal and also on its body weight (small animals have a higher metabolic rate per unit of body weight than larger animals).

base Any compound that will react with an ACID to form a SALT plus water. An example is sodium hydroxide, which reacts with hydrochloric acid to form sodium chloride and water. Most bases are the oxides or hydroxides of metals – ammonia is an important exception. A base that is soluble in water is termed an alkali. Bases can be considered to be any compound which accepts a proton. It is in this sense that the organic bases (PURINES and PYRIMIDINES) are termed bases.

basement membrane A thin layer of proteins secreted by animal EPITHELIAL cells, usually combined with COLLAGEN fibres.

base pair Two NUCLEOTIDE bases in DNA linked by HYDROGEN BONDS between the two strands of the double helix. The pairing is always between a PURINE and a PYRIMIDINE, so the bases ADENINE and THYMINE always link and CYTOSINE and GUANINE always link. Base-pairing also occurs during MESSENGER RNA TRANSCRIPTION and TRANSLATION. URACIL in RNA pairs with adenine.

basidiomycete A member of the phylum BASIDIOMYCOTA.

Basidiomycota A phylum of the kingdom FUNGI. Basidiomycetes are characterized by having septa (partitions) in their HYPHAE and often form large fruiting structures. Examples are the field mushroom (*Agaricus campestris*) and the ink-cap toadstool (*Coprinus*). Reproduction is usually sexual by means of BASIDIOSPORES formed outside basidia (*see* BASIDIUM). Many members of the Basidiomycota are a cause of world-wide crop spoilage.

basidiospore A SPORE produced by the BASIDIOMYCOTA phylum of FUNGI. It is formed by MEIOSIS within a BASIDIUM but is borne outside the basidium.

basidium (*pl. basidia*) A specialized cell of the BASIDIOMYCOTA phylum of FUNGI. It is concerned with the production of BASIDIOSPORES during sexual reproduction.

basophil A type of GRANULOCYTE (blood cell) with cytoplasmic granules that stain with basic dyes.

B cell, *B lymphocyte* A type of LYMPHOCYTE that is formed in BONE MARROW and settles in the SPLEEN or a LYMPH NODE without passing through the THYMUS as T CELLS do. B cells express a specific antibody on their surface and when this binds to a specific ANTIGEN it activates the cell to divide, producing a group of identical cells (*see* CLONE). These are then capable of producing the correct antibody needed to fight the invading antigen.

Following an infection, some B cells do not produce much antibody but instead become 'memory cells' and circulate in the body ready to be reactivated to produce antibody quickly if a second exposure to the initial antigen occurs. Other B cells mature into 'plasma cells', which are specific for one antigen and are the main secretors of antibody if this antigen is encountered again.

See also MAJOR HISTOCOMPATIBILITY COMPLEX.

becquerel (Bq) The SI UNIT of radioactive activity, one becquerel being an activity of one ionizing particle per second.

belt transect *See* TRANSECT.

Benedict's test A test for REDUCING SUGARS based on a modification of FEHLING'S TEST. Only one solution is used, containing sodium citrate, sodium carbonate and copper sulphate in water. This is added to the test sample and if a reducing sugar is present a rust-brown precipitate forms on boiling.

benign (*adj.*) Of a TUMOUR, not MALIGNANT; not usually threatening to life or health.

benthos Animals and plants that live at the bottom of a sea or lake. *Compare* PELAGIC.

beriberi A deficiency of vitamin B_1 (thiamine) that causes inflammation of the nerve endings. It particularly affects the ability to walk. *See also* VITAMIN B.

berry A fleshy FRUIT with an outer skin (exocarp), often brightly coloured to attract birds, a thick fleshy wall (mesocarp) and an inner membrane surrounding many seeds. Examples of berries include the tomato, grape and gooseberry.

beta particles Negatively charged electrons emitted from the nucleus of a radioactive atom during decay. Beta particles are more penetrating than ALPHA PARTICLES but less so than GAMMA RADIATION. They can travel several metres in air and are stopped by 2–3 mm of aluminium.

beta-pleated sheet, *β-sheet* A type of protein structure resulting from HYDROGEN BONDING. Hydrogen atoms from one side of the protein molecule link with the oxygen atoms of the side parallel to it, causing anti-parallel folding of the molecule. *See also* ALPHA HELIX.

β-sheet *See* BETA-PLEATED SHEET.

bicuspid valve, *mitral valve* A flap of tissue in the heart between the left ATRIUM and VENTRICLE. It prevents blood flowing back into the atrium from the ventricle.

biennial plant A plant that completes its life cycle in 2 years and then dies. In the first season, biennial plants store food in underground organs and in the second season they use this energy to produce flowers and seeds. Many root vegetables, including carrot, are biennial. *Compare* ANNUAL PLANT, PERENNIAL PLANT.

big bang theory A theory that the universe began as a small dense mass that exploded, throwing matter out in all directions, which formed the galaxies and stars. *See also* ORIGIN OF LIFE.

bilateral symmetry A type of structure of an organ or organism in which there is only one plane through which the organ or organism can be cut to produce two halves that are mirror images of each other. The plane usually separates the right and left halves. This feature is characteristic of most animals except the CNIDARIANS and ECHINODERMS, which have RADIAL SYMMETRY. Flowers exhibiting bilateral symmetry are termed ZYGOMORPHIC.

bile A brownish fluid made by HEPATOCYTES that assists in the breakdown and absorption of fats in the SMALL INTESTINE. Bile contains BILE SALTS, BILE PIGMENTS, CHOLESTEROL and LECITHIN. Bile is secreted from the liver into a BILE DUCT, which, in humans, then feeds into the GALL BLADDER. It is stored here until it enters the small intestine under hormone regulation. The bile contents that are to be excreted (excess pigments, cholesterol and lecithin) are eliminated with the faeces. In addition to breaking down fat, bile has an alkaline pH, which is important for the functioning of some digesting enzymes.

bile duct A duct from the liver to the DUODENUM. *See* BILE, GALL BLADDER.

bile pigment A breakdown product of old RED BLOOD CELLS. Bile pigments form part of BILE. They are eliminated with the FAECES, which they colour.

bile salt A constituent of BILE that assists in the breakdown and absorption of fats. Examples are sodium chloride and sodium hydrogencarbonate.

binary fission A form of ASEXUAL REPRODUCTION occurring in single-celled organisms, such as bacteria and AMOEBAE. The cell divides into two daughter cells of equal size, each containing half the nuclear material (*see* NUCLEUS) and CYTOPLASM. The division can be transverse or longitudinal.

binomial nomenclature The system used for naming living organisms. It uses a two-part Latin name for every organism. The system was devised by Carolus Linnaeus (1707–78). The first part of the name denotes the GENUS and is written with a capital first letter, and the second part is the SPECIES. Both names are written in italics. The name for humans using this system is *Homo sapiens. See also* CLASSIFICATION.

biochemical evolution theory *See* ORIGIN OF LIFE.

biochemistry The study of the chemistry of living organisms. This includes the study of the structure and functioning of enzymes, proteins and NUCLEIC ACIDS, and molecular biology. The study of biochemistry is fundamental to the understanding of life processes.

biodegradable (*adj.*) Describing any substance that can be broken down by natural biological processes, usually involving bacteria or fungi. Biodegradable substances, such as food and sewage, can therefore be recycled by the ECOSYSTEM. Many substances, such as plastics, are not biodegradable and present a problem of POLLUTION.

biogenesis The accepted theory that all living organisms arose from other living organisms similar to themselves. This theory totally overthrew the SPONTANEOUS GENERATION THEORY of the time.

biological clock *See* BIORHYTHM.

biological control The control of pests by biological means instead of using chemicals. This can be achieved in a number of ways, for example by introducing a pest's predator or a disease-inducing organism, or by breeding resistant crops.

biology The study of life. Biology includes the study of any organism or system to which the term 'living' can be applied. It includes ANATOMY, BIOCHEMISTRY, BIOPHYSICS, BOTANY, CYTOLOGY, ECOLOGY, GENETICS, MICROBIOLOGY, PHYSIOLOGY, VIROLOGY and ZOOLOGY.

bioluminescence The emission of visible light by living organisms, such as the firefly, glowworms, fungi, bacteria and various fish. The light is produced as a result of the oxidation of a substance called luciferin and can serve as a means of protection or as a mating signal or for species identification.

biomagnification The accumulation of harmful substances along the FOOD CHAIN. For example, the insecticide DDT was found in low levels in PLANKTON, but higher levels in fish feeding on the plankton and even higher levels in birds feeding on the fish. *See also* PESTICIDE.

biomass The total mass of living organisms in a given area. This includes the aqueous component; the term 'dry mass' is often used to define the non-aqueous component. Biomass can be specified for a particular species, a general category or global estimates of, for example plant biomass. Measurements of biomass are used to study variations in population sizes or interactions between organisms. *See also* PYRAMID OF BIOMASS.

biome A broad category in a BIOSPHERE, consisting of all the plants and animals in a region subjected to and affected by a common set of climatic conditions, for example, tropical rain forest, desert, tundra (very cold, no trees and low plant growth) or DECIDUOUS forests. A localized area within a biome is called a HABITAT.

biophysics The study of the physics of biological processes. The laws of physics are applied to a biological system, for example the study of crystals of MACROMOLECULES or the use of mechanics to determine the strength of bones.

biorhythm Rhythmic changes in the behavioural or activity patterns of plants and animals, produced by hormones relating to environmental cycles. These patterns can be related to seasonal changes, such as HIBERNATION in winter. In insects and some other invertebrates, a similar period of dormancy called DIAPAUSE can occur at any stage of development (most often the eggs or pupae) but usually only once in a lifetime. Seasonal flowering in plants and breeding in animals is controlled by changes in day length (*see* PHOTOPERIODISM). Bird MIGRATION is another example of a biorhythm.

Some biorhythm patterns are related to short cycles, for example the CIRCADIAN RHYTHM (24 hours) and CIRCALUNAR RHYTHM (28 days), and other patterns are annual. Biorhythms indicate the existence of a 'biological clock', which is probably a series of clocks running simultaneously but starting at different times. The mechanism of these 'clocks' is not understood. In humans, it is suggested that activity is controlled by intellectual, emotional and physical cycles with certain coincidence days being critical.

biosphere The narrow region over the Earth's surface that supports life, including the land (less than 20 km thick, 8 km above sea level) and the water and air around it. The biosphere is either aquatic or terrestrial and can be divided further into BIOMES and HABITATS.

biotechnology The use of living organisms in the large-scale industrial manufacture of food, drugs and other products. For example, yeast is used for FERMENTATION in the baking and brewing industries, and a range of bacteria and fungi are used in the dairy industry for converting milk into cheeses and yoghurts (*see* LACTIC ACID). Other food products that benefit from biotechnology are vinegar, sauerkraut and soy sauce. A number of enzymes made by microorganisms are needed at various stages in food manufacture and are sometimes genetically engineered. Large-scale production of ANTIBIOTICS rely on the activity of various fungi and bacteria.

One of the major biotechnological advances in the medical world has been the manufacture of the hormone INSULIN to treat DIABETES. Using RECOMBINANT DNA technology, insulin can be made on a large scale. Previously, insulin had to be extracted from animal tissues. Human GROWTH HORMONE genetically engineered from micro-organisms has helped in the treatment of human dwarfism, and bovine somatotrophin (BST) is a genetically engineered hormone given to cows to increase milk yield.

See also GENETIC ENGINEERING.

biotic (*adj.*) Referring to the living elements in an ECOSYSTEM. Biotic elements include activities of animal populations, such as competition for food, water, shelter and mates. Predator–prey relationships affect the distribution of species, as do the activities of humans. Many plants depend on animals to disperse them or on insects to pollinate them. These are all biotic elements.

biotin *See* VITAMIN B.

bird A member of the vertebrate class AVES.

bivalve A member of the class PELECYPODA.

bladder In bony fish, amphibians, mammals and some reptiles, a hollow, elastic-walled organ that stores urine before it is discharged. Urine enters the bladder via the URETER, one from each kidney, and in mammals leaves via the URETHRA. In other vertebrates (most reptiles, birds, amphibians and many fish), urine drains from the bladder into the CLOACA. *See also* URINARY SYSTEM.

blastocyst In mammals, the 64-cell stage of a fertilized OVUM. The blastocyst is a hollow cavity surrounded by a ball of cells. There is an inner cell mass that develops into the EMBRYO, and an outer layer of cells called the TROPHOBLAST that develops into the EXTRA-EMBRYONIC MEMBRANES. It is at the blastocyst stage that IMPLANTATION into the wall of the UTERUS occurs. *See also* BLASTULA, EMBRYONIC DEVELOPMENT.

blastula In animals other than mammals, an early stage of development of a fertilized OVUM, when the ovum changes from being a solid mass of cells to a hollow ball of cells (blastomers) with a fluid-filled cavity (blastocoel). It is similar to the BLASTOCYST in mammals. *See also* EMBRYONIC DEVELOPMENT.

blind spot The point in an eye where the OPTIC NERVE leaves the RETINA. At the blind spot, there are no RODS or CONES, so no visual images are transmitted.

blood The liquid circulating in the arteries, veins, capillaries or spaces of animals. It carries oxygen and nutrients to the cells and removes waste products such as carbon dioxide. It also has a role in the IMMUNE RESPONSE and in distributing heat in some animals.

An average human adult has 5.5 litres of blood. It consists of a watery colourless PLASMA that carries a number of blood cells serving different functions. The majority of the cells are RED BLOOD CELLS, which carry oxygen around the body. The main function of WHITE BLOOD CELLS is defence. PLATELETS are important in the BLOOD CLOTTING CASCADE. White blood cells can be divided into two further groups, GRANULO-CYTES and AGRANULOCYTES.

See also BLOOD GROUP SYSTEM, BLOOD VESSEL, CIRCULATORY SYSTEM, HEART, SERUM.

blood clotting cascade A series of events that occurs following injury, in order to prevent excessive bleeding. Small cell fragments called PLATELETS aggregate and release an enzyme called thrombokinase when they come into contact with a damaged blood vessel. The vessel wall also releases this enzyme. In the presence of calcium ions and VITAMIN K, thrombokinase converts the inactive enzyme PROTHROMBIN into the active THROMBIN. This in turn converts the soluble PLASMA PROTEIN fibrinogen into the insoluble FIBRIN, which forms a meshwork of protein fibres and blood cells over the wound. These dry to form a SCAB, which bacteria cannot enter and under which the wound can be repaired. In certain circumstances, clotting has to be prevented and so anticoagulants, such as HEPARIN, which inhibits conversion of prothrombin to thrombin, may be administered. *See also* HAEMOPHILIA.

blood group system The classification of blood into groups according to molecules or ANTIGENS carried on the membrane of RED BLOOD CELLS. There are at least 14 blood group systems in humans, the main ones being the ABO and rhesus systems. In the ABO system, there are two main antigens, A and B, which give rise to four possible blood groups: A (with antigen A only), B (with antigen B only), AB (with both antigens) and O (with neither antigen). In the rhesus system, most individuals carry the RHESUS FACTOR and are called rhesus positive (Rh⁺), while those who do not carry the factor are rhesus negative (Rh⁻). Each of the ABO groups may or may not contain the rhesus factor.

It is important in blood transfusion to match the blood group, since an incompatible donor will cause an antibody response leading to AGGLUTINATION of the recipient's red blood cells, possibly resulting in death. Blood group O is considered to be the universal donor because there are no antigens to cause an antibody response in the recipient. Group AB is the universal recipient because it has both A and B antigens and therefore the recipient can tolerate blood from group A, B or O. The rhesus system can cause a problem during pregnancy if a rhesus negative mother carries a rhesus positive baby, resulting in RHESUS DISEASE. In Britain, 46 per cent of the population are group O, 42 per cent are group A, 9 per cent are group B and 3 per cent are group AB.

blood vessel A specialized tube for carrying blood around the body of multicellular organisms. There are three main types of blood vessels: arteries carry blood away from the heart; veins carry blood to the heart; and capillaries link arteries and veins. All arteries, except the PULMONARY arteries, carry oxygenated blood, and all veins, except the pulmonary vein, carry deoxygenated blood. The main artery leaving the heart is called the AORTA and further away from the heart small arteries are called ARTERI-OLES. The main vein entering the heart is the VENA CAVA and further from the heart small veins are called VENULES. Capillaries are grouped together in capillary beds (or networks) that link arteries and veins and serve as the main site of exchange between blood and body fluids (*see* LYMPH).

blue-green algae The former name of CYANO-BACTERIA.

blue-green bacteria *See* CYANOBACTERIA.

B lymphocyte *See* B CELL.

Bohr effect The effect of carbon dioxide concentration on the association of oxygen with HAEMOGLOBIN. Oxygen is more readily released from haemoglobin in the presence of carbon dioxide (in respiring tissues) and taken up more readily where the carbon dioxide concentration is low (in the lungs). The majority of carbon dioxide in the body is found combined with water as carbonic acid (*see* CARBONIC ANHY-DRASE). This dissociates to release hydrogen ions and it is the presence of these that encourage the dissociation of oxygen from haemoglobin.

The oxygen diffuses into the tissues and haemoglobin combines with the excess hydrogen ions to form HAEMOGLOBINIC ACID.

bolus *See* DIGESTIVE SYSTEM.

bone A hard CONNECTIVE TISSUE forming the skeleton of most vertebrates. Bone consists of COLLAGEN fibres, providing strength, impregnated with mineral ions (calcium phosphate and calcium carbonate), which provide the hardness.

Blood, LYMPH VESSELS and nerves are carried through the bone by small tubes called Haversian canals. The other constituents of bone are arranged in concentric circles or lamellae around these canals. OSTEOCYTES are the bone cells. These are found in spaces in the lamellae known as lacunae, which are linked by fine channels called canaliculi. Inside the long bones of limbs there is a hollow shaft, called the diaphysis, that contains BONE MARROW. Each end of a bone has an expanded head called the EPIPHYSIS. Bone can be a fine spongy network designed to transmit the body's weight, as at the ends of long bones, or compact to resist bending, as in the shaft of long bones.

Bone can develop from CARTILAGE by the process of OSSIFICATION. Bone can also form directly from other connective tissues, and is then plate-like in shape, for example, the skull. Bone shape can be remodelled during growth by OSTEOCLASTS.

bone marrow In vertebrates, a soft material in the centre of some bones that produces blood cells. Yellow marrow is found at the centre of the long bones and consists mainly of fat cells. It can make some WHITE BLOOD CELLS. Red marrow produces RED BLOOD CELLS and is located at the ends of the bone.

boron (B) A non-metallic element, required by plants, but not animals. It is required for the uptake of calcium by plant roots, to aid the germination of pollen grains and for growth of young shoots. Deficiency causes abnormal growth and death of young shoots.

botany The branch of biology that deals with the study of plants. It includes many areas, such as the study of plant structure, function, geographical distribution and classification.

bovine spongiform encephalopathy (BSE) A disease of cows (popularly called 'mad cow disease'), similar to SCRAPIE in sheep and CREUTZFELDT–JAKOB DISEASE (CJD) in humans. All three diseases are characterized by the progressive degeneration of the CENTRAL NERVOUS SYSTEM and the spongy appearance of the brain upon examination after death. The spongy appearance is due to the presence of numerous holes in the tissue. BSE is thought to be caused not by a bacterium or a virus but by a PRION protein or by an agent that can switch on a latent gene to cause production of the prion protein, which infects the brain and SPINAL CORD of its victims.

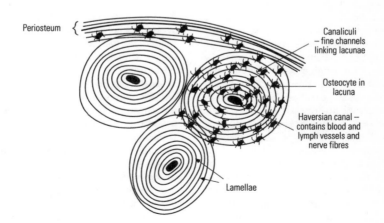

Periosteum

Canaliculi
– fine channels
linking lacunae

Osteocyte in
lacuna

Haversian canal –
contains blood and
lymph vessels and
nerve fibres

Lamellae

Compact bone.

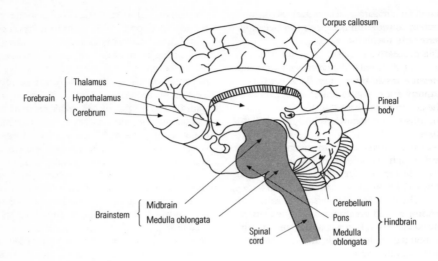

Cross-section through the human brain.

bovine somatotrophin A genetically engineered hormone given to cows to increase milk yield. *See also* BIOTECHNOLOGY.

Bowman's capsule A cup-shaped structure in the KIDNEY that surrounds the GLOMERULUS and forms part of the NEPHRON tubule. Certain smaller substances (of RELATIVE MOLECULAR MASS less than 68,000), such as GLUCOSE, vitamins, amino acids, hormones, UREA, URIC ACID and water, can pass through the wall of the glomerulus and into the Bowman's capsule as a filtrate. Blood cells and larger PLASMA PROTEINS remain in the blood. The cut-off size can be increased by hormonal or nervous signals increasing the blood pressure within the capsule. The filtrate then moves through the tubule, eventually forming urine. There are over a million Bowman's capsules in the outer region of the kidney. *See also* URINARY SYSTEM.

bract A small leaf in whose AXIL (between the bract and stem) a flower or branch develops.

brain In higher animals, a large mass of interconnected NEURONES, forming part of the CENTRAL NERVOUS SYSTEM and controlling the actions of the whole body. It is surrounded by the MENINGES and is protected further by the bones of the skull. Cavities called VENTRICLES are filled with CEREBROSPINAL FLUID, which supplies the brain with nutrients and respiratory gases and removes waste by exchanging these materials with the blood. There are three main regions of the brain: the FOREBRAIN, the HINDBRAIN and the MIDBRAIN. *See also* BRAINSTEM, CEREBELLUM, CEREBRUM, GREY MATTER, HYPOTHALAMUS, MEDULLA OBLONGATA, PINEAL GLAND, SPINAL CORD, THALAMUS, WHITE MATTER.

brainstem The region where the SPINAL CORD merges with the under surface of the brain. It consists of the MEDULLA OBLONGATA and the MIDBRAIN. It controls the body's vital functions, such as breathing, heart rate and blood pressure, and maintains wakefulness by a system of nerve cells which is spread diffusely throughout the brainstem. In many countries, death of the brainstem is considered synonymous with death of the individual, although the heartbeat can be maintained by life-support equipment.

Branchiopoda A subclass of CRUSTACEA that includes water fleas such as Daphnia.

breathing, *external respiration* In terrestrial animals, the process by which air is taken into the LUNGS to provide oxygen and then expelled to release gaseous wastes. It is sometimes called external respiration, to distinguish it from true RESPIRATION, which is internal and is the use of oxygen by cells to make energy.

Air passes through a series of tubes and organs (*see* RESPIRATORY SYSTEM) to reach the lungs, where oxygen diffuses from the air across the thin, moist EPITHELIAL layer of the alveoli (*see* ALVEOLUS), and from here across blood capillaries to combine with HAEMOGLOBIN in the blood for circulation all over the body. Likewise, carbon dioxide diffuses from the blood back across the alveoli for expulsion from the lungs. This process is also called gas exchange.

For air to enter the lungs (inspiration) the pressure inside them must be lower than the atmospheric pressure. This is brought about by the action of the INTERCOSTAL MUSCLES causing the ribs to move upwards and outwards. The DIAPHRAGM also moves and the volume of the THORAX increases so that the lungs fills with air and expands. Breathing out (expiration) occurs when the muscles relax, causing the ribs to move down and so forcing the air out of the lungs.

The rate of breathing is under involuntary control by the breathing centre in the HINDBRAIN and by CHEMORECEPTORS that monitor carbon dioxide levels in the blood. If these levels increase, for example during exercise, then NERVE IMPULSES stimulate the breathing centre in the brain to increase the breathing rate. There is also some voluntary control over breathing.

Human lungs have a total volume of about 5,000 cm^2 of air, but following a forced exhalation 3,500 cm^2 is the most that can be exchanged. This is called the vital capacity. A residual volume of 1,500 cm^2 therefore remains in the lungs, but this is continually circulated as it mixes with fresh air during normal breathing. The volume of air exchanged during normal breathing is called the tidal capacity, and is about 450 cm^2.

breeding The process of bearing offspring. *See also* ARTIFICIAL SELECTION, SEXUAL REPRODUCTION.

bronchiole In vertebrates, one of many small tubes found in the lung that leads off from the larger BRONCHUS. It branches further and finally terminates in the ALVEOLI.

bronchus (*pl. bronchi*) In vertebrates, one of two tubes branching off from the TRACHEA, which carry the air into each of the LUNGS. To prevent their collapse during breathing, the bronchi possess cartilaginous rings for extra rigidity. Glands in the bronchi secrete a sticky liquid called MUCUS to collect dust and other unwanted particles. These are then pushed out towards the mouth for swallowing, aided by CILIA on the walls of the bronchi. The bronchi divide further into small tubes called BRONCHIOLES and then into alveoli (*see* ALVEOLUS). *See also* RESPIRATORY SYSTEM.

brush border The surface of many EUKARYOTIC cells, particularly EPITHELIUM, as it appears under an ELECTRON MICROSCOPE due to the presence of millions of minute projections called MICROVILLI. These increase the surface area of, for example, the SMALL INTESTINE for absorption of food products.

Bryophyta A phylum of the plant kingdom comprising three classes: HEPATICAE (liverworts), MUSCI (mosses) and ANTHOCEROTAE (hornworts). Bryophytes are small terrestrial plants growing in moist habitats, possessing no roots or vascular (conducting) system. Their life cycle shows ALTERNATION OF GENERATIONS. Plants of this phylum consist of simple stems and leaves, or a THALLUS, anchored by RHIZOIDS.

bryophyte A member of the phylum BRYOPHYTA.

BSE *See* BOVINE SPONGIFORM ENCEPHALOPATHY.

buccal cavity The mouth cavity. *See* MOUTH.

bud In plants, a small, undeveloped shoot containing immature leaves or petals. A bud may develop at the tip of a stem or branch, or in the AXILS of a leaf.

budding A method of ASEXUAL REPRODUCTION where a small outgrowth develops from the parent cell and eventually separates, sometimes after becoming more differentiated and forming buds of its own. Yeast and *Hydra* are examples of organisms that can reproduce by budding.

buffer 1. A solution designed to maintain a fairly uniform pH level even when small amounts of acid are added to it This ability is essential in biological systems where a sudden change in pH could adversely affect the functioning of enzymes. Buffer solutions are made in the laboratory by mixing a weak acid or base with a salt of the same acid or base. Natural buffers exist in many biological systems.

bulb A small, underground stem with fleshy leaves, which are a food reserve, and roots growing from its base. Examples include the daffodil and onion. New plants arise from buds (small, undeveloped shoots) that grow between the leaves and then develop into new bulbs. *See also* VEGETATIVE REPRODUCTION.

bulimia nervosa An eating disorder, usually due to psychological problems or stress, characterized by episodes of overeating (binging) followed by self-induced vomiting, fasting, the use of laxatives or excessive exercise.

bundle of His The collective name for the PURKINJE FIBRES.

bundle sheath cell Any one of a layer of cells surrounding the VASCULAR BUNDLE in plants.

C

C₃ plant Any plant that produces a three-carbon intermediate (glycerate 3-phosphate) during the fixation of carbon dioxide in the light-independent stage of PHOTOSYNTHESIS. Most temperate plants and about 85 per cent of all plant species are C₃ plants. *Compare* C₄ PLANT. *See also* CALVIN CYCLE.

C₄ plant Any plant that can produce a four-carbon intermediate (oxaloacetic acid) during the fixation of carbon dioxide in the light-independent stage of PHOTOSYNTHESIS (*see* CALVIN CYCLE). Tropical plants, such as sugar cane, and many cereals are C₄ plants. In these plants, the C₄ system operates in addition to the C₃ PLANT system.

The advantage of the C₄ system is that carbon dioxide can be trapped at lower concentrations than that needed by C₃ plants. This is particularly useful for tropical plants, where carbon dioxide may be in short supply, and the leaves of C₄ plants are modified to suit this method of carbon dioxide fixation. The C₄ system is also more efficient at higher temperatures. PHOTORESPIRATION is avoided in the C₄ system.

caecum In humans, a slight expansion between the small and large intestines, leading to the APPENDIX, that has no function. However, in HERBIVORES it is used (as is the appendix) to assist in the digestion of CELLULOSE in plant cell walls. This requires the presence of micro-organisms with the enzyme cellulase that live in the caecum.

calcitonin A hormone produced by the THYROID GLAND that reduces the levels of calcium ions in the blood. *See* PARATHORMONE, PARATHYROID GLAND.

calcium (Ca) A soft, white metallic element. It is an essential element for living organisms, required for normal growth and development. In animals, it is a major constituent of TEETH and BONES and is required for MUSCULAR CONTRACTION and in the BLOOD CLOTTING CASCADE. Calcium deficiency causes RICKETS and a delay in blood clotting. In plants, calcium is a constituent (as calcium pectate) of the MIDDLE LAMELLA between CELL WALLS, and therefore needed for their proper development. Calcium deficiency in plants, causes death of growing points and thus stunted growth.

calibration 1. The process of using a measuring device to measure known quantities in order to check or adjust the instrument concerned.

2. A marking on the scale of an instrument representing a specified numerical value for the quantity being measured.

calorie (cal) A unit of quantity of heat. One calorie is the amount of heat needed to raise the temperature of one gram of water by 1°C. The calorie has been largely replaced by the JOULE; it is roughly equivalent to 4.2 J.

Calorie (kcal) A unit sometimes used to specify the energy value of foods. One Calorie (capital c) is equivalent to 1,000 CALORIES (small c).

calorific value The energy content of a fuel or food. It is the amount of heat generated by completely burning a given mass of fuel (which can be food) in a piece of apparatus called a bomb CALORIMETER. Calorific value is measured in JOULES per kilogram.

calorimeter A container for performing experiments related to heat transfer and temperature changes, such as the measurement of CALORIFIC VALUE.

Calvin cycle Another name for the light-independent reaction of PHOTOSYNTHESIS, after Melvin Calvin (1911–) who established the details of the reactions occurring. He did this by exposing plants to radioactively labelled carbon dioxide (carbon–14 dioxide), allowing them to photosynthesize and examining the products.

Carbon dioxide from the air diffuses into a leaf through the stomata (*see* STOMA) and then dissolves and diffuses through the CELL MEMBRANE and the CHLOROPLAST membrane into the STROMA of the chloroplast. The light-independent stage of photosynthesis occurs in the stroma. In most plants (C₃ PLANTS), carbon dioxide combines with a five-carbon compound called ribulose bisphosphate to form a 6-carbon intermediate that is unstable and breaks down into two molecules of the three-carbon compound glycerate 3-phosphate (GP). This is then converted, in the presence of ATP and reduced NADP, into TRIOSE phosphate, which combines to form HEXOSE sugars. These polymerize to yield STARCH for the plant to store. Some of the triose phosphate is used to regenerate ribulose bisphosphate.

In C$_4$ PLANTS, carbon dioxide combines with the substance phosphoenolpyruvate (PEP) instead of ribulose bisphosphate and produces the four-carbon oxaloacetic acid instead of GP.

calyx The collective term for the SEPALS of a flower.

cambium In plants, a layer of actively dividing cells in stems and roots that gives rise to the SECONDARY GROWTH occurring in woody PERENNIALS. It is a lateral MERISTEM responsible for increased girth. Vascular cambium is a layer of cells separating the XYLEM and PHLOEM within the VASCULAR BUNDLE, and gives rise to new or secondary xylem and phloem. In trees, the secondary xylem laid down by the vascular cambium forms a new layer of wood annually on the outside of the old wood, and these form the annual rings seen when a tree is felled. Another type of cambium is CORK CAMBIUM, which gives rise to secondary layers of bark and cork.

cAMP See CYCLIC AMP.

canaliculus In zoology, a small channel, such as in bone.

cancer A group of diseases in which certain cells do not show the usual growth restraints and so continue to divide, resulting in unlimited growth of the tissue. Such growth may lead to a swelling or lump called a TUMOUR. If a tumour is cancerous it is called malignant, but tumours can also be benign (non-cancerous). Cancer is the biggest killer in the Western world: there are more than 100 types, named according to the tissue of their origin, for example CARCINOMAS (of EPITHELIAL origin), SARCOMAS (of CONNECTIVE TISSUE) and MYELOMAS (of BONE MARROW).

The causes of cancer are not fully known and are many and complex; there is not usually one direct cause. Some factors (CARCINOGENS) are known to be linked with an increased incidence of cancer; tumour-inducing ONCOGENES have also been identified

in DNA. Some cancers are now treatable and potentially curable. Treatment has historically been by surgery, CHEMOTHERAPY or RADIOTHERAPY, but more recently MONOCLONAL ANTIBODIES attached to CYTOTOXIC drugs to target and kill specific cells have been used with some success. Prevention and early detection still remain the best options.

See also TRANSFORMATION.

canine tooth A sharp, pointed TOOTH in mammals especially developed for catching prey, killing and tearing meat. There are four canine teeth, towards the front of the mouth, one on each side of the upper and lower jaws between the INCISORS and the PREMOLARS. They are highly developed in CARNIVORES, such as dogs and cats, but absent from HERBIVORES, such as giraffes, many rodents and rabbits. In humans they are much reduced in size. In some animals, such as the wild boar, the canine teeth are enlarged as tusks.

cap See DIAPHRAGM.

capacitation The final stage of maturation of a SPERM cell. It takes place in the female tract.

CAPD See CONTINUOUS AMBULATORY PERITONEAL DIALYSIS.

capillary The smallest blood vessel in animals (8–20 μm in diameter), with thin walls only one cell thick containing no muscle or elastic fibres. Capillaries are therefore permeable to nutrients, dissolved gases and waste products. They are grouped together in capillary beds (or networks) that link arteries and veins and serve as the main site of exchange between blood and body fluids (see LYMPH). Blood pressure in the arteries is reduced by capillaries and blood is deoxygenated in capillaries before entering the veins.

capillary effect The effect that causes most liquids, including water, to rise up a glass tube with a narrow bore (capillary tube). The smaller the diameter of the tube, the higher the liquid rises. This is due to the adhesive forces (see ADHESION) between the water and the glass being stronger than the cohesive forces (see COHESION) between the water molecules. It is the capillary effect that causes water to form a curved surface, called a MENISCUS, in a tube.

In plants, XYLEM vessels have a very narrow diameter and therefore considerable capillary forces, which contribute to the movement of water from their roots up into their leaves. This is called the TRANSPIRATION STREAM.

See also OSMOSIS, TRANSLOCATION.

carbaminohaemoglobin The product formed when carbon dioxide combines with HAEMO-GLOBIN. A small proportion, about 10 per cent, of carbon dioxide is carried from the tissues to the respiratory system in this form. The remainder is carried in the form of hydrogen carbonate. *See* CARBONIC ANHYDRASE, CHLORIDE SHIFT.

carbohydrate One of a large group of organic compounds with the general formula $C_x(H_2O)_y$. They are the main energy-providing components of the human diet. There are three main groups of carbohydrates: MONO-SACCHARIDES, DISACCHARIDES and POLYSACCHA-RIDES. Monosaccharides, such as GLUCOSE and FRUCTOSE, are single sugars with the general formula $(CH_2O)_n$ that cannot be split into smaller carbohydrate units. Disaccharides, such as SUCROSE, MALTOSE and LACTOSE, are double sugars, where two monosaccharides are combined, that can be split into their single sugar components. Polysaccharides, such as STARCH and GLYCOGEN, consist of variable numbers of monosaccharides joined together in chains that can be branched or not and can fold for easy storage. They can be broken down into their constituent disaccharides or monosaccharides for use. Some polysaccharides are structural, for example, CELLULOSE and CHITIN. Others are food reserves, for example, starch and glycogen.

Most carbohydrates can form ISOMERS (which have the same chemical formula but a different structure) that give them a different functional property. Enzymes usually only react with one form. Naturally occurring carbohydrates have a D(+) form (*see* DEXTROROTA-TORY).

There are two tests that can be used to identify sugar in a solution: BENEDICT'S TEST and FEHLING'S TEST. Both rely on the ability of monosaccharides and some disaccharides to reduce copper(II) sulphate to copper oxide, causing a colour change on boiling. These sugars can also be classified as reducing or non-reducing. The test for starch is the addition of iodine in potassium iodide, which integrates into the starch polymer causing a colour change from yellow-orange to blue-black.

carbon cycle The constant circulation of carbon between organic and inorganic sources in nature. This recycling maintains the balance between carbon dioxide in the atmosphere and carbon in organisms, and is vital for all forms of life.

Carbon dioxide in the atmosphere provides a major source of carbon that is used by photosynthetic plants and organisms to make CARBOHYDRATES. These photosynthetic plants and organisms then release oxygen back to the atmosphere. HETEROTROPHS (animals and fungi) eat the plants and other photosynthetic organisms, or other heterotrophs (*see* FOOD CHAIN), and through their RESPIRATION return carbon back to the atmosphere as carbon dioxide. Carbon dioxide is also released back to water by some organisms, for example algae, where an equilibrium with atmospheric carbon dioxide is maintained.

Part of any natural cycle includes not only the biological (BIOTIC, or living) component but also a geological (ABIOTIC, or non-living) component. These geological components include rocks and other deposits in the oceans and atmosphere, for example in the form of coal and oil under the ocean, peat in wetlands and limestone rocks, and they provide the largest reservoir of carbon.

The natural processes of photosynthesis and respiration balance each other out, but in recent years human intervention has disturbed this balance. The burning of FOSSIL FUELS and the destruction of large areas of tropical forest has caused the atmospheric levels of carbon dioxide to rise, contributing to the GREENHOUSE EFFECT.

carbon dioxide (CO_2) A colourless, odourless gas present in small quantities in the atmosphere. It is used by plants during PHOTOSYN-THESIS and produced during RESPIRATION. The level of carbon dioxide in the atmosphere has increased over recent years due to the burning of FOSSIL FUELS and the destruction of vast areas of forest. This is thought to be the greatest cause of GLOBAL WARMING.

carbonic anhydrase A zinc-containing enzyme that catalyses the reaction between carbon dioxide and water to form carbonic acid (H_2CO_3). Carbonic acid subsequently dissociates into hydrogen (H^+) and hydrogen carbonate (HCO_3^-) ions. It is in the form of hydrogen carbonate that the majority (85 per cent) of carbon dioxide produced by body tissues is transported to the lungs to be removed as waste. Carbonic anhydrase is found in RED BLOOD CELLS, where this reaction takes place. It is also found in the kidney, where it controls the pH of urine. *See also* CHLORIDE SHIFT.

carbon monoxide (CO) A colourless, odourless gas. It is toxic as it replaces oxygen in HAEMO-GLOBIN in the blood, and so prevents oxygen being transported to the body tissues. *See* CAR-BOXYHAEMOGLOBIN.

carboxyhaemoglobin The product formed when HAEMOGLOBIN combines irreversibly with carbon monoxide. Carbon monoxide competes with oxygen for haemoglobin, but the affinity of haemoglobin for carbon monoxide is far greater than that for oxygen. Carboxy-haemoglobin is therefore very stable. In the presence of carbon monoxide, there is less haemoglobin available for the transport of oxygen. This accounts for the toxic effects of carbon monoxide on the respiratory system.

carboxylic acid Any one of a group of organic compounds that contain the carboxyl group (–COOH) attached to another group, which may be anything from a hydrogen atom to a 24-carbon chain. Ethanoic (acetic) acid is an example of a lower carboxylic acid. Many long chain carboxylic acids occur naturally in animal and vegetable fats and are also known as FATTY ACIDS.

carcinogen Any factor known to be linked with an increased incidence of cancer. Examples are chemicals (including products of smoking and asbestos dust), IONIZING RADIATION, viral infections and dietary or genetic factors. *See also* ONCOGENE, TRANSFORMATION.

carcinoma A MALIGNANT tumour of EPITHELIAL cells, for example of the skin or glandular tissue. *See also* CANCER.

cardiac muscle A specialized network of muscle fibres, found only in the HEART, that is capable of rhythmic contraction and relaxation over a long period. The muscle is said to be MYOGENIC (the contraction is stimulated within the heart itself) and is involuntary. The fibres contain separate cells, each with one nucleus and irregular thickening of their surrounding membrane (SARCOLEMMA), that form intercalated discs. A distinctive feature of this muscle is the branching and rejoining of fibres. *See also* PURKINJE FIBRES.

cardiac sphincter A ring of muscles at the entrance of the stomach that relax and contract to allow food to enter the stomach.

carnivore Any animal, for example a cat, dog, tiger or shark, that eats meat (mainly muscle) from other animals as the main part of its diet. Meat is more easily digested than the CELLU-LOSE in a herbivorous (*see* HERBIVORE) diet, and is richer in nutrients, but it is more difficult to obtain. The main physical adaptation for meat eating is in the jaw, which opens wider, and the teeth, which are very sharp. *See also* OMNIVORE.

carotene A natural CAROTENOID pigment that is responsible for the orange, yellow and red colour of carrots, tomatoes and oranges.

carotenoid Any one of a number of coloured pigments found in many living organisms. Carotenoid pigments are lipids and can be yellow, orange, red or brown. They are frequently found in the CHLOROPLASTS of plants, where they can act as accessory pigments in PHOTO-SYNTHESIS. In some ALGAE, the carotenoid pigments are the main light-absorbing pigments used instead of CHLOROPHYLL. Carotenoids are also found in fruits, roots and petals, giving them their colour, and they provide the autumn colours in leaves. The group includes CAROTENE and the XANTHOPHYLLS.

carotid artery Either one of two main arteries at each side of the neck that supplies the head with blood from the heart. *See also* CIRCU-LATORY SYSTEM.

carpel The essential female reproductive structure in a flowering plant (ANGIOSPERM). In the centre of the carpel is a slender stalk called the STYLE, which supports the STIGMA at the top. The stigma is specialized for receiving pollen; it often has hairs and produces a sticky secretion to trap pollen grains. At the base of the carpel is a hollow called the ovary, with thick walls protecting one or more OVULES, each enclosing the egg nucleus (the female gamete). The ovule is attached to the ovary wall by the funicle and the ovary is attached to the plant by the placenta. The ovary develops into the fruit wall after fertilization, and the ovule into the seed. There may be one or more carpels, which can be fused, as in the tulip, or not fused, as in the buttercup, and collectively the carpels are called the gynoecium. *See also* DOUBLE FERTILIZATION, POLLINATION.

carrying capacity In ecology, the maximum number of individuals of a given species that can be supported by a particular ecological area. When resources, such as food, are exhausted, the population will be reduced by death, emigration or reproductive failure.

cartilage A hard but flexible CONNECTIVE TISSUE that forms the embryonic skeleton and is replaced by bone except in areas of wear and tear, such as the intervertebral discs between the backbones and at bone endings (articular cartilage). In mammals, cartilage is also found in the nose, ear and larynx, and in cartilaginous fish such as sharks it forms the skeleton. Cells called chondrocytes secrete the POLYSACCHARIDE-containing matrix of chondrin. COLLAGEN fibres are embedded within this, giving cartilage its strength. In adults, cartilage contains no blood vessels.

cartilaginous fish See CHONDRICHTHYES.

casein The main protein constituent of milk. Caseinogen is the soluble form, which is precipitated by RENIN to casein. This is particularly important in young animals. Casein is also a major component of cheese.

caseinogen See CASEIN.

Casparian strip A band of SUBERIN found in the walls of endodermal cells (see ENDODERMIS) in plant ROOTS. Since suberin is impermeable to water, it forces water moving up through the plant by the APOPLAST PATHWAY (that is, through the cell walls) to be diverted to the cell cytoplasm. It is thought that this allows salts to be actively secreted into the vascular tissue (XYLEM and PHLOEM) from the endodermal cells. This makes the water potential in the xylem more negative, and in turn draws water along a water potential gradient from the endodermis to the vascular tissue.

catabolism The breaking down of living tissue into energy and waste products. See METABOLISM.

catalase An enzyme that catalyses (see CATALYST) the breakdown of hydrogen peroxide to water and oxygen. It is found in PEROXISOMES.

catalyst Any substance that changes the rate of a chemical reaction without itself being permanently altered chemically by that reaction. The term catalyst usually applies to substances that increase the rate of a reaction (positive catalysts). Those that slow a reaction down are termed negative catalysts.

Many biological reactions rely on complex catalysts, often specific to a particular reaction. These are called ENZYMES.

catecholamine An amino acid derivative that can function as a NEUROTRANSMITTER or a HORMONE, for example, ADRENALINE and NORADRENALINE.

cation A positively charged ION.

cation pump See SODIUM PUMP.

CAT scan, *CT scan, computerized axial tomography* An advanced method of X-RAY imaging. CAT scanning is used in medicine to locate problem areas without the need for exploratory surgery. A CAT scanner is a ring-shaped machine that rotates around the horizontal patient. The CAT scanner passes a narrow fan of X-rays (less than the dose used in routine X-rays) through successive slices of the body area under investigation. These images are converted, by the scanner's own computer, into cross-sectional displays on a viewing screen. By taking views of a series of different angles, a three-dimensional picture of an organ or tissue can be built up and any abnormalities detected. See also TOMOGRAPHY.

caudal (*adj.*) Relating to the tail. The caudal fin is the tail fin of fish and caudal vertebrae are the bones of the tail.

caudal fin The tail FIN of a FISH. It is the main propulsive organ.

caveolae See POTOCYTOSIS.

cDNA See COPY DNA.

cell The smallest mass of self-contained living matter of an animal or plant. Cells normally range in size from 10 to 30 μm. Some organisms consist of only one cell, for example bacteria, PROTOZOA and many other microorganisms, while others, for example humans, are made of billions of cells.

A cell consists of PROTOPLASM, which is the CYTOPLASM and nucleus, surrounded by a CELL MEMBRANE. In plant cells there is also a rigid CELL WALL made of CELLULOSE that gives the cell, and therefore the plant, support. The cell membrane regulates the entry of substances into and out of the cell.

In EUKARYOTES, the DNA is organized into chromosomes and contained within a clearly defined nucleus bounded by a membrane within the cell. Eukaryote cells also contain specialized structures called ORGANELLES (including the nucleus), such as ENDOPLASMIC RETICULUM, GOLGI APPARATUS, LYSOSOMES, MITOCHONDRIA and RIBOSOMES.

In PROKARYOTES, the DNA forms a coiled structure called a NUCLEOID and there is no

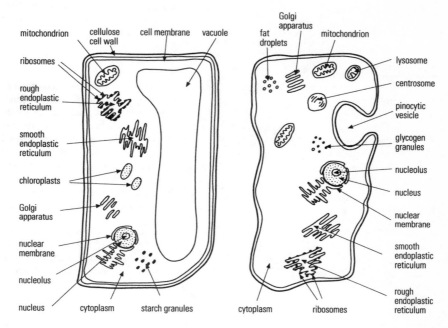

Generalized structure of the eukaryotic cell.
a) Plant cell

b) Animal cell

nucleus and no major organelles. *See also* CELL CYCLE, CELL DIVISION, EXTRACELLULAR MATRIX, INTERPHASE, MEIOSIS, MITOSIS, VACUOLE.

cell adhesion Mechanisms by which cells are able to link with one another or with the EXTRACELLULAR MATRIX. Focal contacts are formed between cells, consisting of a complex arrangement of LIPIDS and proteins linking the extracellular matrix with intracellular ACTIN. The family of INTEGRINS are important in these contacts and other accessory proteins, such as α-actinin, vinculin and talin, are involved. Cell adhesions are fundamentally important in a variety of cellular phenomena such as cell pro-liferation, migration and differentiation. Such interactions are known to be disrupted in many types of tumours.

cell cycle In EUKARYOTES, the regular pattern of events occurring in a dividing CELL. There are two main parts to the cycle: INTERPHASE and MITOSIS (or sometimes MEIOSIS). Interphase can further be divided into three phases: the G_1, S and G_2 phases. During the G_1 phase, biosyn-thesis and growth occur. There is a point in

this phase beyond which the cell is committed to completing the cell cycle. In the S phase, the DNA content of the cell double and the CHRO-MOSOMES replicate. The G_2 phase is one of further growth and preparation for cell and nuclear division in the M phase (mitosis). In rapidly dividing cells, the cell cycle takes about 24 hours and interphase is 90 per cent of this.

The cell cycle can be continuous for some cells (such as single-celled organisms) or it can cease for other cells after a period of time or at a stage of maturity. Once a cell begins on a pathway of differentiation, it has has left the cell cycle. Much research has been focused on understanding the details of the cell cycle in the hope that it will lead to a greater under-standing of cancer, in which the normal growth pattern is changed.

cell division The processes that result in the division of a living CELL into two daughter cells. Cell division is necessary for growth, repair and reproduction of an organism. In EUKARYOTES it is always preceded by division of the nucleus (by MITOSIS or MEIOSIS), followed

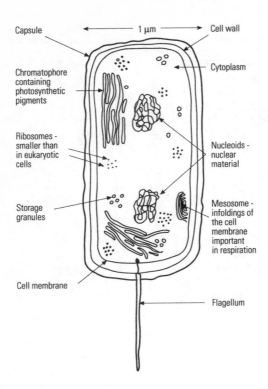

Capsule — 1 μm — Cell wall

Chromatophore containing photosynthetic pigments

Cytoplasm

Ribosomes - smaller than in eukaryotic cells

Nucleoids - nuclear material

Storage granules

Mesosome - infoldings of the cell membrane important in respiration

Cell membrane

Flagellum

Structure of the prokaryotic cell – a generalized bacterial cell.

by duplication of the cell ORGANELLES and the actual splitting of the CELL MEMBRANE. Plant cells divide by forming a cell plate, whereas animal cells divide by constriction. Nuclear division can occur without cell division, giving rise to multinucleate cells. In PROKARYOTES, there is no nucleus and the one event of cell division separates the DNA as well as the cell membrane.

cell-mediated immunity (CMI) The response of an organism to invasion by a foreign object. The cells involved in CMI are PHAGOCYTES, such as MACROPHAGES, or T CELLS, such as CYTOTOXIC T cells. Antibodies play a more minor role in CMI. Foreign ANTIGENS are only recognized by T cells in association with cell antigens of the MAJOR HISTOCOMPATIBILITY COMPLEX. CMI contrasts with HUMORAL IMMUNITY, in which antibodies provide the main line of defence. The distinction between the two is not clear-cut, however, as cells are

needed in the initiation of an antibody response and cell-mediated responses usually also involve the production of antibodies. *See also* IMMUNE RESPONSE, IMMUNITY.

cell membrane, *plasma membrane* A thin layer (10–80 μm thick) of PROTEIN and PHOSPHOLIPID molecules that surrounds CELLS and selectively controls substances passing through it. It is a SEMIPERMEABLE MEMBRANE: small molecules, such as water, GLUCOSE and AMINO ACIDS, can pass through the membrane, whereas large molecules, such as STARCH, cannot. The currently accepted structure of the cell membrane is called the FLUID MOSAIC MODEL.

cell plate A thin partition that forms across the centre of plant cells to effect the division of the CYTOPLASM. *See* CELL DIVISION.

cellulase *See* CELLULOSE.

cellulose A CARBOHYDRATE made of long chains of GLUCOSE that gives strength to plant cell walls. It has many chains running parallel to each other and cross-linked. This provides the strength needed to support plant cell walls. The cellulose in plant cell walls is important in the diet of many animals, although no vertebrate possesses the enzyme cellulase needed to break down cellulose. HERBIVORES (plant-eating animals) digest cellulose (the major part of their diet) by having specialized bacteria in their gut that can make cellulase and therefore digest cellulose. Humans cannot digest cellulose because they do not have the necessary gut micro-organisms or grinding teeth, but it provides a vital source of ROUGHAGE. The strength of cellulose has been utilized by humans in the manufacture of cotton, paper, plastic and cellophane.

cell wall The rigid outer wall of plant, fungal, algal and bacterial cells (but not animal cells). The cell wall provides support and protection for the cell. In most plants and algae it is made from CELLULOSE and may be secondarily thickened by LIGNIN. The secondary thickening in plant cells occurs on the inside of the primary cell wall but some areas remain unthickened. These are called PITS and provide the point at which cell-to-cell contact can be made (since pits of adjacent cells coincide). PLASMODESMATA pass through these pits and water and dissolved minerals are transported between cells in this way. In fungi, the cell wall is made from CHITIN and in bacteria it is made from PEPTIDO-GLYCAN. Some cell walls undergo further modifications, such as the development of a waxy cuticle or the deposition of SUBERIN in cork cells. When a plant cell is turgid (swollen with water) a HYDROSTATIC PRESSURE is exerted on the cell wall that gives the plant the support and rigidity it needs.

Celsius A temperature scale in which the freezing point of water is defined as zero degrees Celsius (0°C), whilst the boiling point of water is 100°C, both temperatures being measured at atmospheric pressure.

centi- Prefix used to denote one hundredth. For example, one centimetre is one hundredth of a metre (0.01 m).

central nervous system (CNS) The part of the NERVOUS SYSTEM that co-ordinates body functions by integrating the SENSORY SYSTEM and the EFFECTOR SYSTEM. The CNS receives information from the sensory NEURONES, interprets these and sends messages to the effector neurones to stimulate the appropriate action.

In vertebrates, the CNS consists of a BRAIN and SPINAL CORD. The spinal cord is enclosed and protected by the spinal column surrounded by three membranes called the MENINGES. In many invertebrates the CNS consists mainly of ganglia (*see* GANGLION), and in some simple invertebrates there is no CNS but instead a simple network of nerve cells called a 'nerve net'. *See also* AUTONOMIC NERVOUS SYSTEM, CEREBROSPINAL FLUID.

centrifuge A machine for separating two different materials on the basis of their relative densities. A centrifuge may be used to separate the components in an EMULSION or a solid suspended in a liquid. The mixture is placed in a tube and rotated very rapidly in a horizontal circle. As the mixture rotates, the denser component is forced outwards along the tube, displacing the less dense component and collecting at the bottom of the tube. Centrifugation is used for separating different cell types and for separating blood PLASMA from the heavier RED BLOOD CELLS. *See also* ULTRA-CENTRIFUGE.

centriole In animal cells, an ORGANELLE that is a hollow cylinder. Centrioles are similar to CILIA. They arise in pairs at the CENTROSOME within the cell CYTOPLASM. During MITOSIS and MEIOSIS they separate and go to opposite poles of the cell, where they give rise to the spindles (*see* MITOSIS). In some higher plant cells a spindle forms even though there are no centrioles.

centromere In EUKARYOTES, the region of a CHROMOSOME at which the two CHROMATIDS join and at which the spindle fibres attach during MITOSIS and MEIOSIS. Under the microscope, it is visible as a constriction in the chromosome. There are no genes at the centromere.

centrosome In animal cells, a distinct region within the CYTOPLASM, situated close to the nucleus. The CENTRIOLES arise from here.

cephalopod A member of the class CEPHALOPODA.

Cephalopoda A class of the phylum MOLLUSCA, including octopus, squid and cuttlefish. Cephalopods are mostly marine and possess a well-developed head surrounded by tentacles. Cephalopods are the most advanced of the molluscs. They have complex eyes, similar to those in vertebrates, and a highly developed nervous system. They possess a rasping tongue or radula for feeding.

cephalothorax The combined head and THORAX of many CRUSTACEANS and ARTHROPODS.

cerebellum Part of the vertebrate HINDBRAIN, overlying the MEDULLA OBLONGATA, that controls the muscle movement needed for posture and locomotion. It is well developed in humans and birds to control the balance needed for walking and flight, respectively, but is smaller in lower animals.

cerebral hemisphere *See* CEREBRUM.

cerebrospinal fluid (CSF) A clear, colourless solution found in the VENTRICLES of the BRAIN and between the membranes of the MENINGES (surrounding and protecting the spinal cord). It acts as a shock-absorber for the CENTRAL NERVOUS SYSTEM. It contains glucose and mineral ions and a few WHITE BLOOD CELLS (but no protein), and so supplies nutrients to the central nervous system. It is secreted continuously by the choroid plexuses (projections of non-nervous EPITHELIUM in the ventricles of the brain) and reabsorbed by veins.

cerebrum Part of the vertebrate FOREBRAIN (the largest part of the human BRAIN) that co-ordinates the body's voluntary activities and some involuntary ones, including the senses and complex activities, such as reasoning, learning and memory. Its surface area is increased by being highly convoluted, allowing greater capacity for more complex activity. In lower animals the size of the cerebrum in relation to the body size decreases.

The cerebrum is divided into two halves, the cerebral hemispheres, which are joined by a strip of WHITE MATTER known as the corpus callosum. The outer layer of the cerebral hemisphere is called the cerebral CORTEX. This is functionally the most important part of the brain and is a particularly large area in humans. The cerebral cortex is made of GREY MATTER on the outside and white matter underneath this. Within this cortex different functional regions have been localized, for example, visual, speech and auditory areas. Linked to these areas are association areas, for example, the visual association area, which allow an individual to interpret information received in relation to previous experience.

In humans, the left cerebral hemisphere is associated with control of the right side of the body, and vice versa, due to the crossing over of nerve fibres as they enter the brain from the body. *See also* THALAMUS.

cervix The neck of the mammalian UTERUS, consisting of a ring of muscle at the base of the uterus opening into the VAGINA. The cervix secretes MUCUS into the vagina.

Cestoda A class of the phylum PLATYHELMINTHES, consisting of flatworms. The Cestoda are all PARASITES and are of great economic importance. An example is the pork tapeworm, *Taenia solium,* which uses the human as its primary host and the pig as an intermediate, and causes anaemia, diarrhoea, weight loss and intestinal blockage and pain in humans.

CFC *See* CHLOROFLUOROCARBON.

chaeta (*pl. chaetae*) A bristle occurring in members of the phylum ANNELIDA. Bristles are made of CHITIN and are present in varying numbers. POLYCHAETA, such as the ragworm, have many chaetae whereas OLIGOCHAETA, which includes the earthworm, have only a few. In earthworms the chaetae project from the skin in each segment and anchor the body while it extends in order to move forwards. In polychaeta the chaetae occur in groups attached to appendages termed parapodia.

Charophyta A phylum of the kingdom PROTOCTISTA that consists of the stoneworts. They grow submerged in hard water and are usually covered in a thick, brittle crust of calcium carbonate.

chemiosmotic theory A theory first proposed by the British biochemist Peter Mitchell (1920–) and now generally accepted, which explains the synthesis of ATP in the ELECTRON TRANSPORT SYSTEM of MITOCHONDRIA. The theory suggests that there is a mechanism within the mitochondrial membrane for actively transporting protons (H^+) from the cell matrix to the space between the inner and outer mitochondrial membranes. There is thus a higher concentration of protons in this space than in the matrix, so creating a gradient of hydrogen ions across the inner membrane. Protons return back across the inner mitochondrial membrane down this gradient through special channels associated with ATP synthetase, the enzyme that catalyses the conversion of ADP to ATP. This movement of protons is thus coupled to the PHOSPHORYLATION of ADP and therefore provides the energy to synthesize ATP. A similar gradient is created across the THYLAKOID membranes of CHLOROPLASTS during the light reaction of PHOTOSYNTHESIS. *See also* PHOTOPHOSPHORYLATION.

chemoautotroph Any organism that is an AUTOTROPH using chemical energy to synthesize organic compounds from inorganic molecules. Since they need no light to function, these organisms can survive where others cannot. The chemical energy comes from the oxidation of inorganic chemicals such as ammonium to nitrite (in the case of *Nitrosomonas*) and sulphur to sulphate (in the case of *Thiobacillus*). *See also* CHEMOSYNTHESIS, PHOTOAUTOTROPH.

chemonasty The NASTIC MOVEMENT of plants in response to chemicals.

chemoreceptor A RECEPTOR cell that responds to chemical changes in the internal or external environment. *See* SENSE ORGAN, TONGUE.

chemosynthesis A method of AUTOTROPHIC NUTRITION (self-feeding) in which chemical energy is used to synthesize organic compounds. It is similar to PHOTOSYNTHESIS, which uses light energy. Certain bacteria oxidize a variety of inorganic chemicals to generate the energy needed for chemosynthesis and are vitally important in the recycling of minerals. Examples include the nitrifying bacteria of the NITROGEN CYCLE, which convert free nitrogen into a form that can be used by plants. *Nitrobacter* converts nitrites (NO^{2-}) to nitrates (NO^{3-}); *Thiobacillus* converts sulphur (S) to sulphate (SO_4^{2-}); and *Ferrobacillus* converts ferric (Fe^{2+}) to ferrous (Fe^{3+}). Such organisms are called chemoautotrophs. As they do not need light to function, they can survive where other organisms cannot.

chemotaxis The directional movement of an organism in response to chemical stimuli. This is common in many bacteria. *See* TAXIS.

chemotherapy The treatment of medical conditions, usually cancer, with chemicals.

chemotropism The directional growth of a plant (or part of it) in response to a chemical stimulus. *See* TROPISM.

chiasma (*pl. chiasmata*) A point where two CHROMATIDS from homologous pairs of CHROMOSOMES join as they wrap around one another during MEIOSIS. It is at chiasmata that chromatids may separate and rejoin during CROSSING-OVER and RECOMBINATION.

Chilopoda A class of the phylum ARTHROPODA, including the centipede.

chi-squared test A statistical test measuring the extent of deviation between an observed result and what was expected. From this, it is possi-

ble to calculate whether the deviation is due to chance (and so is non-significant) or not (and so is significant). Chi is represented by the Greek letter χ^2, shown squared, and can be calculated by the following equation:

$$\chi^2 = \Sigma\,(O - E)^2/E$$

where Σ is 'the sum of', O is the observed result and E is the expected result.

Using a χ^2 table, and relating the χ^2 value to the number of degrees of freedom (which is the number of classes of results minus 1), the probability that the deviation is due to chance alone can be calculated. If the probability is less than 5 per cent then the observed deviation is considered significant. If the probability is more than 5 per cent then the deviation is not significant.

The chi-squared test is useful in genetics. For example, in *Drosophila* there are two types of wings: normal and vestigial. Normal wings are DOMINANT to vestigial wings and if two normal-winged individuals are crossed, the expected ratio of normal to vestigial wings would be 3:1 (*see* MENDEL'S LAWS). In practice the numbers of each wing-type may not correspond exactly to this ratio. Using the chi-squared test it can be calculated whether the actual numbers seen are significantly different from the expected ratio, or whether the difference can be explained by chance.

chitin A structural POLYSACCHARIDE that forms the hard, protective EXOSKELETON of insects and arthropods. It is a complex, long-chain nitrogenous derivative of GLUCOSE. Chitin can also be soft and flexible, as in caterpillars. It is insoluble in water and protects the organism from many solvents, acids and alkalis. In CRUSTACEANS, chitin combines with calcium carbonate to give extra strength. Chitin is also found in the CELL WALLS of FUNGI.

chlorenchyma PARENCHYMA tissue in plants that contains CHLOROPLASTS.

chloride shift The movement of chloride ions (Cl^-) into RED BLOOD CELLS to balance the loss of negatively charged hydrogen carbonate ions, thereby maintaining the neutrality of the red blood cells. Hydrogen carbonate ions are lost as a result of the action of the enzyme CARBONIC ANHYDRASE, which catalyses the reaction between of carbon dioxide and water. Hydrogen carbonate ions diffuse out of the cell into the PLASMA where they associate with sodium ions (from the dissociation of sodium chloride) to form sodium hydrogen carbonate. It is in this form that carbon dioxide is carried to the lungs. The chloride ions from the dissociation of sodium chloride then enter the red blood cell from the plasma to restore the balance.

chlorine (Cl) A greenish, yellow gas with a strong smell. In its pure form, chlorine gas is poisonous. In nature, chlorine is always found in the combined form, for example as dissolved sodium chloride in sea water or as hydrochloric acid in the mammalian stomach. It is used as a bleaching agent and as a germicide in swimming pools and drinking water, and has many applications in organic chemistry. It is an essential MACRONUTRIENT for plants and animals. In animals, chlorine is needed to maintain the electrical, osmotic and cation/anion balance across cell membranes. It is also a component of GASTRIC JUICE as hydrochloric acid and assists in the transport of carbon dioxide by blood (*see* CHLORIDE SHIFT). Deficiency of chlorine may cause muscle cramps. There is never a deficiency in plants since chlorine is readily available in soil.

chlorofluorocarbon (CFC) An inert synthetic chemical used in aerosols as a propellant, in refrigerators and air conditioners as a coolant, and in foam packaging. CFC's can remain in the Earth's atmosphere for more than 100 years and are thought to be responsible for the destruction of the OZONE LAYER. The more destructive CFC's have now been banned and replacements are being developed, and safer methods of disposal are being investigated for existing CFC's. *See also* GREENHOUSE EFFECT.

chlorophyll A green pigment, present in most plants and algae, that is responsible for the capture of light during PHOTOSYNTHESIS, and for the coloration of plants. Chlorophyll is found within the CHLOROPLAST of higher plants and algae. In CYANOBACTERIA, where there is no chloroplast, chlorophyll is found on special photosynthetic membranes (THYLAKOIDS) lying free in the CYTOPLASM. Several chlorophylls exist (a, b, c, d and e); chlorophyll a is common to all plants (and the only one in cyanobacteria). In a few photosynthetic bacteria, another type of chlorophyll, bacteriochlorophyll, occurs. Chlorophyll is similar in structure to HAEMOGLOBIN but contains magnesium instead of iron.

Chlorophyta A phylum from the kingdom PROTOCTISTA that consists of green ALGAE. Green algae can be unicellular or filamentous and contain CHLOROPHYLL a and b, like higher plants. Their CHLOROPLASTS are various shapes: in *Chlamydomonas* the chloroplasts are bowl-shaped and in *Spirogyra* they are spiral. Some green algae have FLAGELLA. Most are found in fresh water. Both ASEXUAL REPRODUCTION and SEXUAL REPRODUCTION can occur.

chloroplast In eukaryotic plant and algal cells, a PLASTID containing the green pigment CHLOROPHYLL that is the site of PHOTOSYNTHESIS. Chloroplasts are therefore found in most of the cells of plants and algae that are exposed to light. They also contain CAROTENOID pigments.

In higher plants, chloroplasts are usually disc-shaped, but change their shape and position in relation to light intensity. They have an outer chloroplast envelope that is made up of two membranes. The outer membrane has a similar structure to the CELL MEMBRANE, but the inner membrane consists of a series of folds. Within the envelope lies the STROMA, which is a colourless, structureless matrix containing a stack of around 50 grana per chloroplast. Each granum is made of a number of closed, flattened sacs called THYLAKOIDS, and inside these are the photosynthetic pigments, such as chlorophyll. Grana can be connected to one another by large intergranal thylakoids. Absorption of light

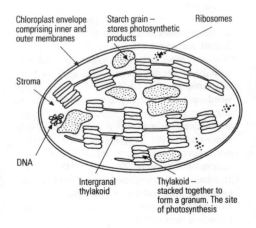

Structure of a chloroplast.

energy occurs in the grana and the subsequent utilization of energy to form CARBOHYDRATES occurs in the stroma. The main food reserve is STARCH, which is stored as grains in the chloroplast. Chloroplasts also contain some DNA and RIBOSOMES.

In algae, chloroplasts can be various shapes and there may be one or more per cell, often associated with PYRENOIDS. Grana do not occur in algae, and thylakoids run across the stroma of the chloroplast as a whole. It is thought that chloroplasts were originally free-living CYANOBACTERIA that invaded larger non-photosynthetic cells and developed a symbiotic relationship (*see* SYMBIOSIS) with them.

choanocyte In sponges, a cell bearing FLAGELLA, which circulate water thorough the body. *See* PORIFERA.

cholecalciserol *See* VITAMIN D.

cholecystokinin A hormone made by the SMALL INTESTINE in response to the presence of acidic, CHYME from the stomach. Cholecystokinin inhibits secretion of GASTRIC JUICE but stimulates production of PANCREATIC JUICE and BILE.

cholesterol A STEROL LIPID that is a component of all CELL MEMBRANES and a precursor of STEROID HORMONES. It is found throughout the body in animals, but is not present in higher plants or bacteria. In the diet, cholesterol is obtained from dairy products and meat. A high level of cholesterol in the blood is thought to contribute to ATHEROSCLEROSIS. It is broken down by the liver into BILE SALTS and any excess is secreted in the BILE.

Chondrichthyes A class of vertebrates, consisting of cartilaginous fish (which have a skeleton made of CARTILAGE). Examples are the dogfish and ray. *Compare* OSTEICHTHYES.

chondrin A firm, elastic substance that forms the matrix of CARTILAGE. Embedded within this matrix are COLLAGEN fibres, which give the cartilage its strength.

chondrocyte A CARTILAGE cell that secretes the matrix of CHONDRIN.

Chordata A phylum of the animal kingdom consisting of animals that at some stage of their lives have a supporting rod of tissue (NOTOCHORD or VERTEBRAL COLUMN) running down their bodies. This phylum includes the vertebrates (*see* VERTEBRATA).

chordate A member of the phylum CHORDATA.

chorion In most mammals, the outermost EXTRAEMBRYONIC MEMBRANE that is next to the uterine walls (*see* UTERUS). VILLI from part of the chorion (chorionic villi) invade the tissue of the mother, forming the TROPHOBLAST that later develops into the PLACENTA. In reptiles and birds, the chorion and ALLANTOIS form a surface for gaseous exchange.

chorionic villus *See* CHORION.

chorionic villus sampling (CVS) In mammals, a procedure performed during early pregnancy to detect chromosomal abnormalities in the foetus. A small biopsy of chorionic villus tissue (*see* CHORION) is taken at 6–10 weeks, which is earlier than when the similar AMNIOCENTESIS analysis is carried out. There is some risk of miscarriage.

choroid A pigmented layer inside the white of the eye that is rich in blood vessels and supplies the RETINA.

choroid plexus A membrane that lines the VENTRICLES of the brain and secretes CEREBROSPINAL FLUID.

chromatid In EUKARYOTES, one of two strands of CHROMATIN that make up chromosomes. Chromatids are joined together at a point called the CENTROMERE. In MITOSIS the chromatids are identical, but in MEIOSIS CROSSING-OVER can occur (exchange of chromosome fragments).

chromatin In EUKARYOTES, the material that CHROMOSOMES are made of. Chromatin consists of DNA and five different HISTONE proteins that are organized into two strands, called CHROMATIDS.

chromatogram The pattern of separated chemicals along a separating medium in CHROMATOGRAPHY.

chromatography An analytical technique used to separate the components of a mixture. The components flow at different rates along some separating medium, which is often packed into a column. The fixed material over which the mixture passes is called the stationary phase and is usually a solid or gel. The moving fluid, often a liquid, but sometimes a gas, is called the moving phase. The mixture is dissolved in a solvent, called an eluent in this context, which then diffuses along the separating medium. The separation of the mixture depends on the competition for molecules of the sample between the moving phase and stationary phase. The liquid (eluate) is usually collected in fractions and the different components may then be identified, and in some cases may be determined quantitatively. *See also* AFFINITY CHROMATOGRAPHY, GEL FILTRATION CHROMATOGRAPHY, ION EXCHANGE CHROMATOGRAPHY, PAPER CHROMATOGRAPHY, THIN-LAYER CHROMATOGRAPHY.

chromatophore 1. In PROKARYOTES, a membrane-bounded vesicle (THYLAKOID) containing photosynthetic pigments. It serves a similar function to the CHLOROPLAST in EUKARYOTES.

2. *See* CHROMOPLAST.

chromoplast, *chromatophore* A PLASTID that may arise from a CHLOROPLAST. It has coloured CAROTENOID pigments but no CHLOROPHYLL.

chromosome A structure in the cell nucleus that carries the genetic material DNA and RNA, as well as proteins. It is only visible during CELL DIVISION. Higher organisms, such as humans, are DIPLOID, which means that there are two copies of each chromosome. Humans have 46 chromosomes, forming 23 pairs. The number of chromosomes differs between species. One set of chromosomes is derived from each parent and carried in the GAMETE. Some organisms have only one set of chromosomes and are called HAPLOID. More than two sets of chromosomes is called polyploidy (*see* POLYPLOID).

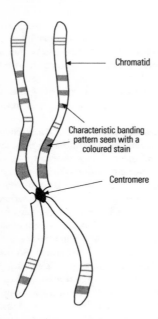

Chromatid

Characteristic banding pattern seen with a coloured stain

Centromere

Structure of a chromosome.

This is rare in animals but spontaneously arises quite frequently in plants.

In EUKARYOTES, chromosomes consist of DNA and HISTONE, which form CHROMATIN. This is organized into two strands (CHROMATIDS) joined together at a point called the CENTROMERE. Each chromosome has a characteristic banding pattern when stained with a routine coloured stain, and varies in size and shape within and between species. PROKARYOTES usually have only one circular chromosome per cell, and may have a small percentage of their DNA in another smaller loop called a PLASMID. In humans, sex is determined by the sex chromosomes, the X-chromosome and the Y-chromosome (*see* SEX DETERMINATION).

See also AUTOSOME, CHROMOSOME MAP, GENE, HETEROSOME, KARYOTYPE, LINKAGE, MEIOSIS, MITOSIS, MUTATION, RECOMBINATION, SEX LINKAGE.

chromosome map A linear map showing the arrangement of GENES on a CHROMOSOME. It is useful for identifying the relative positions of particular genes, and especially for studying abnormal genes. *See also* BACKCROSS.

chromosome mapping The techniques involved in determining a CHROMOSOME MAP. Chromosome mapping involves examining the PHENOTYPES of offspring obtained in appropriate breeding routines and studying LINKAGE. Mapping information can also be obtained by analysis of DNA sequences.

chrysalis *See* PUPA.

Chrysophyta A phylum of the kingdom PROTOCTISTA that consists of the golden-brown ALGAE.

chyme The creamy and partly digested food that is released by the STOMACH into the SMALL INTESTINE.

cilia (*sing.* ***cilium***) Many thread-like structures on the surface of some cells that can contract to produce rhythmic co-ordinated waving movements. Cilia can be used for cell locomotion in single-celled organisms, such as *Paramecium*. In multicellular organisms, they are present as ciliated EPITHELIUM, where they move the surrounding fluid or substances through ducts, such as FALLOPIAN TUBES or TRACHEA, often with the help of MUCUS secreted by single glandular cells or GOBLET CELLS.

Cilia have a characteristic arrangement of internal fibres: two central and nine peripheral pairs. The central fibres can be absent, in which case the structure is non-motile. Cilia are about

0.2 µm in diameter and 10 µm long. They are similar to FLAGELLA, but the latter are longer, fewer and move in a different way.

ciliary muscles The ring of muscles that surrounds the EYE lens. It can contract to squeeze the lens into a fatter shape for viewing nearby objects.

ciliate A member of the phylum CILIOPHORA.

Ciliophora A phylum from the kingdom PROTOCTISTA, consisting of ciliate PROTOZOA. Ciliates possess CILIA, which they use to move and to trap food. Reproduction is by BINARY FISSION or CONJUGATION. They have two nuclei. Examples are *Paramecium* and *Vorticella*.

circadian rhythm A BIORHYTHM found in most organisms that is related to the 24-hour day cycle, such as the sleeping/waking pattern. Other factors, such as hormone concentration and temperature, may also vary throughout the day, affecting behaviour and mood.

circalunar rhythm A BIORHYTHM that follows a 28-day cycle related to the phases of the moon.

circulatory system In an animal, a network of vessels that distributes blood containing essential substances throughout the body. Circulatory system can also refer to other circulating body fluids, such as LYMPH.

In the simplest animals, such as NEMA-TODES, an open blood system exists where there are no vessels, and blood (called haemolymph) moves freely over the tissues through spaces collectively called the haemocoel. Insects have an open system with a heart providing some circulation into the haemocoel. In this open system, blood is moved at low pressure and with little control.

Larger animals, including humans, have a closed blood system in which the blood is contained within the blood vessels and pumped around the body at high pressure by the heart. In humans, other mammals and birds, blood passes from the heart to the lungs and then back to the heart before it is ready for circulation around the body. This is called double circulation and is more efficient than the single circulation of fish, where blood circulates once around the body before returning to the heart and becomes sluggish as the pressure drops further from the heart. In a closed system, blood flows in one direction only, ensured by one-way valves in the heart, arteries and veins.

See also HEART, LYMPHATIC SYSTEM.

cisternum A cavity or vesicle in the cytoplasm of a living cell, formed by membranes of the ENDOPLASMIC RETICULUM or GOLGI APPARATUS.

cistron A region of DNA containing structural GENES responsible for the production of a complete POLYPEPTIDE. In modern terminology, a cistron is considered to be equivalent to a gene. *See also* GENE EXPRESSION, OPERON.

citric acid An organic acid found in many plants. It exists in particularly high concentrations in citrus fruits, such as oranges and lemons. It has a sharp, sour taste. Citric acid is an intermediate in the KREBS CYCLE.

citric acid cycle *See* KREBS CYCLE.

CJD *See* CREUTZFELDT–JAKOB DISEASE.

class One of the subdivisions of a PHYLUM in the CLASSIFICATION of organisms. MAMMALIA is the class consisting of mammals, AVES is the class consisting of birds. Plant class names end in 'idae' and fungi class names end in 'mycetes'. Classes are further subdivided into ORDERS.

classification The organization of all organisms (living and extinct) into groups, for human convenience, based on similarities in their physiological, anatomical and biochemical characteristics. There are many ways of doing this, none of which are perfect.

In the 18th century, Carolus Linnaeus (1707–78) devised a natural classification scheme based on homologous characters (features with a similarity in origin, structure and position) rather than analogous characters (similar functions but with different origins). The divisions (or ranks) in classification used today are based on those of Linnaeus. The first division is a KINGDOM, of which there are usually five: ANIMALIA, PLANTAE, FUNGI, PROTOCTISTA (these four are all EUKARYOTES) and PROKARYOTAE (PROKARYOTES). The divisions below a kingdom are PHYLUM, CLASS, ORDER, FAMILY, GENUS, SPECIES.

The written name given to an organism consists of two parts (BINOMIAL NOMENCLATURE): the genus (with a capital letter) and the species (with a small initial letter) both written in italics. For example, *Homo sapiens* is the name for humans. The species is the lowest level in the classification scheme, but within a species different populations may exist (for example, different breeds of dog).

An artificial classification scheme is based on the differences between organisms and is used to identify species. It uses a series of divisions based on a single character, for example, wings present/wings absent, from which a species can eventually be identified. In this method of classification, groups may contain unrelated forms.

clay A fine-grained deposit, consisting chiefly of silicates of aluminium and/or magnesium. Clay is a constituent of some SOILS in which drainage is slow because of the small particle size.

cleavage In embryology, the first stage of EMBRY-ONIC DEVELOPMENT in which the fertilized egg divides by MITOSIS to form a ball of identical cells, which develops a central cavity to form the BLASTULA or BLASTOCYST. There is no growth during cleavage but the ratio of nuclear material to cytoplasm increases.

climax community A COMMUNITY of organisms that is relatively stable and is in equilibrium with existing natural environmental conditions. A climax community is the result of a series of changes constituting a SUCCESSION. An example of such a community is an oak forest in Britain.

cloaca In birds, reptiles, amphibians, many fish and some marsupials, a chamber containing all excretory products (digestive and urinary) and into which the reproductive tracts enter. Products can be stored in the cloaca before being discharged from the body. Placental mammals do not have a cloaca. *See also* BLAD-DER, URINARY SYSTEM.

clone A group of genetically identical offspring produced by ASEXUAL REPRODUCTION, involving the development of an entire organism from a single cell. This is useful in reproducing certain plants and is theoretically possible in humans (although unlikely to be done). In 1997 a sheep was cloned from one cell. *See also* GENE CLONING.

club moss A member of the phylum LYCO-PODOPHYTA.

CMI *See* CELL-MEDIATED IMMUNITY.

Cnidaria A phylum of mainly marine inverte-brates, commonly known as CNIDARIANS or COELENTERATES. Cnidaria includes *Hydra,* the Portuguese man of war (class HYDROZOA), jellyfish (class SCYPHOZOA), sea anemones and marine coral (class ACTINOZOA). Cnidarians have two structural forms, an attached 'polyp' form with an opening at the opposite end to the attachment, and a free-swimming 'medusa' form that is umbrella-shaped with an opening in the middle of the underside (*see*

CTENOPHORA). One or both of these forms may occur in the life cycle of a cnidarian. *Hydra* (which is a freshwater cnidarian) exists mainly in the polyp phase but is unusual because it moves by extending its body in a series of somersaults. Jellyfish are mostly free-swim-ming medusa. They feed by trapping or sting-ing organisms with their tentacles.

Cnidarians contain specialized stinging cells called cnidoblasts that allow them to adhere to or penetrate their prey. The stings of these animals can be painful and sometimes fatal to humans. Polyps reproduce asexually by BUDDING, but *Hydra* can reproduce sexually in the winter. Other polyp forms produce medusa that can reproduce sexually.

cnidarian Any member of the phylum CNIDARIA, traditionally referred to as COELENTERATES. Cnidarians are aquatic, diploblastic (the body wall is composed of two layers with a jelly-like substance between them), radially symmetri-cal and have one opening to the outside sur-rounded by stinging cells. Cnidarians include *Hydra,* jellyfish, sea anemones and marine coral and the sea-gooseberry (*see* CTENO-PHORA).

cnidoblast A specialized stinging cell found only in CNIDARIA.

CNS *See* CENTRAL NERVOUS SYSTEM.

coagulation The process of particles coming together to form a larger semi-solid mass. ALBUMIN in eggs coagulates on heating, and blood clotting is a coagulation process.

coated pit A specialized region of most EUKARY-OTIC cell membranes that looks like a depres-sion in the membrane, prior to a COATED VESICLE being pinched off. Receptor sites for certain substances occur at some coated pits and the cell membrane invaginates to form a coated vesicle. Once inside the cell, the coated vesicle loses its surrounding protein coat and becomes an ENDOSOME. *See also* ENDOCYTOSIS.

coated vesicle A membranous vesicle budded off from a COATED PIT in the PLASMA MEMBRANE of a cell during receptor-mediated ENDOCYTO-SIS. The vesicle sheds its protein coat after entry into the cell to form an ENDOSOME.

cobalt (Co) A light-grey metallic element. It is a MICRONUTRIENT for plants and animals. In ani-mals, it is a constituent of VITAMIN B_{12} which is important in the synthesis of RNA and in the formation of RED BLOOD CELLS. Deficiency in animals causes PERNICIOUS ANAEMIA.

coccus (*pl. cocci*) A spherical-shaped bacterium (*see* BACTERIA). Examples include *Streptococcus,* which associates in straight chains and is a common cause of sore throats in humans, and *Staphylococcus,* which associates in clusters and is found on the skin and MUCOUS MEMBRANE in humans where it can cause abscesses.

cochlea A spiral-shaped structure in the inner ear concerned with hearing, both sound detection and pitch analysis. The cochlea comprises two narrow canals filled with the fluids perilymph and endolymph, and is coiled to save space. Within the cochlea is a specialized structure, called the organ of Corti, that consists of fine sensory hairs.

Movement of a membrane between the middle and inner ear (the 'oval window'), caused by vibrations of the bone ossicles, causes the perilymph behind the membrane to move, which in turn displaces another membrane (called the 'round window') also between the inner and middle ear. These pressure waves cause movement of the sensory hairs, and this sets up an ACTION POTENTIAL that is transmitted to the brain along the auditory nerve. The pitch is determined by which part of the cochlea is stimulated, and the loudness by how many sensory hairs are stimulated.

codominance In genetics, the expression of a mixture of two ALLELES when neither one of the pair is truly DOMINANT. For example, when HETEROZYGOUS red- and white-flowered snapdragons are crossed, the next generation of snapdragons has pink flowers.

codon In genetics, a triplet of NUCLEOTIDES in DNA or RNA that determines the order in which amino acids are placed during PROTEIN SYNTHESIS. MESSENGER RNA is the template for protein synthesis and the codons are complementary to, and therefore form BASE PAIRS with, ANTICODON triplets in TRANSFER RNA. This leads to the production of a protein with a specific chain of amino acids.

Some codons do not code for an amino acid and are called STOP CODONS. Stop codons, such as UAG, UAA and UGA, terminate protein synthesis. The codon AUG, a START CODON, initiates protein synthesis. There are 64 possible codons, which is more than enough to code for the 20 amino acids, and in fact some amino acids are coded for by several codons.

coelenterate The traditional name for any member of the phyla CNIDARIA and CTENOPHORA (now a class within cnidaria, not a phylum). This collective term is still used in addition to the more modern term of CNIDARIAN.

coeliac disease A rare disorder in which there is poor absorption of essential food from the INTESTINES. The condition is hereditary and is due to an intolerance to GLUTEN, a protein found in cereals such as wheat. Symptoms include wasting and a general weakness, distended abdomen, stools that are bulky, soft and pale and diarrhoea. The disease may subside in children but can recur. Gluten-free food products are available for people with this intolerance and are recommended for all babies.

coelom The main body cavity that separates the DIGESTIVE SYSTEM and associated organs from the body wall. It is a fluid-filled cavity lined with MESODERM and is only absent in the simplest animals.

coenzyme An organic molecule, often a vitamin derivative, that acts as a COFACTOR in an enzyme reaction. It acts without binding, or only temporarily binding, to the enzyme, unlike a PROSTHETIC GROUP. Examples include COENZYME A, coenzyme Q (*see* ELECTRON TRANSPORT SYSTEM), FAD and NAD. Coenzymes are frequently essential for the removal of end-products of enzyme reactions that would otherwise cause inhibition of the enzyme.

coenzyme A A derivative of PANTOTHENIC ACID that is a COENZYME acting as a carrier of ACYL GROUPS in the OXIDATION of FATTY ACIDS (*see* KREBS CYCLE).

coenzyme Q *See* ELECTRON TRANSPORT SYSTEM.

cofactor A non-protein substance that is needed for some enzymes to function efficiently. Cofactors can be COENZYMES, PROSTHETIC GROUPS or activators. Activators are substances other than coenzymes and prosthetic groups that are needed to activate an enzyme, for example, calcium ions are needed to activate THROMBOKINASE (which converts PROTHROMBIN to THROMBIN in blood clotting). An enzyme–cofactor complex is called a haloenzyme and the inactive enzyme on its own is called an apoenzyme.

cohesion The attractive force between molecules of the same type. *See also* ADHESION, CAPILLARY EFFECT.

coil In contraception, *see* INTRAUTERINE DEVICE.

colchicine An ALKALOID that prevents SPINDLE formation during MEIOSIS so that the chromosomes cannot separate, resulting in multiple sets of chromosomes. *See* POLYPLOID.

collagen In vertebrates, a major structural protein found in CONNECTIVE TISSUE. It is made by FIBROBLASTS. Collagen forms fibres that provide strength but little elasticity, and it is the main constituent of LIGAMENTS and TENDONS. It is also found in CARTILAGE and bone.

collenchyma Simple (one cell-type only) plant tissue consisting of elongated cells with cell walls thickened by additional CELLULOSE at the corners, providing extra strength and support. Collenchyma is particularly important in growing stems because the cells can stretch.

colloid A mixture containing small particles of one material suspended in another, often of a different phase. The particles in a colloid have sizes between one and one hundred nanometres, so are larger than the individual molecules that occur in a solution and smaller than the particles that are found in precipitates and can be removed by a filter.

Examples of colloids include aerosols and foams (gas and liquid mixtures), emulsions (liquid mixtures such as milk and paint), and sols and gels (solids dispersed in a liquid).

Particles in a colloid can be separated by passing them through a porous material. The material that forms the separate individual particles is sometimes called the dispersed phase, to distinguish it from the continuous phase, which forms a single connected body of material.

colon Part of the LARGE INTESTINE between the CAECUM and the RECTUM. It is in the colon that any water and minerals in digestive secretions not absorbed by the ILEUM are reabsorbed and the remainder formed into FAECES. The colon also absorbs vitamins, some of which, for example biotin (*see* VITAMIN B) and VITAMIN K, are produced by bacteria, such as *Escherichia coli,* that live in the colon.

colostrum In mammals, the milk produced by the MAMMARY GLANDS during the first few days following the birth of their offspring. It consists of a clear fluid containing water, proteins, vitamins and antibodies, but is low in fat and sugar. Colostrum plays an important role in PASSIVE IMMUNITY. *See also* LACTATION.

column chromatography CHROMATOGRAPHY in which the eluent runs down a glass column containing the STATIONARY PHASE, such as tightly packed powdered alumina.

coma A deep, prolonged unconsciousness in which the person is unable to respond to external stimulus, usually a result of injury or disease.

combined pill *See* PILL.

commensalism An association between two species in which only one partner (the commensal) benefits but the other (the host) is not harmed. Commensalism is a variation of SYMBIOSIS. *See also* MUTUALISM, PARASITISM.

community All of the POPULATIONS of plant or animal species that live together within an ecological area, or HABITAT. A community is often identified by a dominant feature, either a physical feature, such as a swamp, or a dominant species, such as an oak woodland. *See also* SUCCESSION.

companion cell A type of cell found closely associated with SIEVE ELEMENTS in the PHLOEM of flowering plants. The companion cells are responsible for transporting soluble food molecules into and out of the sieve element.

competition The fight between organisms or species for resources that are in short supply. In animals, competition might be for food, water or shelter. In plants, competition might be for light or minerals. Competition results in the survival of or adaptation of some individuals or species instead of others. It is the basis of DARWINISM.

complement (C′) A group of nine serum proteins (C_1–C_9) that are activated in a systematic way and participate in the IMMUNE RESPONSE. The first component is activated by an ANTIGEN–ANTIBODY immune complex and this subsequently triggers the other components, resulting in a complement cascade. The main purpose of the cascade is to assist in the destruction of foreign cells and in the attraction of white blood cells to the site of infection. One of the components, C_3, binds to immune complexes and also to receptors on B CELLS, MACROPHAGES, NEUTROPHILS, EOSINOPHILS and MONOCYTES. This brings together the immune complexes and the cells that are able to ingest and destroy them. The complement cascade is regulated by a complex of inhibitors. The genes encoding complement are located in the MAJOR HISTOCOMPATIBILITY COMPLEX in humans.

complementary DNA *See* COPY DNA.

compound A substance made up of two or more elements that cannot be separated by physical means. In a compound, the quantity of the elements present is fixed and is a simple ratio, though more than one compound may exist containing the same elements, for example, water (H_2O) and hydrogen peroxide (H_2O_2).

compound lens An optical lens made from several pieces of glass.

compound microscope See MICROSCOPE.

computerized axial tomography See CAT SCAN.

concave (*adj.*) Describing an inwardly curving surface, particularly in a lens or mirror.

conditioning A form of LEARNING that involves the association of two stimuli.

An example of conditioning is the conditioned reflex, which involves two stimuli presented together, and is temporary, involuntary and reinforced by repetition. The conditioned reflex was demonstrated by the experiments of the Russian scientist Ivan Pavlov (1849–1936). Dogs learned to associate the ringing of a bell with the arrival of food, and responded to the bell by involuntary salivation even if no food arrived.

Another form of conditioning is trial and error learning, or operant conditioning, as described by Edward Thorndike (1874–1949) and B. Skinner (1903–1990). Here an animal learns a pattern of behaviour based on what occurs after the action. The frequency of a voluntary response is increased by giving a reward for the correct response.

See also HABITUATION, IMPRINTING.

condom A rubber sheath placed over the erect PENIS preventing entry of SPERM into the VAGINA during sexual intercourse. It is used as a method of CONTRACEPTION. It is 97 per cent effective if used with a spermicide (jelly or cream preparations that kill sperm) but only 85 per cent effective if used on its own. The condom also provides protection against sexually transmitted diseases such as AIDS.

conducting vessel A non-living component of XYLEM in flowering plants whose function is in the TRANSLOCATION of water and mineral salts and in providing mechanical support for the plant. The vessels consist of a series of cells arranged end to end which form a tube, often very broad. The side walls are thickened with LIGNIN but have unthickened regions called PITS that enable communication between adjacent vessels. The end walls are perforated to allow connections with the cells above and below, and therefore easy flow of water. Vessels are thought to be more efficient conductors of water than TRACHEIDS because of these perforations.

cone 1. (*botany*) The reproductive structure of CONIFERS. An example is the cone of a pine tree. A cone consists of numerous scales called SPOROPHYLLS that overlap around a central axis. There are usually separate male and female cones. *See also* CONIFEROPHYTA.

2. (*zoology*) Light-sensitive cells in the RETINA of the human eye. Cones are sensitive to colour and are used mostly for day vision.

congenital disease A disease present at birth, which may be a result of genetic factors, injury, infection or environmental factors.

conidium (*pl. conidia*) A non-motile SPORE of some fungi formed during ASEXUAL REPRODUCTION.

conifer A member of the phylum CONIFEROPHYTA.

Coniferophyta A large phylum of plants that have exposed seeds. There are many species, including giant redwoods, firs and pines. Conifers were once the dominant form of vegetation but now the ANGIOSPERMS are. There are fossil remains of conifers that are 350 million years old.

The reproductive structures of conifers are CONES (not flowers as in angiosperms). Female and male cones are usually separate and pollen grains from the male cone are carried by the wind to the female cone. The seeds develop (sometimes over a period of years) on the female cone, and are only released when the scales of the cone open in dry conditions that favour DISPERSAL. Conifers bear no fruit.

Conifers are adapted to dry conditions in many ways, and are often found where water is sparse. Conifers are fast-growing trees and so many species are an important source of timber and wood pulp for paper. They are often planted to prevent land erosion.

conjugated protein A PROTEIN with a non-protein group incorporated into its structure that plays a vital role in the functioning of the protein. For example, HAEMOGLOBIN contains a haem group. *See also* PROSTHETIC GROUP.

conjugation In bacteria, the equivalent of SEXUAL REPRODUCTION. The DNA is exchanged or donated between bacteria by passing through a tube called the pilus.

conjunctiva In vertebrates, an outer protective membrane of the eye that is an extension of the eyelid EPITHELIUM and lies over the CORNEA. The lachrymal glands produce antiseptic tears to nourish and lubricate the cornea and conjunctiva because they have no blood supply. Conjunctivitis is inflammation of the conjunctiva due to infection.

conjunctivitis Inflammation of the CONJUNCTIVA.

connective tissue Animal tissue developed from the embryonic MESODERM (*see* EMBRYONIC DEVELOPMENT). It is made up of a variety of cells and connective tissue fibres (usually COLLAGEN) embedded in the EXTRACELLULAR MATRIX. Connective tissue provides support in cartilage and bone and in the transport system of blood. The strength and elasticity of connective tissue is provided by collagen and ELASTIN fibres, which are produced by FIBROBLASTS. ADIPOSE connective tissue provides insulation. Loose connective tissue (areolar tissue) binds many other tissues together (such as MENINGES of the central nervous system and bone PERIOSTEUM). TENDONS and LIGAMENTS are also connective tissue. As well as providing support, connective tissue provides some defence, due to the presence of tissue MACROPHAGES and MAST CELLS.

conservation The protection of the natural world from the effects of human activities. This includes the protection of endangered species and valuable natural environments, such as the rainforests, protection from pollution, reduction of the GREENHOUSE EFFECT and the recycling of glass, paper, plastics and some metals. Many organizations exist to promote conservation.

   National parks and nature parks have been established to protect particular environments. The first national park was Yellowstone National Park, USA, set up in 1872. The first in the UK was set up in 1949. National parks vary in size and are sites of natural beauty as well as historical or scientific interest. Some national parks are wilderness areas with no traffic or buildings, but others (in the UK) are areas of restricted development. On a smaller scale, SITES OF SPECIAL SCIENTIFIC INTEREST (SSSIs), AREAS OF OUTSTANDING NATURAL BEAUTY (AONBs) and country parks are areas of land worthy of particular protection.

   Some endangered species are protected by law, although this is difficult to enforce. Another way of protecting species is through

commercial farming, for example of mink and deer, to provide sought-after goods without the need to kill wild animals, and zoos provide a safe environment for breeding endangered species for later reintroduction into the wild. In the same way, botanical gardens provide a place where a diverse number of plants, unlikely to be encountered naturally, can be viewed by members of the public. Botanical gardens and other organizations keep seed banks to conserve plants that are rare in the wild.

constitutive (*adj.*) Describing an enzyme that is synthesized all the time, regardless of the availability of SUBSTRATE. *Compare* INDUCIBLE.

contact inhibition A phenomenon seen in TISSUE CULTURES when cells growing in a monolayer stop moving and dividing when they come into contact with other cells. This means that they only fill the available space. Cancer cells lose this regulatory ability. *See also* DENSITY-DEPENDENCE.

contact lens A small MENISCUS lens designed to have the same effect as spectacles but resting directly on the CORNEA (the curved front surface of the eye).

continuous ambulatory peritoneal dialysis (CAPD) A technique for removing the circulating waste products from the blood of patients with kidney failure. It uses the principle of DIALYSIS and is similar to HAEMODIALYSIS except that it uses the membrane enclosing the PERITONEAL CAVITY as the SEMIPERMEABLE MEMBRANE. Dialysis fluid is pumped into the cavity and slowly out again, during which time toxic substances have diffused from the blood to the peritoneal fluid. The patient can move around during dialysis, unlike with haemodialysis.

contraception The deliberate prevention of pregnancy while maintaining a sexual relationship. Contraception can be a simple barrier method, where the GAMETES are prevented from meeting. Barrier methods include the CONDOM and the DIAPHRAGM.

   Some methods of contraception rely on hormonal interference of the female's MENSTRUAL CYCLE, the main one being the oral PILL, which uses synthetic hormones to mimic pregnancy. Other similar preparations include the mini-pill, and the morning-after pill taken after unprotected intercourse (*see* PILL). Hormone implants can be surgically placed under the skin to release hormones over a few months (depo-provera-progestin) or years (norplants release female hormones over 5 years). Another method is the INTRA-UTERINE DEVICE (IUD), which is a plastic or copper coil inserted into the UTERUS. Non-reversible methods of contraception are male vasectomy and female tubal ligation (*see* STERILIZATION).

contractile root A thickened root on the base of CORMS or bulbs that contracts to pull the plant deeper into the ground. Some contractile roots store CARBOHYDRATES for use by the plant.

contractile vacuole A membrane-surrounded cavity in a cell that is important in OSMOREGU-LATION. Contractile vacuoles are common in freshwater organisms such as *Amoeba* and *Paramecium*. They are able to slowly fill with water (an energy-requiring process) and suddenly contract to expel their contents from the cell, thereby preventing excess water building up in the cell.

contrast enhancing medium An X-ray absorbing material introduced into a patient to show up more clearly the various organs or structures during an X-ray.

convex (*adj.*) Describing an outward curving surface.

copper (Cu) A reddish brown metal. It is a MICRONUTRIENT required by living organisms. Copper is a constituent of some enzymes and of the respiratory pigment HAEMOCYANIN. Deficiency in plants causes young shoots to die.

copy DNA, *complementary DNA (cDNA)* DNA formed from an RNA template and therefore complementary to it. It is produced by the action of the enzyme REVERSE TRANSCRIPTASE. Copy DNA is initially single-stranded but can be converted to double-stranded DNA by the action of the enzyme DNA POLYMERASE.

cork A protective tissue forming part of the bark of the stems and roots of most trees and shrubs. Cork is produced by the CORK CAM-BIUM as part of the SECONDARY GROWTH of a plant. It eventually replaces the EPIDERMIS. The inner cells are lined with SUBERIN, which makes them impermeable to water and gases. Some cork cells contain deposits of LIGNIN and others become air-filled. Cork that is used commercially comes from the cork oak tree, which has very thick layers of bark.

cork cambium A type of CAMBIUM in VASCULAR PLANTS that produces secondary tissues resulting in SECONDARY GROWTH of the plant body. This is in contrast to VASCULAR CAMBIUM, which

results in the growth of secondary XYLEM and PHLOEM and usually precedes the formation of cork cambium. Cork cambium gives rise to the outer layers of bark and cork on the stems of woody plants, usually as a complete ring surrounding the inner tissues. The cells of the cork cambium divide to produce an outer layer of CORK (phellem) and an inner cortex (phelloderm). The cork, cork cambium and phelloderm together form the periderm, which provides an outer protective layer for the plant.

corm In plants, a short, thick, round underground stem, surrounded by protective scale-like leaves, that acts as a food store. Gladioli and crocuses, for example, have corms. New buds form in the scale leaves and use the food supply to develop into leafy, flowering shoots. The old corm then withers and a new corm forms at the base of the shoot, on top of the old one (*see* VEGETATIVE REPRODUCTION).

cornea The outer layer of the human eye, lying underneath the protective CONJUNCTIVA. The cornea is curved and acts as a fixed lens carrying out most of the refraction (bending) of the light entering the eye, before it reaches the lens. The cornea receives its nourishment and lubrication from antiseptic tears produced by the lachrymal (tear) glands, because it has no blood vessels. Corneal grafts can be performed in humans to replace diseased or opaque parts that are impairing eyesight.

corolla The collective term for the petals of a flower.

corpus callosum In the brain, the band of WHITE MATTER that joins the two halves of the CERE-BRUM.

corpus luteum, *yellow body* In mammals, a temporary ENDOCRINE GLAND in the OVARY that is formed by the action of LUTEINIZING HOR-MONE on the ruptured GRAAFIAN FOLLICLE after OVULATION (egg release). It secretes the steroid hormone PROGESTERONE that prepares the uterine wall for pregnancy. If pregnancy does not occur, the corpus luteum breaks down. If pregnancy does occur, the life of the corpus luteum is prolonged by the hormone HUMAN CHORIONIC GONADOTROPHIN, until the PLACENTA is fully developed at about 3 months of pregnancy. *See also* MENSTRUAL CYCLE.

cortex 1. In animals, the outer layer of some organs, such as the ADRENAL GLAND, brain and kidney.

2. In plants, the region beneath the outermost EPIDERMIS. *See also* MEDULLA.

corticoid *See* CORTICOSTEROID.

corticosteroid, *corticoid* A collective term for a group of STEROID HORMONES secreted by the cortex of the ADRENAL GLAND. Corticosteroids are either GLUCOCORTICOIDS (such as CORTISOL), concerned with GLUCOSE metabolism, or MINERALOCORTICOIDS (such as ALDOSTERONE), concerned with mineral metabolism. They are produced in response to stress situations under the control of the pituitary ADRENOCORTICOTROPHIC HORMONE. Corticosteroids are widely used as drugs to treat conditions in which inflammation is a problem, or to suppress the IMMUNE RESPONSE. For example, they are used to treat inflammatory bowel disease, asthma and certain skin conditions. Some of these drugs are derived from the natural corticosteroid hormones and some are synthetic variants of these.

corticotrophin *See* ADRENOCORTICOTROPHIC HORMONE.

corticotrophin-releasing factor *See* ADRENOCORTICOTROPHIC HORMONE.

cortisol, *hydrocortisone* A major GLUCOCORTICOID hormone of humans and other mammals, secreted by the cortex of the ADRENAL GLAND in response to internal stress (such as low blood temperature or volume). It raises blood pressure and promotes GLUCONEOGENESIS (the conversion of protein and fat into glucose). It is under the control of the pituitary ADRENOCORTICOTROPHIC HORMONE.

cortisone A GLUCOCORTICOID hormone secreted by the ADRENAL GLAND that promotes the synthesis and storage of glucose. Cortisone is produced as part of the body's normal response to stress. The action of cortisone is similar to CORTISOL and both are used in the treatment of inflammatory conditions such as ARTHRITIS, certain allergies and skin disorders and in ADDISON'S DISEASE where there is adrenal failure.

cotyledon, *seed leaf* A structure in the embryo of a SEED PLANT that may provide food for the growing embryo. The number of cotyledons in an embryo is an important means of classification in flowering plants (ANGIOSPERMS). Most plants have two cotyledons and are called DICOTYLEDONS, but some have a single one and are called MONOCOTYLEDONS. In GYMNOSPERMS there may be up to a dozen cotyledons in each seed.

The cotyledons may remain below ground (hypogeal) following germination, which most monocotyledons do, or may spread out above the soil (epigeal) following germination, forming the first green leaves, as most dicotyledons do. Where an ENDOSPERM (food source) is present in the embryo seed, the cotyledons are thin, but in some plants, for example peas and beans, the cotyledons provide the main food supply and are therefore large. Cotyledons often store LIPIDS, which form a high percentage of the dry weight of a seed, for example in walnuts, coconuts and sunflowers. Less frequently, the cotyledons store protein, for example in LEGUMES and nuts.

countercurrent heat exchangers *See* COUNTERCURRENT SYSTEM.

countercurrent system Any physical or biological system in which two fluids flow in opposite directions along vessels close to one another, so that exchange of heat or contents can occur. The level of substance or heat drops in one fluid and rises in the other.

Some animals living in cold climates reduce heat loss from their feet by countercurrent heat exchangers. A capillary network exists in the extremity of the limb and warm blood entering the limb from the body passes alongside vessels containing cold blood returning to the body from the limb. The heat from the warm blood enters the cold blood as it returns to the body and the limb is kept at a lower temperature than the body, so reducing heat loss from the body.

Other examples are the LOOP OF HENLE in the kidney and the exchange of respiratory gases in the gills of bony fish.

covalent bond A chemical bond in which electrons are shared between two atoms, giving each one a share in the other's electrons. For example, covalent bonds exist between the hydrogen and oxygen atoms in a water molecule (H_2O). Where two atoms share one pair of electrons, a single covalent bond is formed, represented by a single line drawn between the atoms. Where the two atoms share two pairs of electrons, a double bond forms, represented by two lines. A triple bond forms if three pairs of electrons are shared.

Cowper's gland In male mammals, one of a pair of glands below the PROSTATE GLAND. Cowper's glands secrete a sticky fluid into the URETHRA that contributes to SEMEN. *See also* SEXUAL REPRODUCTION.

creation theory One of the theories to explain the ORIGIN OF LIFE. It states that God created all life forms as they are today. This is contradicted by fossil evidence and genetic studies and other theories are therefore favoured.

Creutzfeldt–Jakob disease (CJD) A disease of humans causing progressive degeneration of the CENTRAL NERVOUS SYSTEM. It appears to be caused by a PRION protein, which is found in the brains of affected people after they die. The brains of these people have numerous holes in the tissue, which gives a spongy appearance similar to that seen in SCRAPIE of sheep. The disease usually affects older people because it has a very long incubation period.

During 1996, several cases of younger people dying from a disease with similar symptoms to CJD, but with a different brain appearance, led to a possible link between this new human disease and BOVINE SPONGIFORM ENCEPHALOPATHY.

crista (*pl. cristae*) Foldings of the inner membrane of MITOCHONDRIA. They form extensions into the matrix, some across the entire mitochondria. The role of cristae is to increase the surface area on which respiratory processes take place. The surface of the crista is covered in stalked particles associated with OXIDATIVE PHOSPHORYLATION.

crop In birds, an expanded part of the digestive tract between the OESOPHAGUS and the stomach. It is used to store food, especially seeds. Digestion begins in the crop, with the moistening of the food.

crossing-over A RECOMBINATION process occurring during MEIOSIS that results in pairs of chromosomes twisting around one another and exchanging segments (*see* CHIASMA, CHROMATID). The new combinations are called recombinants. Crossing-over provides the genetic variation that forms the basis of evolution. *See also* CROSS-OVER VALUE.

cross-over value, *recombination frequency* The proportion of recombinants formed in a group of offspring as a result of CROSSING-OVER. This value can be used to determine the distance between genes on a chromosome. The closer together two genes are, the less likely it is that they would be separated by crossing-over, and therefore there would be fewer recombinants.

cross-pollination The transfer of pollen from the male part of one plant to the female part of another plant. *See* POLLINATION.

Crustacea A class of the phylum ARTHROPODA comprising 26,000 species, including crabs, lobsters, prawns, shrimps, woodlice, crayfish, barnacles and the water flea *Daphnia*. All crustaceans possess an EXOSKELETON made of CHITIN impregnated with calcium carbonate, making it hard and impervious to water, acids, alkalis and solvents. The body is segmented, each part bearing appendages that serve a number of purposes. For example, crabs have eight pairs of thoracic appendages for walking and feeding, abdominal appendages for swimming and two pairs of antennae. Some crustaceans, for example *Daphnia*, are filter feeders, filtering particles from the water with hair-like bristles; others use modified appendages as mouth parts to feed on any organic matter; a few are parasitic. Most crustaceans have separate sexes and lay eggs. *See also* BRANCHIOPODA, MALACOSTRACA.

crustacean A member of the class CRUSTACEA.

cryostat An apparatus used to maintain constant low temperatures.

crypt of Lieberkühn One of several intestinal glands between the VILLI of the ILEUM. Crypts of Lieberkühn contain specialized cells (Paneth cells) that release enzymes involved in digestion.

CSF *See* CEREBROSPINAL FLUID.

Ctenophora A class of the phylum CNIDARIA. The body form is neither polyp nor medusa and movement is by means of CILIA fused in rows (combs). Ctenophoras possess specialized sticky cells called lasso cells that are used for capturing prey. They do not possess stinging cells nor do they penetrate their prey as other COELENTERATES do. Examples include comb-jellies and sea-gooseberries.

CT scan *See* CAT SCAN.

cupula A flat gelatinous plate found within the AMPULLA of the inner ear that is concerned with balance (*see* EAR).

curie (Ci) A unit of radioactive activity, now superseded by the BECQUEREL. One curie is equivalent to 3.7×10^{10} Bq.

Cushing's syndrome A rare disorder caused by an over-production of GLUCOCORTICOIDS by the ADRENAL GLANDS. Glucocorticoids regulate the body's metabolism and over-production leads to high blood pressure, high blood sugar, and an obese trunk with spindly limbs and a moon-like face with high colour. The cause is most often a tumour in the PITUITARY GLAND which causes over-production of ADRENOCOR-

TICOTROPHIC HORMONE (ACTH), which stimulates the adrenal cortex. Tumours in other locations can also be a cause, as can long-term hormone treatment for other conditions.

cuticle A tough, non-cellular, outer layer (the waxy layer) of plants that prevents loss of water and provides protection. In higher plants, the cuticle is continuous except for the stomata (*see* STOMA) and LENTICELS. It is secreted by cells of the EPIDERMIS.

In invertebrates, such as insects and other arthropods, the cuticle acts as a protective EXOSKELETON. The insect cuticle often contains additional compounds and is of greater complexity than plant cuticle.

CVS *See* CHORIONIC VILLUS SAMPLING.

cyanide Any salt containing the cyanide ion, CN⁻, such as potassium cyanide, KCN. Cyanides are highly toxic because the ion has the ability to form stable complexes with the iron in HAEMO-GLOBIN, preventing the uptake of oxygen.

cyanobacteria, *blue-green bacteria* A group of single-celled prokaryotic organisms of the kingdom PROKARYOTAE. Formerly called blue-green algae, they are found in damp surfaces of rocks and trees, aquatic habitats and in the soil. They are thought to be the oldest form of life. Many species of cyanobacteria can photosynthesize but their pigments are not contained in CHLOROPLASTS but in THYLAKOID membranes within the cell. Some associate symbiotically with fungi to form LICHENS, and some are important in the NITROGEN CYCLE as they have an ability to fix atmospheric nitrogen.

Cyanobacteria can be a problem in waters polluted with NITRATES and PHOSPHATES from fertilizers and detergents, because they multiply rapidly forming ALGAL BLOOMS that use up all the available oxygen. When the bacteria die after a bloom, they can release toxins poisonous to fish and other animals (*see* POLLUTION).

cyanocobalamin *See* VITAMIN B.

cycads Large palm-like plants, now extinct, previously found in abundance in the tropics and subtropics. They are grouped with the GYMNOSPERMS. *See also* GINKOS.

cyclic AMP (cAMP) A cyclic NUCLEOTIDE produced from ATP by the action of the enzyme ADENYLATE CYCLASE. Cyclic AMP is an important second messenger, and is produced in response to other signals, for example hormones. It determines the rate of many biochemical pathways.

cysteine A sulphur-containing amino acid found in most proteins, especially KERATIN.

cystine The main sulphur-containing amino acid present in proteins.

cytidine In biochemistry, a PYRIMIDINE NUCLEOSIDE consisting of the organic base CYTOSINE and the sugar RIBOSE.

cytochrome A protein forming part of the ELECTRON TRANSPORT SYSTEM. Electrons are transferred to the next cytochrome in a series of electron carriers, resulting in the reduction of oxygen (O_2) to oxygen ions (O^{2-}), which combine with hydrogen ions to form water (H_2O) during aerobic RESPIRATION. The passage of electrons along the carrier chain results in the release of energy, which is used to make ATP. Cytochromes are located in inner mitochondrial membranes, ENDOPLASMIC RETICULUM and THYLAKOIDS of CHLOROPLASTS.

cytochrome oxidase An enzyme involved in the ELECTRON TRANSPORT SYSTEM. Cytochrome oxidase is a complex consisting of the final CYTOCHROMES (a_1 and a_3) of the electron transport system and a copper PROSTHETIC GROUP. It is responsible for combining two hydrogen atoms with an oxygen atom to form water. Cyanide is a respiratory inhibitor, since it can attach to the copper prosthetic group of cytochrome oxidase, so blocking its action.

cytokine A soluble substance produced by cells of the LYMPHOID SYSTEM that is involved in communication between cells of the IMMUNE RESPONSE. There are two categories, LYMPHOKINES, which are secreted by LYMPHOCYTES, and MONOKINES, secreted by MACROPHAGES. However, some cytokines are secreted by both cell types, for example INTERFERON and INTERLEUKIN.

cytokinin A PLANT GROWTH SUBSTANCE that promotes CELL DIVISION in the presence of AUXINS. Cytokinins are derivatives of ADENINE (*see* NUCLEOTIDES) and are found in actively dividing tissues, such as fruits and seeds. They can delay ageing of leaves. They are thought to operate by increasing the metabolism of NUCLEIC ACIDS and PROTEIN SYNTHESIS.

cytology The study of cells and their functions, particularly through MICROSCOPY.

cytolysis The disintegration of cells, usually by destruction or dissolution of their PLASMA MEMBRANES.

cytoplasm All the PROTOPLASM of a living cell, excluding the nucleus. It is a transparent, slightly viscous fluid composed of a soluble jelly-like part (the cytosol) and the ORGANELLES embedded within this.

In many cells, the cytoplasm is made up of two layers: the ectoplasm, a dense gelatinous outer layer containing few granules and associated with cell movement; and the endoplasm (or plasmasol), forming an inner layer more fluid in nature and containing many granules and most of the organelles. In plants, the ectoplasm is equivalent to the CELL MEMBRANE.

cytoplasmic (*adj.*) Relating to CYTOPLASM.

cytoplasmic streaming The directional movement of the CYTOPLASM of certain cells, which allows substances to move through the cell. *See also* HYPHA.

cytosine An organic base called a PYRIMIDINE occurring in NUCLEOTIDES. *See also* DNA, RNA.

cytoskeleton A network of protein filaments (thread-like structures) and MICROTUBULES found within the CYTOSOL of EUKARYOTIC cells, giving the cell its shape and internal organization. The cytoskeleton enables movement of materials throughout the cell and is also concerned with cell locomotion. ACTIN is an important constituent of the cytoskeleton.

cytosol, *hyaloplasm* The liquid part of the CYTOPLASM.

cytotoxic (*adj.*) Capable of killing cells. The term cytotoxic is used particularly to describe drugs that destroy cells, for example in CHEMOTHERAPY, or in reference to a subset of T CELLS (*see* CYTOTOXIC T CELLS).

cytotoxic T cell (Tc) A subset of T CELLS that can recognize and destroy tumour or virus-infected cells. They can only recognize foreign ANTIGEN when it is presented in combination with host cell antigens of the MAJOR HISTOCOMPATIBILITY COMPLEX (*compare* NATURAL KILLER CELLS). Cytotoxic T cells kill cells by releasing proteins that cause perforation of the infected cell membrane, leading to lysis and cell death.

D

Darwinism The theory of EVOLUTION proposed by Charles Darwin (1809–1882). It suggests that new species can arise and old species can become extinct by a process called NATURAL SELECTION. Darwin's work still forms the basis of modern day theories on evolution. *See also* COMPETITION.

Darwin's finches A group of about 14 species of finches, unique to the Galapagos Islands, that were studied by Charles Darwin (1809–1882). Each species had adapted to fill a different ECOLOGICAL NICHE. *See also* ADAPTIVE RADIATION.

DDT, *dichlorodiphenyltrichloroethane* A synthetic PESTICIDE discovered in 1939 that has been used world-wide to kill organisms such as lice, fleas and mosquitoes (to combat malaria). It has caused many problems because it is persistent and accumulates along FOOD CHAINS. The use of DDT is banned in most countries, although it is still in use in developing countries where insect-borne diseases are a problem. Despite being banned, the persistence of DDT means that it is still found in many organisms. Many insects have developed DDT resistance. *See also* BIOMAGNIFICATION.

deamination Removal of the AMINO GROUP, $-NH_2$, from an AMINO ACID to form either ammonia, urea or uric acid, for excretion in the URINE. In vertebrates, this process occurs in the liver, to remove unwanted amino acids.

death phase See BACTERIAL GROWTH CURVE.

deci- Prefix denoting one tenth. For example, a decimetre is one tenth of a metre (0.1 m).

decidua See ENDOMETRIUM.

deciduous (*adj.*) In botany, describing plants that lose their leaves seasonally, for example in the autumn. This is in contrast to evergreen plants, where the leaves are retained all year.

deciduous teeth, *milk teeth* In mammals, the first of the two sets of teeth that mammals have. *See* TOOTH.

decomposer A general term for any organism in the FOOD CHAIN that feeds on dead material or excrement. Decomposers break complex organic compounds into simple organic or inorganic ones, enabling the recycling of nutrients to the soil or atmosphere. Decomposers are mostly saprophytic (*see* SAPROTROPH) bacteria and fungi and play a vital role in the NITROGEN CYCLE, CARBON CYCLE and PHOSPHORUS CYCLE.

decomposition A chemical reaction in which a compound is broken down into its elements or into simpler compounds, usually under the action of heat.

defecation Elimination of FAECES from the body.

deforestation The destruction of forests by humans without replanting new trees or allowing for a cycle of regeneration. Deforestation can occur to provide trees to use as timber or fuel, or to clear land for agricultural purposes or mining. Deforestation of both tropical rainforests and temperate forests is a great ecological problem. More than half of the world's rainforests have been destroyed.

Some of the consequences of deforestation are soil erosion, flooding, drought and GLOBAL WARMING. Also, the salt level in the ground may rise to the surface, making the ground unsuitable for farming. It can lead to DESERTIFICATION, which, along with other factors, such as overgrazing and intensive cultivation, then leads to soil infertility.

dehiscent (*adj.*) Describing a fruit that opens to shed its seeds, for example the poppy and pea.

dehydrate (*vb.*) To remove water from a substance.

deletion A chromosomal MUTATION in which a portion of a CHROMOSOME is lost.

dendrite One of many short projections from the cell body of a NEURONE that conducts NERVE IMPULSES from other neurones towards its own cell body. Impulses are passed to the dendrite of one neurone from the tip of an AXON of another neurone during a SYNAPSE.

dendrochronology The dating of fallen and fossilized trees by comparison of their growth rings. Dendrochronology is used to provide a check on radiocarbon dates, enabling these to be recalibrated to take account of the changes in carbon–14 concentration in the atmosphere. *See also* RADIOCARBON DATING.

dendron An extension from the body of a NEURONE that branches into DENDRITES and conducts NERVE IMPULSES from other neurones towards its own cell body.

denitrification The process by which NITRATES in the soil are converted back to atmospheric nitrogen. Anaerobic bacteria, such as *Pseudomonas denitrificans* and *Thiobacillus denitrificans,* bring about this conversion, particularly in waterlogged soil. Because most organisms cannot utilize atmospheric nitrogen (*see* NITROGEN CYCLE), denitrification reduces soil fertility. To avoid this, soil is ploughed and dug to improve drainage and aeration. *See also* NITRIFICATION.

density-dependence A method by which cells or organisms naturally regulate the size of their population. One or more factor(s) either speeds up the increase in the size of a population when its density is low, or decreases the expansion when the population density is high. The effect of the factor must be proportional to the population density for there to be true density-dependence. Cells that show CONTACT INHIBITION are exhibiting density-dependent inhibition.

dentine The constituent of the main part of a TOOTH, which lies between the hard enamel layer and the pulp cavity. Dentine is secreted by cells called odontoblasts and is similar to BONE in composition, but harder. The living cells of dentine are provided with nutrients and oxygen by nerves and blood vessels within the pulp cavity in the centre of the tooth. These also remove waste products. Elephant tusks are made of ivory which is dentine.

dentition The type and number of teeth in a species. *See* TOOTH.

deoxyribonuclease *See* DNASE.

deoxyribonucleic acid *See* DNA.

deoxyribose A PENTOSE sugar ($C_5H_{10}O_4$) that is a component of DNA.

depolarization A reversal of the charge on a nerve cell (*see* POLARIZATION) that occurs when a NERVE IMPULSE is transmitted along the nerve AXON and an ACTION POTENTIAL is generated. After the impulse has passed, the nerve cell membrane returns to its original polarized state; this is repolarization. Such depolarization is caused by a change in the distribution

of ions across the membrane. Stimulation of a muscle fibre has a similar effect.

depro-provera-progestin A long-lasting contraceptive preparation that is administered as an injection every three months.

dermatitis Inflammation of the skin. *See also* ECZEMA.

dermatology The study of SKIN and its disorders.

dermis The inner layer of the SKIN. The dermis contains blood vessels, hair follicles, nerves (RECEPTORS for touch, pressure, pain and temperature) and SEBACEOUS GLANDS and SWEAT GLANDS embedded in CONNECTIVE TISSUE. *See also* SENSE ORGAN.

desert An arid region capable of supporting only very few life forms.

desertification The creation of deserts as a result of climatic changes or human intervention. Desertification can be caused by DEFORESTATION, overgrazing or intensive cultivation that leads to soil infertility. It can be reversed by replanting trees or grasses and improving soil fertility and water retention.

detritivore Any organism in the FOOD CHAIN that feeds on the organic debris (detritus) from decomposing plants and animals and excrement. Detritivores are usually larger than DECOMPOSERS and digest their food internally. Examples are earthworms, maggots and woodlice. *Compare* SAPROTROPHS, which feed on the excrement or dead bodies of others and digest their food externally.

detritus Organic debris from decomposing plants and animals. *See also* DETRITIVORE.

dextrin A POLYSACCHARIDE formed during the HYDROLYSIS of STARCH.

dextrorotatory (*adj.*) Describing a compound that rotates the plane of polarization of plane-polarized light (light that oscillates in one plane only) to the right (clockwise). This used to be denoted by the prefix *d-* but (+) is now used. This is not to be confused with the prefix D- used to indicate the configuration of CARBOHYDRATES and AMINO ACIDS.

dextrose *See* GLUCOSE.

diabetes A name commonly used to refer to the disease *Diabetes mellitus,* in which there is a failure in the production of the hormone INSULIN by the PANCREAS. Insulin usually regulates blood sugar levels and excess glucose is stored in the liver. In diabetes, excess glucose is not stored so the blood sugar level rises. The kidney cannot absorb all the glucose passing through it, so the excess is secreted in the urine. Diabetes can potentially cause kidney failure, blindness and death in severe cases. Other symptoms include weight loss, thirst and coma.

Treatment of diabetes is by administration of insulin, either orally or by injection. The insulin can be obtained from pigs or calves, or it can be produced synthetically, since its structure is well known. Human insulin can be produced from bacteria by GENETIC ENGINEERING. Mild cases of diabetes can be controlled by diet.

Approximately 4 per cent of the world population have diabetes. The disease can start early in life in which case it is usually insulin-dependent diabetes caused by the autoimmune (self) destruction of certain cells in the pancreas, or can begin later in life in which case it is usually insulin-independent. Onset can also occur in pregnancy, called *Gestational diabetes*, leading to the birth of large babies, but this condition can be temporary.

Another form of diabetes is *Diabetes insipidus*, in which there is a failure to secrete ANTI-DIURETIC HORMONE, resulting in an excess of urine production.

diageotropic The growth of a plant part in response to gravity such that its axis is at right angles to the direction of gravitational force (that is, horizontal). Leaves and RHIZOMES are diageotropic. *See* GEOTROPISM.

dialysate The fluid used in DIALYSIS to surround a semipermeable bag or membrane containing a mixture to be separated. In many cases, the dialysate is water or a BUFFER. Small, unwanted molecules diffuse across the membrane into the dialysate to be disposed of. The term also refers to the dialysis fluid used in HAEMODIALYSIS for patients with kidney failure, into which the toxic substances are filtered out of the blood through a semipermeable membrane.

dialysis The method by which small and large molecules in a mixed solution are separated. The mixed solution is placed inside a semipermeable bag (one that allows molecules of a certain size to pass through its pores) and surrounded by water. Small molecules diffuse out of the bag into the water, which is repeatedly changed, leaving the larger molecules inside the bag. In kidney failure, this principle is used in RENAL DIALYSIS to remove toxic substances from the bloodstream.

In HAEMODIALYSIS, the patient's blood is passed through a pump, where it is separated by a semipermeable membrane from the dialysis fluid (dialysate). Toxic substances, such as urea, are filtered out, but RED BLOOD CELLS and WHITE BLOOD CELLS remain in the blood. Loss of useful substances, such as GLUCOSE and salt is minimized by adding them to the dialysate so that an equilibrium exists between the blood and dialysate.

Another method called CONTINUOUS AMBULATORY PERITONEAL DIALYSIS (CAPD) uses the membrane enclosing the PERITONEAL CAVITY as the semipermeable membrane. The patient on CAPD can move around during dialysis.

Both methods of dialysis are expensive and are not always suitable. Kidney transplants are more effective and desirable in cases of chronic kidney failure.

diapause In insects and some other invertebrates, a period of reduced metabolic activity (*see* METABOLISM) that can occur at any stage of development (usually eggs or pupae) but generally only once in a lifetime. It often coincides with the winter months and increases the organism's chances of survival. *See also* BIORHYTHM.

diaphragm 1. In mammals, a sheet that is made of muscle and TENDON and separates the THORAX from the ABDOMEN. It is arched when resting but flattens during INSPIRATION, so reducing the pressure within the thorax to allow air to be drawn into the lungs.

2. *cap, Dutch cap* A method of CONTRACEPTION that consists of a rubber dome placed over the female CERVIX during sexual intercourse, preventing entry of sperm into the UTERUS. The diaphragm is 97 per cent effective if fitted correctly, used with a spermicide (cream or jelly preparations that kill sperm) and left in place for 6–8 hours after intercourse.

diaphysis The hollow shaft in the centre of a limb bone that contains BONE MARROW.

diatom Any member of the phylum BACILLARIOPHYTA.

dichlorodiphenyltrichloroethane *See* DDT.

dicotyledon A flowering plant that possesses two COTYLEDONS, or seed leaves, in the embryo. Most flowers, vegetables, fruit trees and forest trees (except CONIFERS) are dicotyledons. The cotyledons in dicotyledonous plants are usually surrounded by an ENDOSPERM that provides the food for the embryo, sometimes by passing it on to the cotyledon to be used by the embryo.

Dicotyledons usually have broad leaves, with net-like veins, and can be small or large plants. The flower parts are arranged in groups of five and the VASCULAR BUNDLE forms a regular ring arrangement in the centre of the stem. POLLINATION is usually by insects. After GERMINATION, most dicotyledons spread out above the soil forming the first green leaves, and are called 'epigeal'. *See also* MONOCOTYLEDON.

dictyosome *See* GOLGI APPARATUS.

diet The range of foods eaten by an animal. Diet varies greatly between species. HERBIVORES eat green plants, CARNIVORES eat meat from other animals, and OMNIVORES eat both plants and animal meat. In a mammalian diet, the main constituents are: CARBOHYDRATES and FATS for energy, PROTEINS for growth and repair; VITAMINS and MINERALS for specific functions; water (which generally makes up 70 per cent of the total body weight of the mammal); and ROUGHAGE for efficient digestion.

Humans need nine ESSENTIAL AMINO ACIDS that we cannot make within our bodies and therefore need to take in the form of proteins from meat or vegetables. Minerals needed for various functions are numerous, but include: calcium from dairy foods for bones and teeth; chlorine and sodium in common salt for maintenance of ANION/CATION balance; SULPHATES from dairy foods and meat for proteins; and potassium from meat, fruit and vegetables for nerves and muscle action.

The amount of energy from food required by an individual varies according to gender, age and activity. Excess energy intake leads to obesity (overweight) and too little can lead to a condition called maramus in young children, where the child is irritable, does not grow and becomes thin due to lack of energy from carbohydrates and fat. A lack of protein in the diet of young children results in a condition called Kwashiorkor, where the body is swollen, the hair is soft and changes colour and growth is retarded. A high intake of fat (especially saturated fat; *see* FATTY ACIDS) has been related to heart disease, although other factors, for example, lack of exercise, smoking and stress, play a part.

Anorexia nervosa and bulimia nervosa are self-inflicted eating disorders, seen mostly in teenage girls and young women, that are usually due to psychological problems or stress. Anorexia nervosa is characterized by lack of eating and consequent wasting of the muscles. Bulimia nervosa results in a similar wasting away but the victim overeats (binges) and then causes self-induced vomiting.

dietary fibre *See* ROUGHAGE.

differentiation The process whereby cells develop a specialized function and adopt a particular morphology – the change from simple to more complex forms. Differentiation occurs in EMBRYONIC DEVELOPMENT where cells differentiate into particular tissues and organs. It also occurs in plants in MERISTEM tissue and during regeneration, for example as a result of damage or in plant cuttings. Differentiation can be triggered by a number of factors such as cell density, GROWTH FACTORS or signals from the EXTRACELLULAR MATRIX. When a cell begins to differentiate it usually loses the ability to divide.

diffuse (*vb.*) To spread out by DIFFUSION.

diffusion The spontaneous and random movement of molecules or particles in a fluid (gas or liquid) from a region where they are at a high concentration to one where they are at a low concentration, until a uniform concentration or dynamic equilibrium is achieved. Once at a uniform concentration, the molecules will continue to move in random motion, but there is no net diffusion. The concentration gradient is the difference in concentration of a substance between two regions.

In biological systems, diffusion often occurs across epithelial layers (*see* EPITHELIUM) or CELL MEMBRANES, and since a greater surface area leads to faster diffusion, areas specialized for this purpose often have VILLI to increase their surface area. Sometimes channels occur within a cell membrane or carrier molecules exist to speed up diffusion of specific substances. Diffusion is important in the transport of nutrients, respiratory gases and NEUROTRANSMITTERS within and between cells. *Compare* ACTIVE TRANSPORT.

digestion In animals, the breakdown of food, both physically and chemically (by enzymes), into its basic constituents, ready for absorption to convert to energy. Digestion usually occurs in the STOMACH and INTESTINES. *See also* DIGESTIVE SYSTEM.

digestive system, *alimentary canal, gut* The system of cavities, tubes, organs and glands associated with DIGESTION. In humans the system begins at the MOUTH, where food is mixed

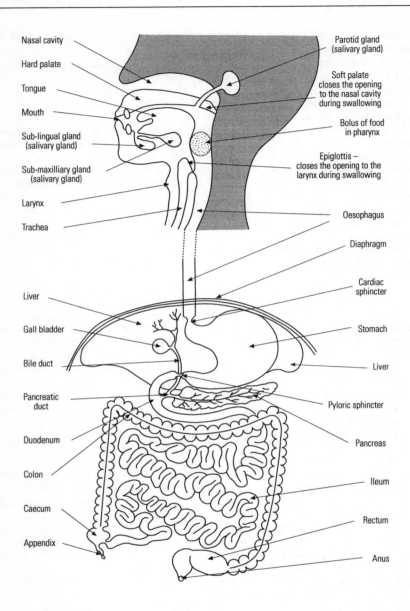

Human digestive system.

Labels (clockwise/as shown):

Nasal cavity

Hard palate

Tongue

Mouth

Sub-lingual gland (salivary gland)

Sub-maxilliary gland (salivary gland)

Larynx

Trachea

Liver

Gall bladder

Bile duct

Pancreatic duct

Duodenum

Colon

Caecum

Appendix

Parotid gland (salivary gland)

Soft palate closes the opening to the nasal cavity during swallowing

Bolus of food in pharynx

Epiglottis – closes the opening to the larynx during swallowing

Oesophagus

Diaphragm

Cardiac sphincter

Stomach

Liver

Pyloric sphincter

Pancreas

Ileum

Rectum

Anus

with saliva and the first stage of digestion occurs. Food is chewed into a round ball called a 'bolus' and this is pushed by the TONGUE to the back of the mouth and to the PHARYNX. In this region, the OESOPHAGUS and TRACHEA meet and the bolus passes down the oesophagus aided by a number of reflexes that prevent it entering the trachea. The food is lubricated by secretions from glands lining the oesophagus, and then passes into the STOMACH. It is here that storage and further digestion is carried out. The resulting fluid (CHYME) passes to the

DUODENUM and further breakdown, and most absorption of nutrients, occurs in the SMALL INTESTINE. What remains passes to the LARGE INTESTINE for water absorption and storage of FAECES before their excretion.

The digestive system has a good blood supply through which nutrients from digested food are carried to the LIVER for use by the body cells.

In simpler organisms, the digestive system can be a cavity with a single opening, for example in jellyfish and *Hydra*, where some of the contents taken in are absorbed and the rest expelled through the same opening.

See also GALL BLADDER, GIZZARD, ILEUM.

digit A finger or toe. *See* PENTADACTYL LIMB.

dihybrid cross In genetics, a cross between two pure-bred individuals of the same species that differ in two characteristics. The characteristics could, for example, be seed colour and shape. The F_1 GENERATION yields offspring identical for those characteristics (e.g. green, round seeds) because the dominant ALLELES are expressed. When the F_1 offspring are crossed with one another, the offspring would show different characteristics (*see* PHENOTYPES) in the ratio of 9:3:3:1, according to MENDEL'S LAWS. In practice, not all dihybrid crosses give this ratio. *See also* MONOHYBRID CROSS.

dioecious The presence of male and female reproductive organs on different individuals. This term is often used to refer to plants that have male and female flowers on separate plants of the same species. This arrangement favours cross-fertilization (*see* FERTILIZATION). *Compare* MONOECIOUS.

diploblastic (*adj.*) Of an animal, having a body that develops from just two of the three GERM LAYERS – the ECTODERM and ENDODERM. COELENTERATES are diploblastic. *Compare* TRIPLOBLASTIC.

diploid (*adj.*) The presence of two sets of CHROMOSOMES in the nucleus of a cell. Most animal cells are diploid, except the GAMETES in sexually reproducing species, which are HAPLOID. *See also* POLYPLOID.

Diplopoda A class of the phylum ARTHROPODA, including the millipede.

disaccharide A double sugar formed by the combination of two MONOSACCHARIDES, with the loss of a water molecule. A disaccharide can be split into its single sugars by the addition of water (a HYDROLYSIS reaction). Di-saccharides are sweet, soluble and crystalline. Examples include SUCROSE (cane sugar or table sugar, $C_{12}H_{22}O_{11}$; GLUCOSE plus FRUCTOSE), maltose (glucose plus glucose) and LACTOSE (glucose plus GALACTOSE). *See also* CARBOHYDRATE, POLYSACCHARIDE.

disease An impairment of the normal functioning of an organism. Causes of disease can be congenital (inborn) or acquired through infection or injury. Many diseases have a known causative agent (a PATHOGEN), such as a virus, bacteria, fungus or PROTOZOA, that results in a characteristic set of symptoms, for example a rash in the case of chickenpox. Some organisms cause disease by releasing toxins or venoms. For some diseases, such as cancer, arthritis and multiple sclerosis, the cause is complex or even unknown.

If many people suffer from the same disease at the same time it is called an epidemic. Some diseases only occur in certain parts of the world (endemic), other diseases occur all over the world (pandemic). Sporadic refers to irregular outbreaks of diseases in particular places.

dispersal The movement of seeds, eggs and offspring away from their parents in order to prevent unfavourable competition or overcrowding. There are various methods of dispersal, for example by insects, wind, animals (that eat or carry seeds) and locomotion (of animal offspring themselves).

DNA, *deoxyribonucleic acid* A complex, very large molecule that contains all the information for building and controlling a living organism. It is a double-stranded NUCLEIC ACID made of NUCLEOTIDES with the bases ADENINE, GUANINE, CYTOSINE and THYMINE (never URACIL, which is found only in RNA) and a PENTOSE sugar that is always DEOXYRIBOSE. Except in bacteria, DNA is found in the nuclei of cells, arranged in CHROMOSOMES. The molecular structure of DNA was first proposed by James Watson (1928–) and Francis Crick (1916–) in 1953, for which they were awarded the Nobel prize for medicine and physiology.

The DNA molecule consists of two polynucleotide strands (each of millions of nucleotides) linked to each other by base pairing (*see* BASE PAIR) and HYDROGEN BONDS. The bases adenine and thymine always link and cytosine and guanine always link. The linking of one PURINE and one PYRIMIDINE in this way

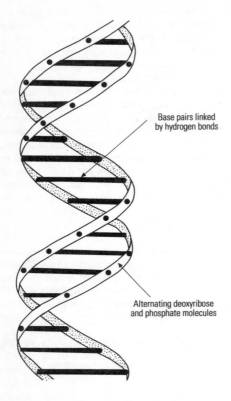

Base pairs linked
by hydrogen bonds

Alternating deoxyribose
and phosphate molecules

The DNA structure is like a ladder with alternating deoxyribose and phosphate molecules forming the uprights and base pairs forming the rungs. The ladder is twisted to form a double helix. The uprights run in opposite directions to each other. Base pairings are always adenine-thymine and cytosine-guanine.

allows the same spacing between the strands throughout the length of the molecule. So the DNA is like a ladder, with the base pairs forming the rungs and the deoxyribose and PHOSPHATE groups forming the uprights. In addition, the two chains forming the upright run in opposite directions and are called antiparallel. The ladder is then twisted into a double helix.

The hereditary information is stored as a specific sequence of bases. Individual AMINO ACIDS are coded for by a set of three bases called a CODON. The precise sequence of bases therefore determines the amino acids that are made and therefore the PROTEINS that are

produced by the cell. This sequence is called the GENETIC CODE and, because of the importance of the proteins it codes for, it controls the whole organism. In order for the genetic information to be passed on from cell to cell and generation to generation, DNA must be able to replicate. It does this by SEMI-CONSERVATIVE REPLICATION.

See also COPY DNA, GENE, GENETIC ENGINEERING, PROTEIN SYNTHESIS, RECOMBINANT DNA.

DNA fingerprinting *See* GENETIC FINGERPRINTING.

DNA ligase *See* LIGASE.

DNA polymerase A POLYMERASE enzyme that catalyses the elongation of a new DNA strand during SEMI-CONSERVATIVE REPLICATION.

DNAse, deoxyribonuclease One of many enzymes that hydrolyse (*see* HYDROLYSIS) DNA by breaking down the sugar–phosphate bonds. *See* RESTRICTION ENDONUCLEASE.

DNA sequencing A method used to determine the sequence of NUCLEOTIDES in a length of DNA. Two methods are used: the Maxam-Gilbert (base-destruction) method and the Sanger (dideoxy chain terminator) method, the latter being more commonly used. Both use single stranded DNA labelled with radioactive 32P-phosphate. The methods use different ways of obtaining DNA fragments of different lengths, cleaved at specific nucleotide bases. Both use POLYACRYLAMIDE GEL ELECTROPHORESIS and AUTORADIOGRAPHY to analyse the order of nucleotides, which can be read directly from the four tracks on the autoradiograph, each representing one of the four nucleotide bases (A,C,G or T).

dominant In genetics, an ALLELE that is expressed in preference to another in the HETEROZYGOUS form. The allele that is masked, and therefore not expressed, is RECESSIVE. *See also* CODOMINANCE.

dopamine ($C_8H_{11}NO_2$) In biochemistry, a NEUROTRANSMITTER that is an intermediate in the synthesis of ADRENALINE. There are special areas in the brain that particularly use dopamine to transmit nerve impulses. Patients with the tremors of Parkinson's disease show degeneration of such areas.

Doppler effect The apparent observed change in observed frequency (or wavelength) of a wave due to the relative motion between the source and the observer. An example is a police car siren that increases in pitch (frequency) as it moves towards a stationary observer and decreases in pitch as it moves away.

dormancy In plants, a phase of reduced activity shown by seeds, spores and some buds that often aids survival of the plant during unfavourable conditions. Some seeds will remain dormant for many years, even during apparently favourable conditions. Particular requirements, such as a period of cold (to prevent GERMINATION until after winter), a period of light, or a period of 'after-ripening' (for internal changes to take place), are needed for dormancy to be broken. Dormancy can often be broken artificially. *See also* BIORHYTHM.

dorsal (*adj.*) **1.** Of an animal, relating to the back or spine.

    2. Of a plant, relating to the back of an organ or part. *Compare* VENTRAL.

dorsal fin The FIN on the back of a fish that controls balance.

dorsal root The uppermost pair of spinal nerves along the length of the SPINAL CORD which carry only sensory nerves (*see* SENSORY SYSTEM). The cell bodies of these nerves form the dorsal root ganglion, a swelling within the dorsal root lying just outside the spinal cord. *Compare* VENTRAL ROOT.

double circulation *See* CIRCULATORY SYSTEM.

double fertilization An event unique to flowering plants, in which two male nuclei fuse with two female nuclei. When a POLLEN grain lands on a STIGMA of a flowering plant, it absorbs water and an outgrowth called a POLLEN TUBE grows inwards down the STYLE towards the OVULE (*see* CARPEL). The tube transports two male nuclei to the ovule, and when it enters the EMBRYO SAC (usually through a small hole called the MICROPYLE) it disintegrates and the nuclei are released. One nucleus fuses with the female egg nucleus to give rise to a DIPLOID ZYGOTE, and the other fuses with two polar nuclei (*see* POLAR NUCLEUS) to form a TRIPLOID PRIMARY ENDOSPERM NUCLEUS. The latter provides nourishment for the zygote and the

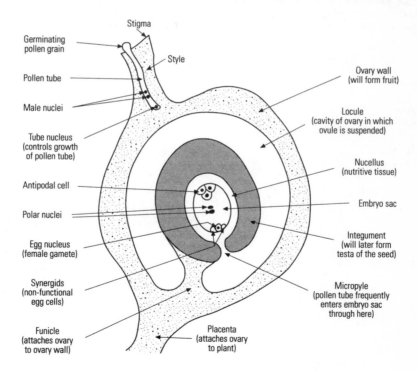

Mature carpel during double fertilization.

process seems to be a way of ensuring that nourishment is only provided if there is a zygote to use it. *See also* FERTILIZATION, GERMINATION, POLLINATION.

Down's syndrome A genetic disorder characterized by mental retardation, altered facial features and reduced life expectancy. It is caused by the presence of three copies of chromosome 21, which occurs when the pair of chromosomes does not separate at MEIOSIS, resulting in one GAMETE with 24 chromosomes and one with 22, instead of 23 each. This is called non-disjunction. Down's syndrome occurs in 1:700 births, and the incidence increases with maternal age.

drug Any substance administered to animals or humans that alters a biological function. Most drugs are used in medicine, for example antibiotics, immunosuppressives, sedatives and pain-relievers. Other drugs are used socially, for example nicotine in tobacco, alcohol and heroin. All drugs are potentially harmful.

drupe, *stone fruit* A fleshy fruit similar to a BERRY, with an outer skin (exocarp) and a fleshy middle layer (mesocarp) but containing a single seed (berries have many seeds) that is surrounded by a hard woody layer (endocarp). Examples include the cherry, plum, coconut (the outer wall is dry when mature) and blackberry (which comprises many drupes together).

duodenum In vertebrates, a short length of SMALL INTESTINE immediately after the stomach, where most digestion and some absorption of food occurs. The inner lining of the walls of the duodenum, like the rest of the small intestine, consists of projections called VILLI, which are folds to increase its surface area. Food enters the duodenum through a ring of muscle at the base of the stomach called the pyloric sphincter, and mixes with BILE juice from the liver and PANCREATIC JUICES, both of which neutralize the acid CHYME entering from the stomach and aid the digestion of fat.

duplication A chromosomal MUTATION in which a portion of, or an entire CHROMOSOME is copied.

dura mater The outermost of the three membranes that cover the brain and spinal cord. *See* MENINGES.

Dutch cap *See* DIAPHRAGM.

E

E102 *See* TARTRAZINE.

ear In animals, the organ of hearing and balance. The ear consists of three parts. The outer ear is a flap of cartilage (the pinna) that collects, amplifies and focuses the air vibrations that constitute sound, and directs them along the AUDITORY CANAL to a membrane called the EARDRUM. Sound causes the eardrum to vibrate. The middle ear is an air-filled cavity containing small bones called EAR OSSICLES. Vibration of the eardrum causes movement of the ossicles, which in turn causes a second membrane called the 'oval window' to vibrate. The pressure inside the middle ear cavity is controlled by the EUSTACHIAN TUBE. The inner ear is a fluid-filled region consisting of a COCHLEA, which is concerned with hearing, and the SEMI-CIRCULAR CANALS, which are concerned with balance. Amphibians and some reptiles have no outer ear (the eardrum is in the skin) and fish have no ear. *See also* SENSE ORGAN.

ear canal *See* AUDITORY CANAL.

eardrum, *tympanic membrane* A thin membrane separating the outer EAR from the middle ear. Sound causes the eardrum to vibrate and this movement is detected by the bone EAR OSSICLES of the middle ear.

ear ossicle One of three small bones in the middle EAR. The three ear ossicles are called the

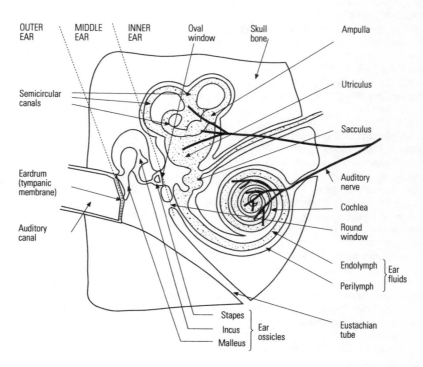

The human ear.

malleus (hammer), incus (anvil) and stapes (stirrup), connected to one another and held in place by muscles. Vibrations of the EARDRUM are detected by the ossicles, which in turn vibrate and cause movement of a membrane called the 'oval window', which separates the middle and inner ear. Sound vibrations are amplified here more than 20 times. Birds, reptiles and many amphibians have only one ear ossicle.

ecdysis The periodic shedding of the EXOSKELE-TON of arthropods or the outermost layer of skin of reptiles to allow new growth. *See also* ARTHROPODA.

Echinoderm A member of the phylum ECHINO-DERMATA.

Echinodermata A phylum of spiny-skinned marine invertebrates with a 5-way radial symmetry. Examples include the starfish, sea urchins and sea cucumbers. Echinoderms have a water vascular (conducting) system with an opening for transporting substances around the body. They have small water-filled sacs called tube-feet, which may have suckers on their ends, by which they move. Most feed on MOLLUSCS, CRUSTACEANS and ANNELIDS. Some can reproduce asexually, but others have separate sexes and fertilization is external. *See also* ECHINOIDIA, STELLEROIDIA.

Echinoidia A class of the phylum ECHINODER-MATA which consists of the sea urchins. Members are globular with no arms. *Compare* STELLEROIDIA.

ECM *See* EXTRACELLULAR MATRIX.

ecological niche The position any one species occupies within an ecological HABITAT, including not just the physical environment but all interactions between living, non-living and behavioural components.

ecological pyramid A diagrammatic representation of the levels in a FOOD CHAIN. *See* PYRAMID OF BIOMASS, PYRAMID OF ENERGY, PYRAMID OF NUMBERS.

ecology (Greek *oikos* = house) The study of the relationship between organisms and the environment in which they live (the study of ECOSYSTEMS). Ecology includes the study of living (BIOTIC) elements and non-living (ABIOTIC) elements and is concerned with factors such as feeding habits, behaviour, competition between individuals or between whole COMMUNITIES, and can include CONSERVATION and management of POLLUTION.

ecosystem The interactions between the living (BIOTIC) and non-living (ABIOTIC) elements within a biological system, such as a lake or a forest, forming an ecological unit. The living elements constitute a FOOD CHAIN. The balance in an ecosystem is delicate and removal of one species can destroy the balance. Because energy is lost at each stage of the food chain, the flow of energy through an ecosystem is in one direction only (unlike minerals or elements such as carbon, nitrogen and phosphorous, which are recycled).

ectoderm In animals, the outermost GERM LAYER of a developing embryo. The ectoderm develops into a neural tube, which is the precursor of the CENTRAL NERVOUS SYSTEM, and also forms EPIDERMIS (e.g. skin and hair). Failure of the neural tube to develop normally leads to a condition called spina bifida. *See also* ENDODERM, MESODERM.

ectoparasite A PARASITE that lives outside its host.

ectopic pregnancy In humans, the IMPLANTATION of a fertilized egg outside the UTERUS (for example, in the FALLOPIAN TUBE or, more rarely, within the abdomen). This condition is life-threatening and the pregnancy must be terminated.

ectoplasm, *plasmagel* The dense, outer layer of CYTOPLASM that is concerned with movement.

ectotherm *See* POIKILOTHERM.

ectothermy *See* POIKILOTHERMY.

eczema A condition of the skin characterized by redness, itching and blistering. Once blisters burst, they may become infected, but eventually dry out with scabs and crusts. If eczema becomes chronic, the skin remains inflamed and flaky and easily bleeds. The mechanisms causing eczema are complex and not fully understood.

There are several types of eczema. Contact eczema is an ALLERGIC REACTION to particular irritants such as surgical tape or perfume. Atopic eczema is the type most commonly found in children. It affects those with a family history of ASTHMA or HAYFEVER and usually improves with age. Seborrhagic eczema develops where SEBACEOUS GLANDS are numerous, for example behind the ears, or as cradle cap in babies. Other types include discoid and varicose eczema. Treatment is usually by steroid skin preparations.

edaphic factor Any factor relating to the physical or chemical composition of SOIL in a particular area, especially in relation to the ECOSYSTEM in that area.

effector A cell or organ by which an animal responds to internal or external stimuli received via the NERVOUS SYSTEM. An effector can be a MUSCLE or a GLAND.

effector neurone A NEURONE that transmits NERVE IMPULSES from the CENTRAL NERVOUS SYSTEM towards an EFFECTOR.

effector system, *motor system* The part of the NERVOUS SYSTEM that carries NERVE IMPULSES away from the CENTRAL NERVOUS SYSTEM to a body muscle or gland, called an EFFECTOR, which then acts upon the impulse. The effector system has two parts: the AUTONOMIC NERVOUS SYSTEM (self-governing), which activates involuntary responses; and the somatic nervous system, which activates voluntary responses.

egestion, *elimination* The removal of undigested material from the DIGESTIVE SYSTEM in the form of FAECES. This is different from EXCRETION because the material has never been inside the body cells. In most animals, faeces are removed through the ANUS.

egg The female GAMETE. After FERTILIZATION by a male gamete, the egg divides to form an EMBRYO. In animals, the eggs can develop inside the female (VIVIPARY or OVOVIVIPARY) or can be deposited by her (OVIPARY). In birds and reptiles, the eggs are protected by a shell and the embryo is nourished by a YOLK. *See* OVULE, OVUM.

egg membrane *See* EXTRAEMBRYONIC MEMBRANE.

elaioplast A type of LEUCOPLAST that stores oils in plants.

elastin A protein of animal CONNECTIVE TISSUE that is very elastic. It is found in LIGAMENTS, arteries and lungs.

electromagnetic radiation RADIATION resulting from the acceleration of electric charge. Examples of this type of radiation include visible light, microwaves, radio waves, INFRARED RADIATION, ULTRAVIOLET RADIATION, X-RAYS and GAMMA RADIATION. Electromagnetic radiation is usually only harmful if it is a high-energy emission, such as close to powerful radio transmitters. This can cause some cancers, particularly leukaemia, loss of hearing, cataracts, organ damage or other more minor problems such as depression or headaches. *See also* IONIZING RADIATION.

electron A negatively charged elementary particle that occurs in all atoms. Atoms consist of a central nucleus surrounded by orbiting electrons. The electron structure is responsible for the chemical properties of the atom. In a neutral atom the number of electrons is equal to the number of PROTONS in the nucleus. An electron that has become detached from an atom is known as a free electron. The electron is the basic particle of electricity.

electron microscope An instrument, developed in 1933, that magnifies objects using a beam of electrons. The electrons are accelerated by a high voltage through the object (held in a vacuum) and focused by powerful electromagnets, instead of optical lenses, onto a fluorescent screen for viewing. A camera may be built in to record what is seen on the screen. A beam of electrons of 0.005 nm is used, compared to the wavelength of about 500 nm of the light rays used in a light microscope. Therefore the RESOLVING POWER of the electron microscope is vastly improved to about 1 nm, and objects can be magnified more than 500,000 times.

The sample to be viewed is fixed and embedded in a special resin called Araldite, which enables ultra-thin sections to be cut. Living specimens cannot be viewed. There are several types of electron microscope.

See also ELECTRON-PROBE MICROANALYSER, SCANNING ELECTRON MICROSCOPE, SCANNING TRANSMISSION ELECTRON MICROSCOPE, SCANNING TUNNELLING MICROSCOPE, TRANSMISSION ELECTRON MICROSCOPE, ULTRAMICROTOME.

electron-probe microanalyser A modified ELECTRON MICROSCOPE in which the object emits X-rays when it is hit by electrons, the different intensities of which indicate the presence of different chemicals. The specimens can be examined without being destroyed.

electron transport system, *respiratory chain* The third stage of cellular RESPIRATION in which hydrogen atoms from the KREBS CYCLE are converted to ATP, so providing energy. The hydrogen atoms are carried from the Krebs cycle by the electron carriers NAD or FAD, and transferred to a series of other carriers at lower energy levels. At each transfer the energy released is used to produce ATP. The other carriers are COENZYME Q and a series of CYTOCHROMES. The hydrogen atoms are split into their protons and electrons and the electrons are passed along the carrier chain and recombine with hydrogen at the end, which links

Summary of the electron transport system.

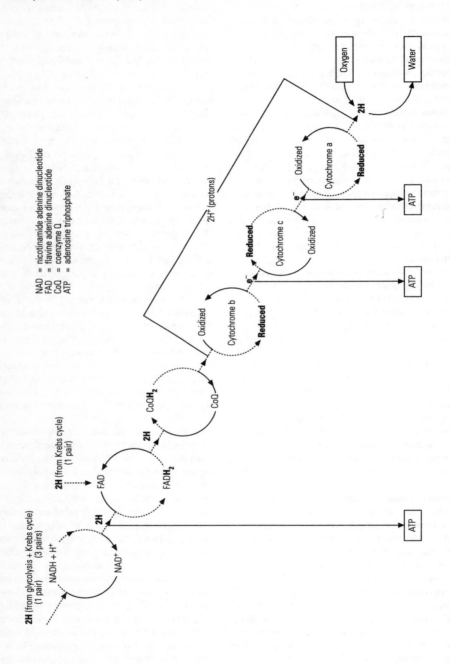

with oxygen to form water. The whole process of forming ATP through oxidation of hydrogen atoms is called OXIDATIVE PHOSPHORYLATION and occurs in the MITOCHONDRIA.

It is thought that electron carriers and their enzymes attach, in a precise sequence, to particles within the inner mitochondrial membrane. In addition, within this membrane there is a mechanism for actively transporting protons from the cell matrix to the space between the mitochondrial membranes, creating a gradient of hydrogen ions across the membrane that provides the energy to synthesize ATP. This is called the CHEMIOSMOTIC THEORY. The respiratory inhibitor cyanide inhibits the enzyme (CYTOCHROME OXIDASE) that links the hydrogen atoms and oxygen to form water, and therefore hydrogen atoms build up and aerobic respiration ceases.

See also GLYCOLYSIS.

electrophoresis, *cataphoresis* A technique used for separating molecules on the basis of their charge. The mixture to be separated is placed on a gel or paper in a BUFFER solution at a given pH, and an electric current is applied through the buffer. Substances of different sizes can be separated because they diffuse at different rates, and can be compared with known standards. Electrophoresis is widely used in biology as an analytical technique, to separate proteins or NUCLEIC ACIDS, for example. *See also* IMMUNOELECTROPHORESIS, POLYACRYLAMIDE GEL ELECTROPHORESIS.

electroreceptor A RECEPTOR cell that detects electric fields. *See* SENSE ORGAN.

elimination *See* EGESTION.

ELISA *See* ENZYME-LINKED IMMUNOSORBANT ASSAY.

embryo In mammals, the name given to a developing fertilized egg before it becomes a FOETUS at 7 weeks of pregnancy. In plants, the young developing plant after FERTILIZATION of an egg cell (*see* OVULE, SEED). *See also* BLASTOCYST, EMBRYONIC DEVELOPMENT, SEXUAL REPRODUCTION.

embryology The study of EMBRYONIC DEVELOPMENT.

embryonic development In an animal, the growth and development of a fertilized egg to become a fully formed FOETUS ready for birth. There are three stages of development: cleavage, gastrulation and organogenesis. Cleavage is the mitotic division (*see* MITOSIS) of the

fertilized egg to form a ball of identical cells. In gastrulation, the ball of cells arranges into definite layers called GERM LAYERS, of which most animals have three. These are the outer ECTODERM, middle MESODERM and inner ENDODERM. In organogenesis, the cells of the germ layers differentiate.

Some cell types are derived from more than one lineage. External factors can have an effect on embryo development, for example the rubella (German measles) virus causes brain damage and deafness, while smoking, alcohol and nutrition have less defined effects.

See also PREGNANCY.

embryo sac A large cell within the OVULE of a SEED PLANT within which DOUBLE FERTILIZATION occurs and the embryo subsequently develops. The embryo sac is formed from one of the cells of the NUCELLUS and its nucleus divides by MITOSIS until there are typically eight nuclei. One of these nuclei develops into the egg cell, which is the female GAMETE that fuses with the male gamete at fertilization. Three other nuclei become ANTIPODAL CELLS and two are POLAR NUCLEI that fuse to form the PRIMARY ENDOSPERM NUCLEUS. Two others degenerate.

embryo transfer *See* IN VITRO FERTILIZATION.

emulsion A mixture of two liquids, with one liquid forming small droplets suspended in the other. *See also* COLLOID.

emulsion test A laboratory test for detecting LIPIDS. The test solution is mixed with alcohol and then with water. The presence of a lipid is shown by the solution turning milky. The lipid is soluble in the alcohol and the water can mix with the alcohol, but the lipid is not miscible with water and gives the milky appearance.

encephalin One of a variety of PEPTIDES produced by nerve cells in the brain that acts as a natural pain-killer in a similar way to OPIATE drugs. However, encephalin is not addictive as it is quickly degraded by the body. *See also* ENDORPHINS.

endemic (*adj*) **1.** A disease or pest that is peculiar to and recurring in a particular locality. For example MALARIA is endemic in parts of Africa.

2. A plant or animal that is native or confined to a particular locality.

endocarp The innermost layer of the wall of a fruit. The endocarp may be hard and woody, as in a DRUPE.

endocrine gland A ductless GLAND that secretes its product (such as a hormone or enzyme) directly into the blood by DIFFUSION. Such glands include the PANCREAS, PITUITARY GLAND, ADRENAL GLAND, THYROID GLAND, OVARY and TESTIS. *Compare* EXOCRINE GLAND.

endocrine system One of the two major co-ordinating systems of animals (the other being the NERVOUS SYSTEM), consisting of a series of ENDOCRINE GLANDS that secrete products such as hormones and enzymes directly into the bloodstream. Compared to the speed of nerve impulses, the endocrine system is a slower means of co-ordination. It controls growth, breeding and maintenance of body fluid composition, for which a slower rate is appropriate. The study of the endocrine system is called endocrinology.

In some instances, there is an overlap between the nervous and endocrine systems shown by neurosecretory (neuroendocrine) cells. These are cells of the nervous system (for example in the brain and HYPOTHALAMUS) that can both conduct nerve impulses and secrete hormones (which travel along the AXON and then in the blood to target cells). ADRENA-LINE is another example of overlap because it can act as a hormone and as a NEURO-TRANSMITTER.

endocrinology The study of the ENDOCRINE SYSTEM.

endocytosis A term used to describe the bulk intake of material into a cell. It is often used as a collective term for PHAGOCYTOSIS and PINOCY-TOSIS. In endocytosis, the cell membrane folds around the material outside the cell, forming a vesicle. The vesicle is then pinched off the cell membrane so that it lies inside the cell. The material is thus taken into the cell without passing through the cell membrane. *See also* COATED PIT, COATED VESICLE, ENDOSOME. *Compare* EXOCYTOSIS.

endoderm The innermost GERM LAYER of the developing animal embryo. It gives rise to the digestive system, bladder, lungs, liver, pancreas and thyroid gland. *See also* ECTODERM, MESO-DERM.

endodermis A layer of single cells surrounding the central column of vascular tissue (XYLEM and PHLOEM) in plant ROOTS and some stems. Part of the cell wall is impregnated with SUBERIN, which forms a characteristic band called the CASPARIAN STRIP. The endodermis

controls the passage of materials into and out of the vascular system.

endolymph Fluid found in the inner ear. *See* COCHLEA, SEMI-CIRCULAR CANAL.

endometrium The inner lining of the UTERUS. The endometrium is a glandular tissue that changes structure in response to hormone stimulation during the MENSTRUAL CYCLE and pregnancy. During the menstrual cycle, the endometrium thickens and its blood supply increases in preparation for pregnancy. If FER-TILIZATION of the OVUM (egg) does not occur, the endometrium is shed and rebuilds again during the next cycle. If fertilization does occur, the ZYGOTE (fertilized ovum) implants into the endometrium, which then thickens further and becomes known as the decidua. In early pregnancy, the decidua produces many hormones and proteins thought to be impor-tant in early embryo survival. Part of the decidua becomes the PLACENTA that supports the foetus during pregnancy.

endoparasite A PARASITE that lives inside its host.

endoplasm, *plasmasol* The inner layer of CYTO-PLASM containing many granules and most of the cell ORGANELLES.

endoplasmic reticulum (ER) A complex system of membranous stacks (cisternae) and tubes within the CYTOPLASM of EUKARYOTIC cells. It is an extension from and is continuous with the nuclear membrane, and is only visible under the ELECTRON MICROSCOPE. Some membranes are called rough ER (rER). These are most abundant in rapidly growing cells, are lined with RIBOSOMES and are the site of PROTEIN SYN-THESIS. Smooth ER (sER) has no associated ribosomes, is concerned with LIPID synthesis and is present in cells producing lipid-related secretions. The membranes of the ER are thought to regulate the exchange of materials passing through them. The ER stores proteins needed by the cell (in the LUMEN) and organizes them into transport vesicles that bud from the main membrane to be carried elsewhere in the cell. A specialized stack of membranous sacs similar in structure to smooth ER but more compact is the GOLGI APPARATUS.

endorphin A PEPTIDE produced by the PITUITARY GLAND and HYPOTHALAMUS. Endorphins in-clude some amino acid sequences found in hormones. They are considered to be natural morphine-like pain-killers and operate in ways similar to both hormones and NEUROTRANS-MITTERS (for example in the relief of pain, reduction of breathing and heart rate and water conservation). Endorphins reduce the perception of pain by reducing the transmission of nerve impulses. They also affect the release of sex hormones from the pituitary gland. Their effects are short-lived because they are rapidly degraded by the body.

endoscope A device for viewing the inside of otherwise inaccessible structures, such as the inside of the human body, by sending light along a bundle of OPTICAL FIBRES.

endoskeleton An internal SKELETON of vertebrates consisting of bone or cartilage. It provides support, protection and a system of levers to which muscles are attached to enable movement. *Compare* EXOSKELETON.

endosome A type of vesicle formed in a cell during some cases of ENDOCYTOSIS, prior to one cell becoming a LYSOSOME. Certain substances bind to RECEPTOR SITES on the cell surface and the PLASMA MEMBRANE invaginates. A vesicle is pinched off that contains the substance and its receptor. This vesicle, the endosome fuses with other vesicles that have budded from the GOLGI APPARATUS to form lysosomes. The receptor is returned to the cell surface in a vesicle that buds from the endosome. *See also* COATED PIT, COATED VESICLE.

endosperm The food supply in the seeds of most flowering plants, containing starch, fat and protein. The endosperm is a mass of triploid (three sets of chromosomes) cells resulting from mitotic divisions (*see* MITOSIS) of the PRIMARY ENDOSPERM NUCLEUS. Some seeds have no endosperm because the food has been used up by the embryo before GERMINATION.

endothelial (*adj.*) Relating to ENDOTHELIUM.

endothelium (*pl. endothelia*) A single layer of flattened cells derived from embryonic MESODERM. Endothelia line blood vessels, lymph vessels and the heart. It is similar to EPITHELIUM. *Compare* MESOTHELIUM.

endotherm *See* HOMEOTHERM.

endothermic (*adj.*) 1. Describing a chemical reaction in which heat energy is taken in.
2. Describing an ENDOTHERM.

endothermy *See* HOMEOTHERMY.

enterokinase An enzyme in the SMALL INTESTINE that activates TRYPSIN from its inactive form trypsinogen.

entomology The study of insects.

entomophily POLLINATION by insects.

E number One of a group of ADDITIVES (not including flavourings) approved by the European Community. E numbers do not have to be listed with the ingredients of a product. Some E numbers are more often referred to by their name, for example E102 is TARTRAZINE. Some can cause side-effects in certain people, for example tartrazine, used to provide an orange colour, is known to cause hyperactivity and worsen asthma.

environment The sum of conditions in which an organism lives. This includes all the BIOTIC (living) and ABIOTIC (non-living) factors. The term 'internal environment' is sometimes used to refer to an organism's internal conditions. Environment is also used to mean the total global environment, rather than that relating to an individual.

enzyme A biological molecule that alters the rate of a reaction (usually speeds it up) without undergoing a chemical change itself (and can therefore be used repeatedly). An enzyme is a natural CATALYST that does not alter the final equilibrium of a reaction, only the speed at which it is achieved. Enzyme reactions are always reversible.

Most enzymes are GLOBULAR PROTEINS with a three-dimensional structure that provides a specific ACTIVE SITE where the SUBSTRATE molecule it acts upon fits, like a LOCK-AND-KEY MECHANISM. Modern interpretation of this theory suggests that the three-dimensional shape of the active site changes as the substrate binds. This is called 'induced fit'. An enzyme–substrate complex forms until the substrate is altered or split. Its shape then changes and it no longer fits into the active site, so the enzyme falls away. All enzymes act on specific substrates but some will bind a variety of similar substrates, while others are very specific and bind only one.

Enzymes need different and precise conditions in which to function. Deviations from the optimum pH and temperature will cause the shape of the enzyme to change, which will eventually make it non-functional. The concentration of both enzyme and substrate has an effect on enzyme activity. A very low concentration of enzyme is needed for a reaction, and if there is an excess of substrate the rate of the reaction is proportional to the enzyme concentration.

The affinity of the enzyme for the substrate is variable. Some enzymes are CONSTITUTIVE while others are INDUCIBLE, dependent on whether or not the CISTRON(S) encoding the enzyme is expressed (*see* GENE EXPRESSION). Some enzymes need a COFACTOR for them to function. There are six recognized categories of enzymes based on their functions: OXIDOREDUCTASES, TRANSFERASES, HYDROLASES, LYASES, ISOMERASES and LIGASES. The nomenclature often gives information regarding an enzyme's activity, for example peptidases break down PEPTIDES.

Enzymes have many uses outside the body, in medicine and industry and as research tools in molecular biology, and they can be extracted from bacteria and even modified by GENETIC ENGINEERING for a particular purpose. *See also* COENZYME, PROSTHETIC GROUP.

enzyme inhibition The reduction of the rate of the reaction of an ENZYME by an INHIBITOR, which can be reversible or non-reversible. Heavy metal ions, for example mercury and silver, cause non-reversible inhibition and therefore permanent damage to enzymes by altering their shape when they break the SULPHIDE bonds. Reversible inhibition is temporary and the enzyme function returns once the inhibitor is removed.

Competitive inhibitors function by having a structure similar to the SUBSTRATE, and they bind and remain in the ACTIVE SITE. Non-competitive inhibitors attach elsewhere on the enzyme molecule and change the enzyme's shape so that the substrate can no longer bind. Cyanide is a non-competitive inhibitor that attaches to the copper PROSTHETIC GROUP of the enzyme cytochrome oxidase, inhibiting RESPIRATION. In many metabolic reactions the end-product of a pathway may inhibit the enzyme at the start of the pathway; this is an example of NEGATIVE FEEDBACK. *See also* HOMEOSTASIS.

enzyme-linked immunosorbant assay (ELISA) A laboratory technique using MONOCLONAL ANTIBODIES that determines the amount of ANTIGEN in a given sample. The test has many uses, such as detecting drugs in athletes' urine, pregnancy testing and the AIDS test. The technique immobilizes a monoclonal antibody on a laboratory dish and passes a test solution over it. If the appropriate antigen is present it will bind. Then a second antibody with an enzyme attached is added that will bind only to the original antibody bound to the antigen. A SUBSTRATE is added that changes colour when reacted with the enzyme, which can be detected and measured.

eosinophil A type of GRANULOCYTE (blood cell) with cytoplasmic granules that stain with acid dyes. Eosinophils increase in certain parasitic infections and allergies.

epidemic (*adj.*) Describing a disease, particularly an infectious one, that spreads rapidly throughout a population, affecting a large number of individuals.

epidermal growth factor (EGF) A GROWTH FACTOR that stimulates many cells to divide. It is important in new cell proliferation.

epidermis The outermost layer of cells covering an organism's body, providing protection and prevention of water loss. Plants and many invertebrates have an epidermis that is a single layer of cells, often with a non-cellular, tough outer CUTICLE that prevents desiccation (*see* PARENCHYMA). In vertebrates, the epidermis consists of a surface cornified layer (stratum corneum), a tough waterproof layer of dead cells impregnated with KERATIN that provides protection from bacteria and prevention of loss of water. The cornified layer is constantly worn away and replaced by the living cells of the granular layer (stratum granulosum) below. As cells move through the granular layer they incorporate keratin and eventually die. The deepest layer of the epidermis is the MALPIGHIAN LAYER, from which actively dividing cells move to the granular layer. There are no blood vessels in the epidermis. *See also* SKIN.

epididymis A 6-metre long coiled tube in the TESTIS of male vertebrates. SPERM pass into the epididymis from the SEMINIFEROUS TUBULES, where they are made, to be stored for 18 hours, during which time they gain their motility. If the sperm are not used they are reabsorbed by the epididymis (after 4 weeks). Sperm are passed from the epididymis during sexual intercourse into a muscular tube called the VAS DEFERENS. *See also* SEXUAL REPRODUCTION.

epigeal (*adj.*) Describing seed GERMINATION whereby the COTYLEDONS emerge from the soil following germination and form the first green leaves.

epiglottis A small flap of cartilage at the back of the mouth in the throat that moves during swallowing to prevent food entering the

TRACHEA (windpipe), which would cause choking. The epiglottis is necessary because the trachea and OESOPHAGUS (the tube carrying food to the stomach) both meet at the PHARYNX. The movement of the epiglottis is complex.

epinephrine *See* ADRENALINE.

epiphysis 1. *See* PINEAL GLAND.

2. The end of a long bone. The epiphysis is initially separated from the shaft of the bone by a section of CARTILAGE that eventually ossifies (see OSSIFICATION), so that the two portions merge together.

epithelial (*adj.*) Relating to EPITHELIUM.

epithelium (*pl. epithelia*) Animal tissue composed of cells firmly held together in single or compound sheets or tubes by a minimal amount of intercellular substance. Epithelia line cavities, tubes and exposed surfaces of the body, and provide a protective function. Most epithelial tissue is derived from embryonic ECTODERM. One surface of epithelia is always attached to a BASEMENT MEMBRANE, while the other is free.

Epithelia are classified as cubicle, columnar, and squamous (flattened), according to the height of the cell relative to its breadth, and according to whether the sheet is one cell thick (simple) or many cells thick (stratified). Most epithelia are specialized for absorption or secretion. Columnar epithelial cells often have minute finger-like projections on their free surface called MICROVILLI, which increase their surface area for absorption. Further fine cytoplasmic projections called CILIA can be found on these cells, which aid movement of fluid or substances through ducts. Stratified epithelium consists of a germinative layer attached to the basement membrane that undergoes MITOSIS to create new cells. The new cells push old ones to the surface, where they become flattened and are shed. Stratified epithelium may be thickened with KERATIN, as in skin, to provide protection. A specialized type of stratified epithelium called transitional epithelium is found in structures that need to stretch, such as the bladder. Transitional epithelium consists of three to four layers, flattened at the surface, that are not shed. *See also* MUCOUS MEMBRANE, SEROUS MEMBRANE. *Compare* ENDOTHELIUM, MESOTHELIUM.

ER *See* ENDOPLASMIC RETICULUM.

erythrocyte Another name for a RED BLOOD CELL.

essential fatty acid One of several FATTY ACIDS that are required in the diet of humans and certain other animals for normal growth. They include LINOLEIC and LINOLENIC and ARACHIDONIC ACIDS. The latter is a precursor of PROSTAGLANDINS. Deficiency of essential fatty acids can lead to weight loss, scaly skin, hair loss and eventually death.

essential amino acid Any AMINO ACID that is needed by humans but cannot be made by them and must therefore be included in their diet. The nine essential amino acids are HISTIDINE, ISOLEUCINE, LEUCINE, LYSINE, METHIONINE, PHENYLALANINE, THREONINE, TRYPTOPHAN and VALINE.

essential oil A natural oil with a pleasant odour obtained from plants and used in perfumes and flavourings.

ester An organic compound formed when an acid and alcohol react together with the elimination of water. Esters occur naturally in fruit and many have a characteristic fruity odour. Esters are used as food flavourings and to provide the scent in perfumes. LIPIDS are all esters of FATTY ACIDS. Molecules containing three ester groups are called triesters and are found naturally in oils and fats. *See also* GLYCERIDE, TRIGLYCERIDE.

ethene, *ethylene* (C_2H_4) A colourless, gaseous HYDROCARBON. It is a by-product of plant METABOLISM. Ethene acts as a PLANT GROWTH SUBSTANCE, stimulating the ripening of fruit. It is therefore useful commercially as a spray to ripen fruit, such as tomatoes and grapes. Ethene also stimulates ABSCISSION (drop) of leaves, fruit and flowers and is used commercially to promote fruit loosening.

Ethene is also widely used in industry, for example in the manufacture of plastics, detergents, paints and pharmaceuticals.

ethylene *See* ETHENE.

eugenics The study of the ways in which the human race can be improved, especially by ARTIFICIAL SELECTION.

Euglenophyta A phylum from the kingdom PROTOCTISTA. Members of this phylum are characterized by their possession of CHLOROPHYLL a and b (which enable photosynthesis) and FLAGELLA. The products of photosynthesis are stored as a carbohydrate called paramylon, which is not found in other organisms. Reproduction is by BINARY FISSION. *Euglena* is the best example of an organism in this phylum. *Euglena* can undergo a flowing movement because of strips of protein called the pellicle inside the cell membrane (there is no cell wall) that allow a change in shape.

eukaryote (*Eu* = true, *karyo* = nucleus) An organism possessing a clearly defined membrane-bound NUCLEUS in its cells, and other cell ORGANELLES, such as MITOCHONDRIA and CHLOROPLASTS, which are lacking in the simple cells of PROKARYOTES. All organisms are eukaryotes except BACTERIA and CYANOBACTERIA. The genetic material within the nucleus is arranged in CHROMOSOMES, which are not seen in prokaryotes. The RIBOSOMES of eukaryotic cells are larger and denser than prokaryotes, and certain proteins exist (such as HISTONES, ACTIN, MYOSIN) that are not found in prokaryotes.

eukaryotic (*adj.*) Describing an organism that is a EUKARYOTE.

Eustachian tube A narrow tube connecting the PHARYNX to the middle ear. On swallowing, air can enter or leave the middle ear via this tube. This ensures that the pressure inside the middle ear cavity remains the same as that of the atmosphere, so avoiding stretching of the EARDRUM and sound impairment.

Eutheria The placental MAMMALS. These are the most advanced of the MAMMALIA. The young develop within a UTERUS inside their mother nourished by a PLACENTA. *Compare* METATHERIA, PROTOTHERIA.

eutrophication Excessive enrichment of lakes and rivers by NITRATES and PHOSPHATES. Eutrophication can occur naturally or as a result of human activities. The natural accumulation of salts into a body of water is slow and is usually counter-balanced by loss of salts through natural drainage.

The more serious cause of eutrophication is artificial. Artificial eutrophication can be caused by the addition of nitrates and phosphates from fertilizers (washed from the soil by rain), sewage and detergents. This enrichment causes the growth of ALGAL BLOOMS, which prevent light reaching deeper regions of the water and so aquatic plants die because they are unable to photosynthesize. The dead plants and algae are decomposed by saprophytic bacteria (*see* SAPROPHYTE), which use all the available oxygen in the water, leading to the death of other species, such as fish.

See also POLLUTION.

evergreen In botany, plants that retain their leaves all year round. *Compare* DECIDUOUS.

evolution The slow, continuous process by which changes in life forms take place. Such changes can be in the appearance of organisms (microevolution) or in the origin and extinction of species (macroevolution). The emergence of new species is called SPECIATION. There are many theories regarding the ORIGIN OF LIFE, but the process of evolution is the one most widely accepted by scientists today.

Two main theories of evolution were put forward in the 19th century, the first by the French naturalist Jean-Baptiste Lamarck (1744–1829). He suggested that useful characteristics acquired by an organism during its lifetime would be inherited by its offspring, while disuse of other characteristics would result in their eventual disappearance from the species. This theory influenced the more widely believed theory of the English naturalists Charles Darwin (1809–1882) and Alfred Wallace (1823–1913), who independently suggested that evolution occurs through NATURAL SELECTION. A combination of Darwin's theory and Mendel's theories (*see* MENDEL'S LAWS) on GENETICS gives us today's theory of evolution, which is called neo-Darwinism.

Factors other than natural selection may play a role in evolution, such as sexual selection (the choosing of sexual partners by characteristics considered to be attractive, therefore increasing their frequency of inheritance) and GENETIC DRIFT.

The process of evolution does not seem to be constant but occurs in periods of rapid change and relative stability. Much of the evidence to support the theory of evolution comes from the examination of fossils, which appear to show a gradual and progressive change from simple to complex forms, although some intermediate forms are yet to be found. Comparative molecular biology (e.g. amino acid sequences) and biochemical studies also support the theory of evolution. Other evidence for evolution is found in comparative anatomy and embryology (some different adult species have similarities in their embryos, suggesting common ancestry).

See also ADAPTIVE RADIATION, LAMARCKISM, VARIATION.

excretion The removal of waste metabolic products from an organism. This is different from

EGESTION, which is the removal of waste that has never been inside body cells. Excretion involves the removal of, for example, carbon dioxide from the lungs and nitrogenous compounds and water in the form of urine from the kidney. In plants and simple animals, excretion is by DIFFUSION rather than by specialized organs. *Compare* SECRETION.

exocarp The outer skin of a fruit.

exocrine gland A gland of EPITHELIAL origin that transports its secretion(s) to an epithelial surface either directly or, more usually, via ducts. Exocrine glands can be classified according to their type of duct, which may be a simple tube or compound with many branches. Exocrine glands can also be classified according to whether or not their cells break down after secreting their contents. Merocrine glands (such as the SALIVARY GLANDS) remain intact after secreting their product, but holocrine gland (for example sebaceous or SWEAT GLANDS) cells are destroyed with the discharge of their secretion. In apocrine glands (such as the MAMMARY GLANDS) the apical part of the cell breaks down during secretion. *Compare* ENDOCRINE GLAND.

exocytosis A term used to describe the bulk removal of materials from a cell. A vesicle (such as a secretory vesicle) bud from the ENDOPLASMIC RETICULUM or the GOLGI APPARATUS, fuses with the PLASMA MEMBRANE of the cell and releases its contents to the exterior. Exocytosis is used in the secretion of substances and to remove waste products from the cell. *Compare* ENDOCYTOSIS.

exon A region of DNA in a GENE that codes for all or part of the gene product. In EUKARYOTES, exons are separated by non-coding sequences called INTRONS.

exoskeleton A hard, external skeleton of ARTHROPODS, such as crabs and insects. *See also* CHITIN, CRUSTACEAN.

experiment A set of measurements or observations, often performed on equipment designed specifically for this purpose, and designed to suggest or to test a theory. If the theory satisfies a sufficient range of experimental tests, it may then be used to predict what will happen in similar situations. If an experiment provides results that genuinely contradict the theory, the theory must be revised. In the design of any experiment, it is important to ensure that only those factors being studied can change,

and that all other factors remain constant throughout the experiment.

expiration Breathing out. *See* BREATHING.

exponential growth Growth of cells or populations in which the rate of growth is dependent upon the numbers of individuals present and their reproductive rate, with no other limiting factors such as food or space. Exponential growth is characteristic of the initial rapid growth phase in a population before competition for limiting factors occurs. If a population number is plotted on a graph against time, a J-shaped curve is obtained. For most populations, this levels off to give an S-shaped curve as factors become restrictive. For others, growth may end abruptly as a result of factors such as disease. If an exponential growth curve is plotted on a logarithmic scale a straight line is obtained. The human population growth is at present exponential, in other words it is increasing unchecked. Thomas Malthus suggested as far back as 1798 that future food supplies would not support population size – this problem is still true today. *See also* BACTERIAL GROWTH CURVE, GROWTH, POPULATION GROWTH.

external auditory meatus *See* AUDITORY CANAL.

external respiration *See* BREATHING.

extracellular matrix (ECM) A network of proteins and POLYSACCHARIDES in which animal and plant cells are embedded and form TISSUES. ECM forms part of CONNECTIVE TISSUE, plant and bacterial CELL WALLS and the EXOSKELETON of ARTHROPODS. Structural proteins, such as COLLAGEN and ELASTIN, and cell adhesion molecules, such as FIBRONECTIN, are components of ECM. ECM also contains signalling molecules that play an important role in the regulation of cell functions, for example growth, differentiation, division and cell death.

extraembryonic membrane In mammals, one of the membranes derived from and surrounding the EMBRYO during pregnancy. The extraembryonic membranes are the CHORION, AMNION, ALLANTOIS and YOLK SAC. *See also* EMBRYONIC DEVELOPMENT, TROPHOBLAST.

eye The SENSE ORGAN responding to light. The human eye is roughly spherical and protected by bony sockets in the skull (the orbits) and attached to the skull by the rectus muscles, which allow the eyeball to rotate. The whole eyeball (except the CORNEA) is covered by the sclerotic coat (sclera, the white of the eye),

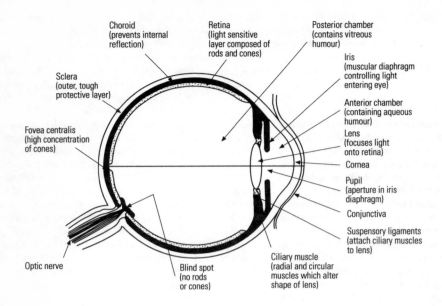

The human eye.

which is made of COLLAGEN fibres and so protects and maintains the shape of the eyeball.

Light passes through the external layers of the eye, the CONJUNCTIVA and the cornea, and then enters the eye through an aperture called the PUPIL at the centre of the coloured diaphragm, the IRIS. Most of the refraction (bending) of the light entering the eye is carried out by the cornea, which is curved and acts as a fixed lens. Behind the pupil is the lens, which carries out the final focusing of the light entering the eye. The lens can alter its shape, due to the action of CILIARY MUSCLES (radial and circular muscles) that surround it. These alter the tension of suspensory ligaments supporting the lens and so allow it to change shape (fatter for near objects, thinner for distant ones). This allows objects at different distances to be focused, an ability called 'accommodation'. In front of the lens is an ANTERIOR chamber containing a clear liquid called AQUEOUS HUMOUR,

and behind the lens is a larger posterior chamber containing clear, jelly-like VITREOUS HUMOUR, which helps to maintain the shape of the eyeball. The lens focuses the light entering the eye onto a layer called the RETINA. Light-sensitive cells here (RODS and CONES) convert the light they receive into NERVE IMPULSES that pass along the OPTIC NERVE to the brain. Internal reflection of light is prevented by a pigmented layer inside the sclera called the choroid, which is rich in blood vessels and supplies the retina. In some nocturnal animals, light is reflected by a layer called the TAPETUM.

Most invertebrates have simple eyes with no lens. Insects have compound eyes made of many OMMATIDIA.

eyepiece In any optical instrument, such as a microscope or telescope, the lens placed closest to the eye. Generally, this acts as a magnifying glass, producing an enlarged image of the image formed in the instrument.

F

F₁ generation, *first filial generation* In genetics, the first generation of a cross between pure breeding parents.

F₂ generation, *second filial generation* In genetics, the offspring obtained by crossing members of the F₁ GENERATION.

F-actin The form of ACTIN (termed filamentous actin) that exists as long fibrous molecules consisting of polymerized G-ACTIN monomers. Two chains of F-actin twist around one another to form the thin filaments characteristic of MUSCLE MYOFIBRILS.

facultative parasite A PARASITE that is able to survive without its host.

FAD (flavine adenine dinucleotide) A derivative of the vitamin riboflavin (*see* VITAMIN B) that acts as an electron carrier in the ELECTRON TRANSPORT SYSTEM. It is also a PROSTHETIC GROUP in some enzymes.

faeces The remains of food and other waste products from the DIGESTIVE SYSTEM of animals. Faeces consist mostly of residual material from digestive juices, some undigested ROUGHAGE, cells from the intestinal walls, bacteria and water. They are passed to the outside via the ANUS.

Fahrenheit A temperature scale obsolete in science but still in everyday use in the UK and US. The fixed points of this scale are taken as the freezing point of water, 32°F, and the boiling point of water, 212°F, both at atmospheric pressure.

Fallopian tube, *oviduct* In female mammals, one of two tubes leading from the OVARY to the UTERUS, down which the ova (*see* OVUM) travel, aided by the muscular movements of the tube and CILIA lining it. The ovum is often fertilized in the Fallopian tube and sometimes implants there instead of the uterus, causing an ECTOPIC PREGNANCY. *See also* FERTILIZATION.

family One of the subdivisions of ORDERS in the CLASSIFICATION of organisms. Family names end in '-idae' for animals and '-aceae' for plants and fungi. Families are groups of related genera (*see* GENUS).

fat A LIPID mixture consisting of GLYCEROL and FATTY ACIDS, mostly TRIGLYCERIDES, which are solid at room temperature – unlike OILS, which are liquid fats. In many animals, fats are stored to provide an energy reserve and also to give some protection and insulation to the animal and its internal organs. Fats are an essential constituent of an animal diet, but too much fat has been linked to heart disease in humans.

fatigue A reduction in the responsiveness of cells or tissues to nervous stimulation that occurs after prolonged stimulation. An example is muscle fatigue after prolonged exercise.

fatty acid An organic compound (a CARBOXYLIC ACID) made with a straight HYDROCARBON chain of up to 24 carbon atoms with a carboxyl group (–COOH) at one end. Carbon atoms can be joined by double or single bonds (*see* COVALENT BOND). If a fatty acid chain has a double bond, it is said to be unsaturated. Oleic acid is an example of an unsaturated fatty acid. Polyunsaturates, for example LINOLEIC ACID, have two or more double bonds. Saturated fatty acids, for example, palmitic and stearic acids, have a single bond between the carbon atoms. The carbon atoms therefore carry all the hydrogen atoms possible. The more double bonds the fatty acid chains contain, the lower the melting point of the fat; for example, oil has many double bonds and lard has none. Polyunsaturates (such as in margarine) are thought to be less likely to contribute to cardiovascular disease than saturated fats (butter has both). Fatty acids usually combine with GLYCEROL to form LIPIDS. They are also constituents of PHOSPHOLIPIDS, STEROLS and WAXES. *See also* ESSENTIAL FATTY ACIDS.

fatty acid oxidation, β-oxidation The pathway by which fats are metabolized to release energy. Most energy comes from the breakdown of CARBOHYDRATES, but when these are exhausted, for example during starvation or dieting, fats and proteins are used. Fats are broken down to their constituents, GLYCEROL and FATTY ACIDS. Glycerol is phosphorylated and enters GLYCOLYSIS as glyceraldehyde-3-phosphate. Fatty acids are progressively broken down into 2-carbon fragments, which combine with COENZYME A to form ACETYL COENZYME A, which then enters the KREBS CYCLE. This fatty acid oxidation occurs in the MITOCHONDRIA and is accompanied by the production of reduced NAD, which enters the ELECTRON TRANSPORT SYSTEM where it yields ATP. More than twice as much energy is liberated by fats compared to the same quantity of carbohydrates. *See also* RESPIRATION.

fatty tissue *See* ADIPOSE TISSUE.

Fehling's test A test used on organic substances to determine which are REDUCING AGENTS. It is usually used to detect REDUCING SUGARS and ALDEHYDES. The test involves heating the sample with a fresh solution of copper(II) sulphate, sodium hydroxide and sodium potassium tartrate. The presence of a reducing sugar is indicated by the production of a red precipitate. *See also* BENEDICT'S TEST.

fermentation The process by which sugars are broken down by bacteria or yeasts, in the absence of oxygen. This process is considered to be analogous to ANAEROBIC RESPIRATION. In nature, the role of fermentation is to remove the hydrogen ions formed at the end of GLYCOLYSIS. This allows the process to continue, since in the absence of oxygen the hydrogen ions cannot be used further. The products of fermentation can be alcohol and carbon dioxide (alcoholic fermentation) or LACTIC ACID (lactate fermentation).

The process of fermentation has been utilized by humans in the baking and brewing industries, in cheese and yoghurt manufacture and also in the production of some antibiotics. In brewing, the alcohol ethanol is the important product, and in baking the carbon dioxide is more important, the bubbles causing bread to rise.

In animals, lactate fermentation is more common since lactic acid can be used to release energy if oxygen becomes available again, which allows the animal to withstand short periods without oxygen. A common situation where lactate fermentation occurs is in the muscles during strenuous exercise, where the build up of lactic acid causes cramps and prevents the muscle operating (*see* OXYGEN DEBT).

fern A plant of the phylum FILICINOPHYTA.

fertilization The fusion of two GAMETES in SEXUAL REPRODUCTION to form a single cell (or ZYGOTE) combining the genetic material of both gametes.

Some organisms are DIOECIOUS, the OVUM (egg) coming from the female and SPERM coming from the male individual. In mammals, the ovum is released from the OVARY and begins to move down the FALLOPIAN TUBE and the sperm swim towards it through the UTERUS. A small proportion of the sperm released by the male reaches the tubes and only one fertilizes the egg. This is achieved by the ACROSOME REACTION, in which the outer membrane of the egg is softened so that the sperm can penetrate it. The outer membrane of the egg then thickens to form a fertilization membrane, which prevents entry of a second sperm. The head and middle piece of the sperm enter the ovum (the tail separates), the two nuclei fuse and CELL DIVISION begins immediately. If the ovum is not fertilized within 24 hours it dies and is shed in the menstrual flow (*see* MENSTRUAL CYCLE). In some species, the female stores the sperm as a thick mucus plug and fertilization is delayed until a time when survival of offspring is optimum.

Some organisms are MONOECIOUS (or HERMAPHRODITE), where both sex organs are carried by one individual. In this case, either self-fertilization (the gametes come from the same individual) or cross-fertilization (the gametes come from two individuals) can occur. Self-fertilization occurs in some plants but rarely in animals. Fertilization in some plant species is preceded by POLLINATION.

External fertilization is common in aquatic vertebrates, where both sexes release their gametes into the water. *See also* DOUBLE FERTILIZATION, IMPLANTATION, PARTHENOGENESIS.

fibre A thin elongated cell, such as nerve or muscle fibre. Fibre also refers to the structure of molecules such as COLLAGEN and ELASTIN. *See also* ROUGHAGE, SCLERENCHYMA.

fibrin An insoluble blood protein, formed following injury, that prevents excessive bleeding. It is formed as part of the BLOOD CLOTTING CASCADE from the soluble PLASMA PROTEIN fibrinogen by the action of the enzyme THROMBIN. Fibrin forms a meshwork of protein fibres and blood cells over the wound that dry to form a scab under which the wound can be repaired and bacteria cannot enter.

fibrinogen A soluble PLASMA PROTEIN made by the vertebrate liver that is involved in the BLOOD CLOTTING CASCADE. *See also* FIBRIN.

fibroblast A spindle-shaped cell characteristic of CONNECTIVE TISSUE. It produces fibres of the proteins COLLAGEN and ELASTIN to provide strength and elasticity.

fibronectin A GLYCOPROTEIN found in the EXTRA-CELLULAR MATRIX that is involved in CELL ADHESION. Cells often interact with such extra-cellular matrix glycoproteins via a family of INTEGRINS located across the cell membrane.

fibrous protein An insoluble PROTEIN that consists of long coiled strands or flat sheets. This structure confers strength and elasticity and so fibrous proteins provide structural roles. Examples include COLLAGEN, ACTIN, MYOSIN and FIBRIN.

fibrous root A type of root system in plants formed from ADVENTITIOUS ROOTS that consist of many branching roots (no main root) growing from the base of a stem. MONOCOTY-LEDONS have this type of root.

filament 1. In flowering plants, the stalk of the STAMEN.

2. In fungi, the thread-like structures that form HYPHAE.

3. A general term used to describe long thread-like structures or molecules, for example, ACTIN and MYOSIN.

Filicinophyta A phylum of the plant kingdom consisting of the ferns. Fern leaves are called 'fronds', and ferns can be various shapes and sizes. More than 7,000 species exist. There are many small ferns that provide ground cover in moist areas, but there are some very tall ferns, such as tropical tree ferns.

The life cycle of ferns shows ALTERNATION OF GENERATIONS. For most of its life, a fern is a short stem (RHIZOME) with roots and leaves growing from it. This is the SPOROPHYTE generation. For the short-lived GAMETOPHYTE generation, a fern is a small heart-shaped plant. SPORES are carried on the underside of the leaves in sacs that split to release their contents. The spores develop into the prothallus, and during this gametophyte stage GAMETES fuse to produce a fertilized egg that develops into a new frond and root. Ferns are VASCULAR PLANTS.

fin In FISH, a projection from the body that provides stability and locomotion. The term usually refers to fins of bony and cartilaginous fish (*see* FISH). There are several types of fin. Pectoral fins occur as a pair attached at the shoulder girdle and there is a pair of pelvic fins attached at the pelvic (hip) girdle. These are for steering and breaking and can be considered to be equivalent to the limbs of tetrapods. There is usually one or more dorsal and ventral fins (which may be continuous), an anal and a caudal fin, all of which control balance or rolling. The caudal, or tail fin, is also the main propulsive organ. Fins are strengthened by cartilaginous or bony fin rays, but retain flexibility.

first filial generation *See* F_1 GENERATION.

fish A general term for an aquatic vertebrate of fresh or sea water that obtains oxygen through GILLS and uses FINS as a means of locomotion. There are three groups: bony fish (OSTE-ICHTHYES), for example cod and tuna; cartilaginous fish (CHONDRICHTHYES), for example sharks and rays; and jawless fish (AGNATHA), for example lampreys and hagfish.

Bony fish constitute the majority of fish (20,000 species), and have a skeleton of bone and a body covered in scales. They have a number of mobile fins that control their movements (dorsal, pectoral, pelvic and caudal). In many bony fish their buoyancy is adjusted by a swim bladder. Most bony fish lay eggs; some retain the eggs inside their body and give birth to live young.

Cartilaginous fish have a CARTILAGE skeleton, a large sensitive nose and a series of open gill slits along the neck region. There are about 600 species. Jawless fish have a NOTOCHORD instead of a backbone and resemble primitive vertebrates before the true fishes with jaws evolved.

flagella (*sing. flagellum*) A long thread-like structure used for locomotion of unicellular organisms or individual cells, such as the sperm of multicellular organisms. Flagella are very similar to CILIA but are longer (100 μm), occur in fewer numbers and their action of movement is different. Unlike cilia, flagella can function independently, usually moving in one plane or occasionally in a spiral motion. Often a single flagellum is found at the rear of a cell that is propelled forwards by it, although sometimes (such as in bacteria) flagella are found in groups called tufts. The external structure of flagella in EUKARYOTIC cells is as found in cilia – two central fibres and nine peripheral pairs – but this arrangement is not found in PROKARYOTIC cells.

flagellate A member of the phylum ZOOMASTI-GINA. Euglenoid flagellates are members of the phylum EUGLENOPHYTA.

flatworm A member of the phylum PLATY-HELMINTHES.

flavine adenine dinucleotide *See* FAD.

floret A small flower, usually making up part of a composite flower head.

flower The reproductive structure in flowering plants (ANGIOSPERMS). A flower consists of four sets of modified leaves, SEPALS, PETALS, STAMENS and CARPELS, attached to a RECEPTACLE, which is the modified end of the stem (PEDUNCLE). Flowers differ in their colour, size, number and arrangements of their parts depending on the method of POLLINATION. Insect-pollinated flowers usually have brightly coloured petals, whereas wind-pollinated flowers may have small or no petals and long STIGMAS to collect POLLEN. Flowers can contain male and female organs together (*see* HERMAPHRODITE), or the male and female organs occur in separate flowers (*see* MONOECIOUS) or on separate plants (*see* DIOECIOUS). The sepals and petals together form the PERIANTH, which protects the reproductive organs and attracts pollinators.

The structure of a flower can be represented symbolically in a floral formula using letters for parts of the flower, for example *K* to represent the CALYX and *P* perianth, and numbers to indicate how many of a part are present. A floral diagram can be used with this formula to show the arrangement of the parts.

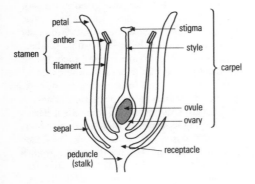

Cross-section of a generalized flower.

A flower with RADIAL SYMMETRY is called ACTINOMORPHIC, for example the buttercup (the petals and sepals are of similar size), and a flower with BILATERAL SYMMETRY is called ZYGOMORPHIC, for example the white dead nettle (unequal sepals and petals of different shapes).

A floret is a small flower, usually making up part of a composite flower head, and there are often two types of florets in one flower, for example the daisy has yellow and white florets.

Although the flower is thought of as the main reproductive structure, leaves, stems and roots can also carry out this function.

See also DOUBLE FERTILIZATION, INFLORES-CENCE.

fluid mosaic model The currently accepted structure of a CELL MEMBRANE. A cell membrane has two main components, PROTEIN and PHOSPHOLIPID. A phospholipid molecule is composed of a hydrophobic (water-repelling) tail end and a hydrophilic (water-loving) head end. In the fluid mosaic model, the phospholipids form a bimolecular layer in which the hydrophobic tail ends associate together at the centre of the membrane and the hydrophilic heads extend towards the surface. The proteins are arranged in a mosaic fashion dotted throughout the membrane: some on the surface, some extending into the phospholipid layer and some extending across it. Since the phospholipid layer seems to be capable of movement, the term 'fluid' was added to the model. In this model it is suggested that the proteins provide structural support and also give specificity to the cell (for recognition by antibodies, hormones, etc.). The proteins also assist ACTIVE TRANSPORT across the cell membrane.

fluorescence An effect in which ultraviolet light is absorbed and then re-emitted immediately as visible light. *See also* BIOLUMINESCENCE.

fluorescence microscopy A modification of the use of a light microscope in which the tissue to be viewed is stained with a fluorescent dye that is used as the light source. This dye can be added to a marker, such as an antibody, to highlight specific cells or proteins.

fluorine (F) A pale yellow, highly reactive poisonous gas with a strong odour. Fluorine combines with many elements and as such has many uses. It is required by animals as a MICRONUTRIENT since it is a component of

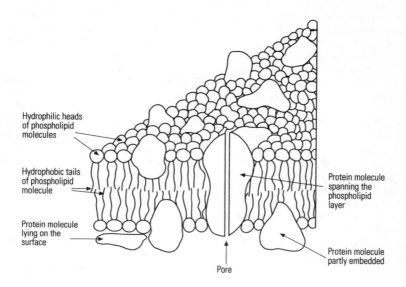

Hydrophilic heads of phospholipid molecules

Hydrophobic tails of phospholipid molecule

Protein molecule lying on the surface

Protein molecule spanning the phospholipid layer

Pore

Protein molecule partly embedded

The fluid mosaic model of the cell membrane.

bones and teeth. Fluoride combines with calcium to form calcium fluoride, which strengthens teeth and helps to prevent decay. Small quantities of fluorine salts, such as sodium fluoride, are added to the drinking water in many countries to help prevent tooth decay. Fluorine is not required by most plants.

focal contact See CELL ADHESION.

foetus The name given to the unborn young of mammals after the first bone cells appear in the cartilage. In humans this is at about 7 weeks of pregnancy, by which time the foetus is almost fully formed (internal organs formed, limbs, fingers, toes, eyelids). *See also* EMBRYO.

folic acid *See* VITAMIN B.

follicle Any cluster of cells that protects and nourishes an enclosed cell or structure. For example, a hair follicle surrounds the root of a hair. *See also* GRAAFIAN FOLLICLE.

follicle-stimulating hormone (FSH) A gonadotrophic GLYCOPROTEIN produced by the PITUITARY GLAND. In females it stimulates development of the GRAAFIAN FOLLICLE within the OVARY. Mature ova, or egg cells, are released from the Graafian follicle under the further control of another hormone, called LUTEINIZING HORMONE. In males, FSH controls the production of SPERM by the TESTIS. FSH also stimulates OESTROGEN production. It is inhibited by PROGESTERONE.

food chain A sequence of organisms showing the order in which energy and nutrients pass through an ECOSYSTEM. Each organism in the chain feeds on, and is dependent on, the one preceding it, and is then itself eaten by the succeeding organism. There are usually three or four, and possibly up to six, links in a food chain.

The primary producers in a food chain are the photosynthetic organisms: green plants and some bacteria that are autotrophic (*see* AUTOTROPH) and manufacture sugars from raw materials using energy from the sun. All other organisms in the chain are consumers (HETEROTROPHS) that obtain their energy by consuming other organisms. Primary consumers are those that feed on the primary producers, for example HERBIVORES and some plant parasites. Secondary consumers are CARNIVORES that feed on herbivores, and tertiary consumers feed on other carnivores. Each of these levels in the food chain is called a trophic level and only a small proportion of energy passes from one level to the next because a lot of energy is lost during RESPIRATION. When producers and consumers die, organisms called DECOMPOSERS and

DETRITIVORES break down the dead material, enabling nutrients to be recycled to the soil or atmosphere.

The food chain is an over-simplification because many organisms have more varied diets than it suggests (*see* OMNIVORE). All the food chains in a COMMUNITY may interact (*see* FOOD WEB).

See also BIOMAGNIFICATION, PRODUCTIVITY, PYRAMID OF BIOMASS, PYRAMID OF ENERGY, PYRAMID OF NUMBERS.

food preservation Various methods used to prevent the spoilage of food by micro-organisms (such as bacteria, mould or yeast), oxidation of fats (making them rancid) or the action of enzymes. Food preservation can be achieved by the use of modern chemical ADDITIVES, but traditional methods are still used. Methods include canning, which is the sealing of food after destroying micro-organisms and enzymes with high temperatures; pickling, which uses vinegar (ethanoic acid) to stop growth of mould; curing or salting (soaking in salt) of meats; preserving in sugar; drying or freeze-drying (the latter is less damaging to the food); heat treatment (*see* PASTEURIZATION); refrigeration (below 3°C) for cooked foods, which slows down spoilage; and deep freezing (−18°C), which stops most spoilage.

food web The interaction of all of the FOOD CHAINS in a COMMUNITY. One organism may occur in several chains at different TROPHIC LEVELS, creating a complex feeding system.

forebrain The largest of the three parts of the human BRAIN. The forebrain contains the CEREBRUM, which co-ordinates all of the body's senses and higher mental activities such as reasoning, learning and memory, and the THALAMUS, which is a relay centre for other regions of the brain and is associated with pain and pleasure. *See also* HINDBRAIN, MIDBRAIN.

foreskin A loose retractable skin covering the end of the PENIS. It is sometimes surgically removed for medical or religious reasons in a small operation called circumcision.

formaldehyde, *methanal* A gas with a pungent odour which is used dissolved in water as FORMALIN.

formalin A solution of FORMALDEHYDE (methanal) in water which is used as a preservative for biological specimens.

fossil The remains of an animal or plant preserved in some way, usually in rocks but also in ice, peat and tar. Fossils are generally found in the layers of rock that form from the slow deposition of mud and silt; the oldest fossils are found in the lower layers. The scientific study of fossils is called PALAEONTOLOGY.

fossil fuels Petroleum oil, coal and natural gas. Petroleum and natural gas are produced in rocks from the decay of marine life. Coal forms as a result of similar geological processes compressing decayed forests. There is increasing pressure to reduce the consumption of fossil fuels as they are NON-RENEWABLE. The burning of such fuels also releases carbon dioxide into the atmosphere, which is believed to be responsible for the GREENHOUSE EFFECT.

fovea centralis The region of the RETINA of the eye with the greatest concentration of CONES.

fragmentation A method of ASEXUAL REPRODUCTION in some relatively undifferentiated organisms, such as algae, sponges and worms, that involves dividing into sections and separating to form new individuals. It is a form of regeneration in some organisms, where as a result of injury the separated fragments will regenerate parts.

frameshift mutation An alteration in the reading frame of DNA bases (*see* NUCLEOTIDE) read as triplets (CODONS) during TRANSCRIPTION. Frameshift mutations arise as a result of the deletion or addition of a single base. This gives rise to an abnormal triplet and each subsequent triplet is altered. Thus the MESSENGER RNA produced carries a different sequence, which is translated (see TRANSLATION) into a different protein. *See also* MUTATION.

fructose ($C_6H_{12}O_6$) A MONOSACCHARIDE that combines with GLUCOSE to form SUCROSE. Fructose contains a KETONE group and is therefore called a ketose sugar. It is a component of fruit and NECTAR, which plants use to attract animals to assist in seed dispersal.

fruit The ripened OVARY in flowering plants containing and protecting one or more SEEDS. Fruits are often edible, which aids DISPERSAL of seeds because the seeds pass through the guts of animals undigested. Simple fruits, such as the peach, are formed from one ovary, whereas multiple fruits, such as the blackberry, are formed from the ovaries of several flowers. A fruit consists of a wall or PERICARP that is divided into layers. Fruits may open to shed their seeds (DEHISCENT) or remain unopened (INDEHISCENT). Fruits are classified as either fleshy or dry. Examples of fleshy fruits are the BERRY or DRUPE. There are many examples of dry fruits, including LEGUMES, NUTS and ACHENES. A soft fruit, or PSEUDOCARP, can be formed from parts other than the ovary. For example the strawberry develops from the RECEPTACLE and its true fruits are the pips on the outer surface.

FSH *See* FOLLICLE-STIMULATING HORMONE.

Fungi A large kingdom of EUKARYOTIC organisms that have no CHLOROPHYLL, and many reproduce by SPORES. The kingdom Fungi includes the phyla ZYGOMYCOTA, ASCOMYCOTA and BASIDIOMYCOTA. Fungi (*sing.* fungus) were once classified with plants, but they have no leaves or roots and have the POLYSACCHARIDE CHITIN in their cell walls, which is never found in plants.

Fungi usually have no distinct cell boundaries and the main body of a fungus, called a MYCELIUM, consists of a mass of threadlike HYPHAE. The spores by which a fungus reproduces are contained within a structure called a SPORANGIUM, which grows at the ends of the hyphae. ASEXUAL REPRODUCTION or SEXUAL REPRODUCTION can take place. The lack of chlorophyll means fungi cannot photosynthesize and so obtain their nutrients as SAPROTROPHS or PARASITES.

Fungi are important to humans, for example in the production of ANTIBIOTICS (such as PENICILLIN from the mould *Penicillium*), in the production of alcohol and bread (YEAST), and in the decomposition of sewage and soil materials. However, they are also a cause of economic loss, causing decay of stored foods (for example in moulds on bread and fruit) and deterioration of materials such as leather and wood.

Fungi also cause a number of plant (and to a lesser extent animal) diseases, for example, Dutch elm disease and powdery mildew, which attacks cereal crops.

See also MYCORRHIZA.

fungicide A PESTICIDE that kills unwanted fungi.

fungus (*pl. fungi*) Any member of the kingdom FUNGI.

funicle The stalk that attaches the OVULE to the ovary wall in the CARPEL of a flowering plant. *See also* DOUBLE FERTILIZATION.

G

G-actin The form of ACTIN that exists as GLOBU-LAR PROTEIN monomers. These monomers join together to form long fibrous molecules of filamentous actin, or F-ACTIN, which are the thin filaments characteristic of muscle MYOFIBRILS.

galactose ($C_6H_{12}O_6$) A MONOSACCHARIDE that combines with GLUCOSE to form LACTOSE. Galactose contains an ALDEHYDE group and is therefore called an ALDOSE sugar.

gall bladder A small muscular sac forming part of the DIGESTIVE SYSTEM. Not all vertebrates have a gall bladder. In humans, the gall bladder is located under the liver and is connected to the SMALL INTESTINE by the BILE DUCT. It receives and stores BILE from the liver until its release, under intestinal hormone control, into the small intestine. GALLSTONES can occur if the CHOLESTEROL content of bile is too great; gall stones block the bile duct and often need surgical removal.

gallstone A small, hard concretion, mainly of CHOLESTEROL and mineral salts, that sometimes forms in the GALL BLADDER or one of its ducts.

gamete A cell of male or female origin that is involved in SEXUAL REPRODUCTION. A gamete of one sex combines with the gamete of the opposite sex to produce the beginnings of a new organism. Most gametes are HAPLOID (have one set of chromosomes), so when two gametes fuse a full set (two sets) of chromosomes is restored in the DIPLOID offspring. In higher organisms, such as humans, the female gamete is called an OVUM (or egg) and is produced in the OVARY. The male gamete is called a SPERM and is produced in the TESTIS. *See also* FERTILIZATION.

gametophyte One form of a plant that shows ALTERNATION OF GENERATIONS. The gametophyte is the HAPLOID generation that produces GAMETES by MITOSIS. *See also* SPOROPHYTE.

gamma radiation High frequency ELECTROMAGNETIC RADIATION similar to X-RAYS but with a shorter wavelength. Gamma radiation is emitted by the nuclei of atoms during radioactive decay. It is only weakly ionizing (*see* IONIZING RADIATION), less so than ALPHA or BETA PARTICLES, but is highly penetrating and therefore extremely dangerous. A dense material such as lead will reduce the intensity of gamma-rays, provided a thickness of several centimetres is used. Gamma radiation is used to kill microorganisms and to sterilize medical and laboratory equipment.

gamma ray *See* GAMMA RADIATION.

ganglion (*pl. ganglia*) A solid cluster of nervous tissue made up of NERVE CELL bodies and SYNAPSES enclosed in a sheath. The CENTRAL NERVOUS SYSTEM of many invertebrates, such as insects, crustaceans and worms, consists of these well-developed ganglia in the head that are analogous to the vertebrate brain. Ganglia are also found in vertebrates, but mainly outside the central nervous system.

gastric juice An acidic product of secretory cells that occurs in gastric pits lining the wall of the STOMACH. Gastric juice consists of mostly water mixed with hydrochloric acid and provides the acid environment needed for the digestive enzymes to function and to kill bacteria brought in with the food. *See also* SECRETIN.

gastric pit *See* STOMACH.

gastrin In mammals, a hormone secreted by the STOMACH and DUODENUM that stimulates the production of GASTRIC JUICE.

gastrocoel *See* ARCHENTERON.

Gastropod A member of the class GASTROPODA.

Gastropoda A class of the phylum MOLLUSCA, including snails, slugs and winkles. Gastropods often have a single shell (univalve) that is usually coiled. The head is distinct with tentacles and eyes. They possess a rasping tongue (radula) for feeding. *Compare* PELECYPODA.

gastrula The stage of EMBRYONIC DEVELOPMENT following on from the BLASTULA. The GERM LAYERS are laid down during this stage. *See also* GASTRULATION.

gastrulation In embryology, the second stage of EMBRYONIC DEVELOPMENT during which the GERM LAYERS are laid down. During gastrulation, the embryo becomes a cup-shaped structure containing a cavity called the ARCHENTERON.

Gaussian curve *See* NORMAL DISTRIBUTION.

Geiger counter A device for detecting RADIATION.

gel A semi-solid COLLOID. A gel is similar to a SOL, except that both the continuous and dispersed phases of the colloid contribute to the three-dimensional network of the material. This results in a jelly-like mass, such as GELATIN, or a more rigid structure, such as silica gel.

gelatin, *gelatine* One of several glutinous substances obtained by boiling COLLAGEN in skin, bones and connective tissue of animals in dilute acid. Gelatins swell in hot water to form a viscous solution that sets on cooling to form a jelly. Gelatins are widely used in the food industry and in photography. Gelatin is rich in the amino acids GLYCINE and LYSINE.

gelatinous (*adj.*) Describing a GEL.

gel filtration chromatography A type of CHROMATOGRAPHY that separates molecules according to their size. A mixture of liquids is passed down a gel in a column. Small molecules are able to enter the pores in the gel, whereas larger molecules, which cannot enter the pores, pass straight through the column. Fractions are collected which contain a gradation of large to small molecules. The gel can be chosen to achieve the desired separation. This technique is useful for separating proteins.

gene The basic unit of inheritance encoded by a specific length of DNA controlling one particular function or characteristic, for example eye colour. It was thought that one gene encoded one ENZYME, but it is now known that this is not always the case because proteins are made up of several POLYPEPTIDES each encoded by a separate gene. Today, a gene is considered to be a length of DNA encoding any molecular cell product. In higher organisms, genes are located on CHROMOSOMES.

The position of a gene within a DNA molecule is called the LOCUS and a gene may have two (or sometimes more) variants called ALLELES, each specifying a particular form of the characteristic, for example blue or brown eyes. Genes can undergo MUTATION and RECOMBINATION. The full set of genes carried by a cell or an an individual or the range of genes carried by a particular species is called the GENOME.

See also CISTRON, GENETIC CODE, GENETIC ENGINEERING, GENE REPLACEMENT THERAPY, LINKAGE, SEX LINKAGE, POLYGENE.

gene amplification The repeated duplication of a specific length of DNA to produce sufficient amounts for genetic analysis.

The main technique used for gene amplification is the polymerase chain reaction (PCR). A small sample of DNA is heated and cooled in the presence of POLYMERASE enzymes (which allow it to replicate) and an excess of NUCLEOTIDES. The DNA strands separate, new paired strands are formed, and the DNA is reassembled. The process is repeated until enough DNA is present to analyse. PCR typically takes about 20 cycles of 30 minutes. The technique allows very small amounts of DNA to be examined, for example to test for genetic defects in a single cell taken from an embryo, which would otherwise be impossible to analyse.

See also GENETIC ENGINEERING.

gene bank, *gene library* A stored collection of genetic materials, such as frozen sperm, frozen OVA, seeds or bacterial cultures. The materials are collected for future use in, for example, medicine, breeding, agriculture and re-stocking extinct species. In GENETIC ENGINEERING, a gene bank is a collection of stored DNA fragments obtained by digestion of a GENOME with RESTRICTION ENZYMES. The fragments can be cloned (*see* GENE CLONING) and stored frozen in host cells. Such gene banks can then be used in genetic engineering to investigate a specific GENE using an appropriate probe in, for example, SOUTHERN BLOTTING.

gene cloning The production of exact copies (clones) of one or more GENES using GENETIC ENGINEERING techniques.

gene expression, *gene regulation* The mechanisms controlling whether or not a particular GENE is transcribed (used to make a POLYPEPTIDE; *see* TRANSCRIPTION). It is now accepted that one gene codes for one polypeptide, but the genes need to be switched on and off as necessary. The JACOB–MONOD THEORY puts forward the concept of an OPERON, a group of adjacent genes that act together. This theory explains the control of some enzymes that are only produced when needed, and enzymes where the production is switched off when necessary. The theory does not fully explain all eukaryotic gene expression, which is known to be more complicated.

gene flow The spread of GENES through populations. Genetic material is exchanged through breeding or the spread of reproductive structures, such as spores and seeds. Gene flow is also affected by NATURAL SELECTION, GENETIC DRIFT and the movement of individuals.

gene frequency The frequency of a GENE in a population. This is dependent on MUTATION, NATURAL SELECTION and GENETIC DRIFT. *See* HARDY–WEINBERG PRINCIPLE.

gene pool All the different ALLELES possessed by all the members of a given species or population at any one time.

gene probe A single-stranded DNA or RNA fragment used in GENETIC ENGINEERING to search for a specific DNA sequence. Probes can be constructed in the laboratory to have a base sequence that is complimentary to the target sequence. The probe will then attach to the target DNA by base-pairing (*see* BASE PAIR). By labelling the probe with a radioactive isotope, it can be identified after separation. Gene probes are used in SOUTHERN and NORTHERN BLOTTING. Probes are also used to identify embryos carrying genes for human genetic disorders, such as SICKLE-CELL DISEASE, using CHORIONIC VILLUS SAMPLING.

gene regulation *See* GENE EXPRESSION.

gene replacement therapy The treatment of diseases caused by GENE defects by manipulating the gene outside the body and replacing the functional (repaired) gene into the individual. This technique has been carried out successfully in animals and some trials have begun in humans as a treatment for cancer, cystic fibrosis and IMMUNODEFICIENCY caused by an enzyme defect in LYMPHOCYTES. Blood disorders such as THALASSAEMIA, in which the cause is a gene defect, are most likely to be curable by this technique because the cells are easily obtainable and have been grown successfully in the laboratory. There are stringent requirements that have to be met before a trial can be carried out, one of which is to ensure that the replaced genes are not transmitted to the GERM LINE (which would then interfere with evolution) *See also* GENETIC ENGINEERING.

gene splicing A stage in the processing of MESSENGER RNA (*see* TRANSCRIPTION). Non-coding INTRON sequences are cut from the primary mRNA transcript and the coding EXON sequences are spliced together to form the mRNA to be translated into protein (*see* TRANSLATION). Gene splicing takes place only in EUKARYOTIC cells.

It is catalysed by a complex of small RNA particles and protein called splicosomes. Sometimes 'alternative splicing' occurs in which the primary RNA transcript is spliced in different ways (for example one of two exons may be cut out). This results in a single gene producing different forms of the same protein at translation, in different cell types or at different developmental stages.

genetic code The way in which the information contained within an organism's genetic material (DNA) is stored. The code is in the sequence of bases on the NUCLEOTIDE strands of DNA. Each AMINO ACID is coded for by a set of three bases called a CODON. The same codons code for the same amino acids in almost all organisms. The genetic code is said to be degenerate because several codons can code for the same amino acid. Three codons, UAA, UAG and UGA (the abbreviations stand for the names of the bases, for example U is URACIL), are called nonsense or STOP CODONS because they do not code for an amino acid and cause PROTEIN SYNTHESIS to come to an end. Another codon, AUG, is called the START CODON because this is where protein synthesis begins. The code is non-overlapping, so that a sequence GAUCAGUGA would be read as GAU-CAG-UGA and not GAU-AUC-UCA, etc. *See also* ALLELE.

genetic drift A change in the GENE frequencies in a population resulting from factors other than NATURAL SELECTION, emigration or immigration. For example, the accidental death of a disproportionate number of speckled moths would reduce the frequency of that ALLELE controlling colour within the population as a whole. Genetic drift applies to small populations because the change in allele frequency would be insignificant in a larger population. In some cases, an allele could be eliminated by genetic drift, for example if only one speckled moth survived in a population of ten moths and it did not lay eggs. *See also* EVOLUTION.

genetic engineering The deliberate biochemical manipulation of GENES, or DNA itself, so that the genes are spliced (divided), altered, added or removed or transferred from one organism to another. Genetic engineering has enabled large-scale production of useful chemicals such as ANTIBIOTICS, INTERFERON and INSULIN.

The manipulation of DNA involves cutting and rejoining it with special enzymes called RESTRICTION ENDONUCLEASES and DNA LIGASE respectively. Any combined DNA from two different organisms is called RECOMBINANT DNA. Using recombinant technology, fragments of human DNA can be inserted into bacterial PLASMID DNA. By growing cultures of this bacteria, many copies of the foreign DNA can be obtained. This is the way large amounts of human insulin are made; the DNA fragment joined to the plasmid contains the gene for insulin production, and replication within the bacterial plasmid allows large-scale production of the gene. If bacteria are used that secrete insulin, then large-scale production of a pure protein is possible. There is no danger of contamination, for example with HIV, as is possible when extracts from other human donors are used, nor does insulin produced in this way contain foreign material likely to cause immune rejection. If the required portion of DNA is not available, then COPY DNA can be made from MESSENGER RNA, taken from cells known to synthesize the protein in question, by using the enzyme REVERSE TRANSCRIPTASE.

Plants and bacteria can be modified by genetic engineering to achieve practical ends, such as the production of drugs by bacteria, better growth of plants or disease resistance. In the future, it is likely that some genetic diseases will be curable by genetic engineering, for example by GENE REPLACEMENT THERAPY. However, as well as the benefits of genetic engineering, there are potential hazards and ethical issues. Organisms with altered genes could be dangerous if they spread away from their intended location. Gene manipulation could affect the natural balance of evolution and could potentially be dangerous.

See also GENE AMPLIFICATION, BIOTECHNOLOGY, TRANSLATION.

genetic fingerprinting, *DNA fingerprinting* A method for revealing the pattern of sequences of the genetic material, DNA, that are unique to an individual (except identical twins). This technique can then be used, like fingerprinting, to identify individuals. Genetic fingerprinting was discovered by Alex Jeffreys (1950–) of Leicester University, England, and has been of great importance in identifying criminals, including rapists, and establishing paternity. The material needed for testing can be a spot of blood, a sample of skin, a few sperm or a hair.

The method involves separating DNA from the sample, breaking it into smaller fragments with enzymes, and comparing the banding patterns of the component chemicals between individuals. Much of the DNA is similar between individuals but some regions are unique and it is these that are compared. The patterns obtained can be expressed in digital code, which provides greater accuracy, and the information is accepted as a legal means of identification. Other uses of the technique include detecting some inherited diseases and monitoring BONE MARROW transplants.

genetic mapping *See* CHROMOSOME MAPPING, RESTRICTION MAPPING.

genetics The study of inheritance, GENES and their effects. It includes the manipulation of genes in GENETIC ENGINEERING. The work of Gregor Mendel (1822–1884) forms the basis of genetics. *See* MENDEL'S LAWS.

gene tracking A method used to trace the inheritance of a particular GENE through a family. It is used to diagnose and predict genetic disorders, such as cystic fibrosis and Huntingdon's chorea. GENE PROBES are used to detect RESTRICTION FRAGMENT LENGTH POLYMORPHISMS (RFLPs) close to the locus of interest. These RFLPs can then be used as markers and traced through the family or used to detect presence of the disease in the unborn foetus.

genitalia In mammals, the external sexual reproductive organs. In males, the genitalia are the PENIS and testes (*see* TESTIS); in females they are the VULVA and CLITORIS.

genome The full set of GENES carried by a cell or an individual, or the range of genes carried by a particular species. In humans, the genome is about 3×10^9 BASE PAIRS. An international effort, called the Human Genome Project, to map all the genes is expected to be completed by 2005.

genotype The genetic constitution of an organism, or the set of ALLELES inherited by it. *Compare* PHENOTYPE. *See also* VARIATION.

genus One of the subdivisions of families (*see* FAMILY) in the CLASSIFICATION of organisms. The subdivision genus comes between family and SPECIES. Members of a genus are thought to have a common ancestral origin and share many characteristics. *See also* BINOMIAL NOMENCLATURE.

geotropism The response of a plant to gravity. Shoots are negatively geotropic (they grow upwards) and roots are positively geotropic (they grow downwards). Leaves grow at right angles and are said to be diageotropic. *See also* TROPISM.

germination The initial stages of development and growth in a SEED, SPORE or POLLEN grain. Germination begins when the right conditions of water, oxygen, temperature, light and other factors needed to break DORMANCY are met. The uptake of water by the seed (through a small hole called the MICROPYLE) initiates germination. The embryo plant differentiates into the RADICLE (young root), PLUMULE (young shoot) and COTYLEDON (seed leaves). Food is provided for the growing embryo by the ENDOSPERM or cotyledon. The radicle appears first; its apical MERISTEM pushes through the soil and may develop into the entire root system (or ADVENTITIOUS ROOTS may develop).

The cotyledons may remain below ground (hypogeal) or spread out above the soil (epigeal). In epigeal germination, the plumule develops after the cotyledons have grown above ground, and forms the shoot bearing the first true leaves. In hypogeal germination, for example in the broad bean, the leaves develop on the plumule in the seed and the cotyledons remain below ground when the plumule grows up through the soil. Germination is considered to have ended when the first true leaves appear.

germinative layer *See* MALPIGHIAN LAYER.

germ layer In animals, one of the main recognizable layers of cells to appear in a developing embryo after gastrulation (*see* EMBRYONIC DEVELOPMENT). There are three germ layers in most animals: the ECTODERM (outermost layer), the MESODERM (middle layer; not present in more primitive animals) and the ENDO-DERM (inner layer).

germ line The cell line in the development of many animals that goes on to form the GAMETES.

gestation In most mammals, the period between conception and birth. The gestation period varies between species: in humans it is 9 months, in elephants it is 18–22 months and in cats it is 60 days. In humans the gestation period is commonly called PREGNANCY.

gibberellin A PLANT GROWTH SUBSTANCE that promotes cell elongation and therefore growth. Gibberellin was originally isolated from a fungus and over 50 types are now known. Unlike AUXINS, gibberellins can stimulate growth in dwarf varieties and restore them to normal size, largely by causing elongation of the stem. They can also break the DORMANCY in certain buds and seeds and induce flowering. They have no effect on TROPISMS (the directional growth of a plant in response to an external stimulus). Gibberellin and auxin have some SYNERGISTIC effects but also some ANTAGONISTIC ones.

gill The main respiratory organ of fish. Water passes over the gills, providing oxygen that diffuses across a gill membrane to the circulation, and carbon dioxide passes out into the water.

ginkos A group of GYMNOSPERM plants that were once widespread. The maidenhair tree is the sole survivor. *See also* CYCADS.

gizzard In some animals, for example birds and reptiles, a muscular grinding organ that forms part of the DIGESTIVE SYSTEM. Food is ground in the gizzard before passing to the stomach for the main digestion. The wall is very tough and hardened with a layer of the protein KERATIN to prevent damage during the grinding process. Grit or stones may be swallowed to aid grinding in the gizzard.

gland An organ (or sometimes cells within an organ) specialized for producing and secreting HORMONES, ENZYMES or other chemicals. Glands can vary in size, for example tear glands are very small and the PANCREAS is large. In animals there are two types of glands: EXOCRINE GLANDS and ENDOCRINE GLANDS. Together the glands form the ENDOCRINE SYSTEM, which is one of the two major co-ordinating systems of animals (the other being the NERVOUS SYSTEM). Glands exist in plants, where they are always small, sometimes comprising only a single cell.

glans penis The expanded region at the end of the PENIS. It is highly sensitive and covered by a loose retractable FORESKIN.

glial cell A non-conducting nerve cell that provides support and protection for NEURONES. SCHWANN CELLS, astrocytes (which attach neurones to blood vessels), microglial cells (PHAGOCYTES) and OLIGODENDROCYTES are all examples of glial cells.

global warming An increase in the overall temperature of the Earth, believed to be caused by the increasing level of GREENHOUSE GASES, particularly carbon dioxide, in the atmosphere. *See* GREENHOUSE EFFECT.

globular protein A PROTEIN that is compact and rounded in structure and usually soluble. These proteins provide a metabolic role rather than the structural role provided by the FIBROUS PROTEINS. ENZYMES are globular proteins as are certain HORMONES, including INSULIN. Other examples include CASEIN and ALBUMIN.

globulin Any one of a group of GLOBULAR PROTEINS characterized by a spherical shape, solubility in weak salt solutions and a metabolic (*see* METABOLISM) rather than structural function. ENZYMES, some HORMONES and HAEMOGLOBIN are all globular proteins. In animals globulins are found in blood PLASMA. They are also found in plant seeds.

glomerulus A tight knot of blood capillaries forming part of a NEPHRON in the KIDNEY. Blood enters the glomerulus at high pressure, from ARTERIOLES of the RENAL artery, and certain substances, including water, amino acids and sugar, are forced out through the capillary wall to form a filtrate that enters the BOWMAN'S CAPSULE. The filtrate then moves through the tubule of the nephron. Much of the water and sugars are reabsorbed by the LOOP OF HENLE and the remaining waste materials remain in the fluid that becomes URINE.

glucagon A POLYPEPTIDE hormone (29 amino acids), produced by the PANCREAS, that is involved in the regulation of blood sugar levels. Glucagon is produced by the α-cells of the ISLETS OF LANGERHANS and it converts GLYCOGEN, stored in the liver, to GLUCOSE in response to a drop in the concentration of glucose in the blood. This process is called GLYCOGENOLYSIS. Glucagon acts antagonistically with INSULIN.

glucocorticoid Any one of a group of CORTICOSTEROID hormones that are concerned with GLUCOSE metabolism. They are secreted by the CORTEX of the ADRENAL GLAND, and examples include CORTISOL and CORTISONE. Glucocorticoids raise blood pressure and blood sugar levels. They raise blood sugar levels by increasing formation of glucose from fat and proteins, by inhibiting INSULIN and also by increasing the rate of glycogen formation in the liver.

Too little glucocorticoid results in ADDISON'S DISEASE, characterized by low blood sugar levels, low blood pressure and fatigue. Overproduction of glucocorticoid results in CUSHING'S SYNDROME, characterized by high blood sugar levels due to excessive break down of protein (and therefore tissue and muscle wasting) and high blood pressure.

gluconeogenesis The conversion of proteins and fats into GLUCOSE by the LIVER. *See also* GLYCOGENESIS, GLYCOGENOLYSIS, PANCREAS.

glucose, *dextrose* ($C_6H_{12}O_6$) A MONOSACCHARIDE. In humans, glucose is found in the blood and is a source of energy for the body because it is used to generate ATP. Glucose is made by the HYDROLYSIS of STARCH or SUCROSE. Levels of glucose in the blood are regulated by the LIVER and the pancreatic hormones GLUCAGON and INSULIN. *See also* GLUCONEOGENESIS, GLYCOGEN, GLYCOGENESIS, PANCREAS.

glutamic acid A non-essential acidic AMINO ACID, commonly called glutamate. GLUTAMINE is a derivative. Glutamic acid is found widely in plant and animal tissue. It is important in nitrogen metabolism and also acts as a NEUROTRANSMITTER. Its sodium salt, MONOSODIUM GLUTAMATE is widely used in the food industry.

glutamine A non-essential AMINO ACID that is a derivative of GLUTAMIC ACID. It is found in plant and animal tissue and used in medicine and research.

gluten A protein found in wheat, barley, rye and possibly oats. Some people have an intolerance to gluten, which causes COELIAC DISEASE. Gluten-free products are available for such people. Babies should not be given gluten in their diet until the age of 6 months, since an undetected allergy before this could be dangerous.

glycerate 3-phosphate A substance produced in C_3 PLANTS as an intermediate in the light-independent stage of PHOTOSYNTHESIS. *See* CALVIN CYCLE.

glyceride A LIPID that is formed by the combination of a FATTY ACID with GLYCEROL. Glycerides can be mono, di or tri, depending on how many of the three hydroxyl groups on glycerol have combined with the fatty acids. If one of the hydroxyl groups on glycerol combines with a phosphate group, a PHOSPHOLIPID. A GLYCOLIPID results from combination with sugar. TRIGLYCERIDES are found naturally as the main constituents of fats and oils. See also ESTER.

glycerine See GLYCEROL.

glycerol, *glycerine, propan-1,2,3-triol* ($HOCH_2CH(OH)CH_2OH$) A sweet, colourless liquid ALCOHOL extracted from animal and vegetable oils. Glycerol reacts with FATTY ACIDS to form LIPIDS. It is used in antifreeze solutions, explosives and cosmetics.

glycine The simplest AMINO ACID and the main one in sugar cane. Glycine is a sweet, crystalline amino acid, prepared commercially from GELATIN for use in medicine and research.

glycogen A POLYSACCHARIDE (made of branched chains) of the sugar GLUCOSE that is stored in the liver as a carbohydrate source until needed. Glycogen is the animal equivalent of STARCH. It is converted back to glucose by the pancreatic hormone GLUCAGON (a process known as GLYCOGENOLYSIS), and glucose is converted to glycogen by INSULIN (GLYCOGENESIS). These two hormones therefore act antagonistically to regulate blood sugar levels.

glycogenesis The conversion of GLUCOSE in the blood to GLYCOGEN by the LIVER to be stored until needed, when blood glucose levels fall. *See also* GLUCONEOGENESIS, GLYCOGENOLYSIS, PANCREAS.

glycogenolysis The breakdown of GLYCOGEN in the liver to GLUCOSE by GLUCAGON. Glucagon is a hormone produced by the PANCREAS in response to a fall in blood sugar levels and is secreted directly into the blood, and so to the liver. *See also* GLUCONEOGENESIS, GLYCOGENESIS.

glycol *See* ETHANE-1,2-DIOL.

glycolipid A LIPID that is covalently linked to a CARBOHYDRATE. Glycolipids are found in CELL MEMBRANES. There is a wide variation in the composition and complexity of glycolipids.

glycolysis (*Glycol* = sugar, *Lysol* = breakdown) The series of reactions resulting in the breakdown of the six-carbon sugar GLUCOSE to two molecules of the three-carbon compound pyruvate with the release of usable energy in the form of ATP. The process requires no

Summary of glycolysis.

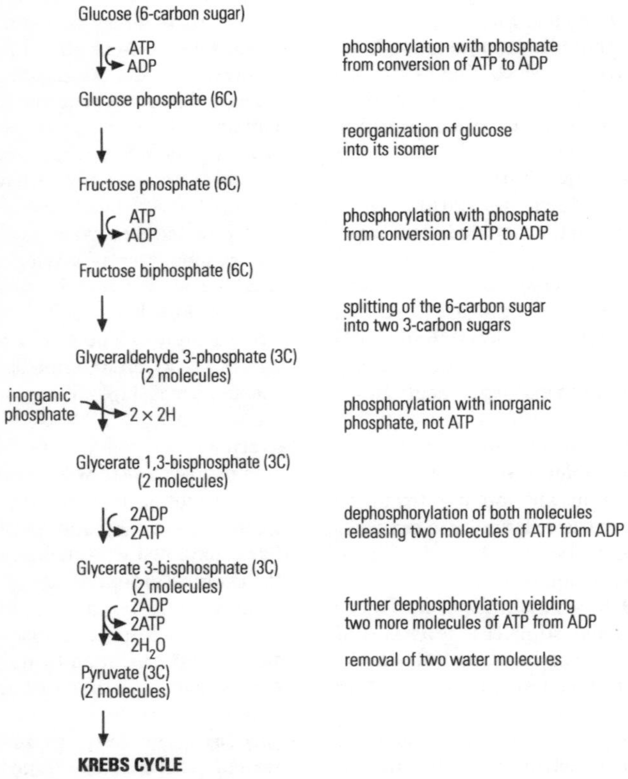

Glucose (6-carbon sugar)

$\downarrow$ ATP → ADP

phosphorylation with phosphate from conversion of ATP to ADP

Glucose phosphate (6C)

$\downarrow$

reorganization of glucose into its isomer

Fructose phosphate (6C)

$\downarrow$ ATP → ADP

phosphorylation with phosphate from conversion of ATP to ADP

Fructose biphosphate (6C)

$\downarrow$

splitting of the 6-carbon sugar into two 3-carbon sugars

Glyceraldehyde 3-phosphate (3C) (2 molecules)

inorganic phosphate → $2 \times 2H$

phosphorylation with inorganic phosphate, not ATP

Glycerate 1,3-bisphosphate (3C) (2 molecules)

$\downarrow$ 2ADP → 2ATP

dephosphorylation of both molecules releasing two molecules of ATP from ADP

Glycerate 3-bisphosphate (3C) (2 molecules)

$\downarrow$ 2ADP → 2ATP → $2H_2O$

further dephosphorylation yielding two more molecules of ATP from ADP

removal of two water molecules

Pyruvate (3C) (2 molecules)

$\downarrow$

KREBS CYCLE

oxygen and is the only form of RESPIRATION and ATP synthesis in anaerobic organisms (*see* ANAEROBE). In AEROBES, it is the first stage of cellular respiration and the pyruvate formed then enters the KREBS CYCLE, which requires oxygen. Glycolysis occurs in the CYTOPLASM of cells, usually in the MITOCHONDRIA. There is an overall gain of two molecules of ATP for each glucose molecule broken down, and two pairs of hydrogen atoms either go into the ELECTRON TRANSPORT SYSTEM to generate a further six molecules of ATP if oxygen is present, or are removed by the FERMENTATION process if no oxygen is present.

glycoprotein A protein covalently linked to a CARBOHYDRATE. The protein can form the bulk of the molecule, as in cell-surface glycoproteins, or the carbohydrate can represent the major part. The addition of sugar residues to proteins is called glycosylation and occurs in the GOLGI APPARATUS or ENDOPLASMIC RETICULUM of a cell.

glycosylation The addition of sugar residues to protein. *See* GLYCOPROTEIN.

glyoxylate cycle A modified form of the KREBS CYCLE that occurs in plants and micro-organisms. The glyoxylate cycle uses fat as the carbon source instead of carbohydrates and enables carbohydrates to be synthesized from FATTY ACIDS. This is achieved by avoiding the carbon dioxide releasing-steps of the Krebs cycle and using the enzymes isocitrate lyase and malate synthase. The glyoxylate cycle occurs in fat-rich tissues, such as germinating seeds. The enzymes are found in organelles called glyoxysomes.

goblet cell A cell that is specialized for the production of MUCUS. Goblet cells are found in some epithelia (*see* EPITHELIUM), for example in the intestines.

goitre A swelling of the THYROID GLAND.

Golgi apparatus, *Golgi body, dictyosome* A stack of membranous sacs found in the CYTOPLASM of a EUKARYOTIC cell, similar in structure to smooth ENDOPLASMIC RETICULUM (ER). The Golgi apparatus is named after its discoverer, Camillo Golgi (1843–1926).

The Golgi apparatus is well developed in secretory cells and NEURONES, and is small in muscle cells. It is thought to play some role in the production of secretory material. Many molecules travel through the Golgi apparatus and are modified or assembled inside the sacs on their way to other organelles or ER. For example, carbohydrate may be added to protein to form GLYCOPROTEINS such as MUCIN. With an ELECTRON MICROSCOPE, vesicles can be seen budding from the Golgi apparatus.

In cells of vertebrates, there is usually only one Golgi apparatus, but in invertebrates and plants there may be several, called dictyosomes.

Golgi body *See* GOLGI APPARATUS.

gonad In animals, the sex organ producing the GAMETES required for SEXUAL REPRODUCTION. A male gonad is the TESTIS and a female gonad is the OVARY.

gonadotrophic (*adj.*) Describing a HORMONE that stimulates reproductive activity of the GONADS (the TESTIS or the OVARY).

gout A disease characterized by painful inflammation of the joints, caused by crystalline deposits of URIC ACID.

Graafian follicle In mammals, a fluid-filled structure in the mammalian OVARY, up to 1 cm in size within which the OVUM (egg cell) develops until it matures. The development of the ovum is under the control of the FOLLICLE-STIMULATING HORMONE (FSH). The mature ovum is released (OVULATION) when the follicle ruptures under the control of the LUTEINIZING HORMONE (LH). The Graafian follicle is then transformed, again under LH control, into the CORPUS LUTEUM. FSH is inhibited by PROGESTERONE, secreted by the corpus luteum, which therefore prevents the maturation of further Graafian follicles.

grafting The TRANSPLANTATION of a small portion of living tissue onto another tissue on either the same or a different organism.

gram A unit of mass, nowadays defined as one thousandth of a KILOGRAM.

Gram's stain A method of staining, devised by Christian Gram (1855–1938) in 1884, that differentiates BACTERIA into those that take up the stain and those that do not. This is of great value in the differentiation of otherwise similar bacteria. The procedure involves staining a smear of bacteria with a crystal violet solution, rinsing this off with Gram's iodine solution and applying 95 per cent ethanol until most of the dye has been removed. Gram-positive bacteria retain the dye while Gram-negative bacteria do not. A counterstain is applied that only stains the Gram-negative bacteria. The staining reflects differences in the bacterial cell membrane, but there are other differences between the two groups. *See also* PEPTIDOGLY-CAN.

granulocyte, *polymorphonuclear leucocyte* One of the two types of WHITE BLOOD CELLS in humans. The other type of white blood cell is the AGRANULOCYTE. Granulocytes are characterized by their granular CYTOPLASM and lobed, darkly staining NUCLEUS. They are about 8 μm in diameter and act as PHAGOCYTES. They form about 60–70 per cent of human white blood cells. In humans, the cytoplasmic granules of most granulocytes do not take up acid or basic dyes and are called NEUTROPHILS. Others stain strongly with basic dyes (BASOPHILS) or acid dyes (EOSINOPHILS).

granum (*pl.* **grana**) A structure within the CHLOROPLAST consisting of a stack of flattened sacs containing CHLOROPHYLL. The light-dependent stage of PHOTOSYNTHESIS occurs in the grana.

graph A way of showing how one variable behaves as a function of another. Points are plotted on a two-dimensional surface according to the values of the two variables. The variables are usually referred to as x, measured along a horizontal AXIS, and y, measured up a vertical axis. The independent variable is generally plotted along the x-axis, with the dependent variable up the y-axis. If statistical or experimental data is plotted, a series of measurements will build up a scatter diagram. A mathematical function may be represented by a line or curve on the graph.

graticule A small measuring device, like a ruler, placed in the eyepiece lens of a light microscope, that can be used to determine the exact size of a feature being viewed.

greenhouse effect A supposed increase in the average surface temperature of the Earth as a result of changes in the composition of its atmosphere. Radiation from the Sun, mostly visible light, is able to penetrate the atmosphere and warm up the Earth's surface. The radiation given off by the Earth is in the infrared part of the electromagnetic spectrum and can be absorbed by gases in the atmosphere, particularly carbon dioxide and water vapour. The concentration of such gases (called GREENHOUSE GASES) has a marked effect on the surface temperature of the Earth. This effect is called the greenhouse effect since the Earth's atmosphere behaves in a way similar to

the glass in a greenhouse, which also allows in visible radiation but absorbs infrared.

Fears have been expressed over the increase in carbon dioxide in the atmosphere caused by the burning of increasing quantities of FOSSIL FUELS over the last century, and attempts are being made to reduce carbon dioxide emission by increasing use of renewable energy sources. The systematic destruction of vast areas of forest has also contributed to the increase of carbon dioxide levels. Other greenhouse gases include methane (a by-product from agriculture), water vapour (as a by-product from industry) and CHLOROFLUO-ROCARBONS (CFCs; from refrigerators, aerosol sprays and polystyrene).

The problem with continued GLOBAL WARMING is that it will cause the polar ice caps to melt, resulting in a rise in sea levels and consequent flooding of low-lying land, which could include whole countries and many world capital cities. A change in the climate would also affect crop growth. It is not clear exactly what or how rapid the consequences of the greenhouse effect will be, because many effects will interact with one another, but the issue raises great concerns about the future of humankind.

greenhouse gas Any gas, such as carbon dioxide or methane, which contributes to the GREEN-HOUSE EFFECT in the Earth's atmosphere by absorbing infrared radiation. Other greenhouse gases are CHLOROFLUOROCARBONS, water vapour, ozone and nitrogen oxides.

grey matter Tissue of the vertebrate BRAIN and SPINAL CORD, so called because of its appearance. It consists of many cell bodies, DENDRITES, SYNAPSES, GLIAL CELLS and blood vessels. It forms an inner layer in the spinal cord and some regions of the brain, but is the outer layer of the cerebral cortex of higher primates (*see* CEREBRUM). *See also* WHITE MATTER.

growth The increase in size and weight of an organism during development. More accurately, growth is the increase in dry, and not fresh mass (which includes water), but in practice the former is difficult to measure because the organism would have to be destroyed. If the growth of most populations, organisms or organs is plotted against time on a graph then a S-shaped growth curve is produced. This represents initial slow growth (lag phase), then fast growth (EXPONENTIAL GROWTH) and then non-existent growth (and even negative growth) towards death.

The growth of a population occurs in several stages: the lag phase is where the growth of a small number of individuals is slow; the exponential phase is where enough individuals are available for reproduction and growth is rapid; the stationary phase is where factors within a given area become limiting and growth of the population stabilizes; and the death phase is where the numbers of individuals dying exceeds the numbers of growing individuals.

The rapid growth of an organism is usually limited by genetic and environmental factors, although some plants show unlimited growth during their lives. Factors that may limit growth include the supply of food, water, oxygen or light, disease, predation or the accumulation of toxic waste. In some populations, growth is density-dependent, so that at a certain population density all the available resources are used up. Growth may be a modification of the S-curve, for example in humans the rapid growth phases are in the early years and at adolescence.

Some populations show a J-shaped growth curve, where the initial density increases very rapidly in an exponential manner and then stops suddenly due to environmental resistance or other limiting factors. Some organisms, for example ARTHROPODS, show intermittent growth because their EXOSKELETON cannot expand; they have to moult periodically during growth.

See also ALLOMETRIC GROWTH, BACTERIAL GROWTH CURVE, ISOMETRIC GROWTH, POPULATION GROWTH.

growth factor A general term referring to a group of substances that affect the growth of cells. Growth factors can signal a variety of effects, such as CELL DIVISION, differentiation, locomotion or survival. All growth factors bind RECEPTORS on the cell surface and are linked to intracellular EFFECTOR molecules, such as enzymes, which set off a cascade of signals. Examples include T-cell growth factor (TCGF), which is required for the proliferation of T CELLS, and epidermal growth factor (EGF), which stimulates many cells to divide.

growth hormone, *somatotrophin* A hormone produced by the anterior PITUITARY GLAND that controls body METABOLISM and growth generally. The production of growth hormone is itself regulated by the HYPOTHALAMUS. Low levels in humans results in dwarfism and high levels in gigantism.

guanine An organic base called a PURINE that occurs in NUCLEOTIDES. *See also* DNA, RNA.

guanosine A PURINE NUCLEOSIDE, consisting of the organic base GUANINE and the sugar RIBOSE.

guard cell A specialized cell in plants that controls gas exchange and water loss. Guard cells contain CHLOROPLASTS and are found in pairs at intervals along the underside of leaves. They are crescent-shaped and surround pores called stomata (*see* STOMA) that are the main route of water loss in a plant and the site of carbon dioxide and oxygen exchange. The pores can be opened or closed by the guard cells changing shape (by changes in turgidity). The inner wall of the guard cell is thicker and less flexible than the outer wall, so when water is taken up by OSMOSIS the cells become turgid and kidney-shaped, causing the stomata to open. When they are less turgid the stomata close. In this way water loss can be adjusted, for example they can be closed during warm weather, to prevent water loss through evaporation, and at night when photosynthesis cannot take place.

gut *See* DIGESTIVE SYSTEM.

gymnosperm A plant of the phylum CONIFEROPHYTA, whose seeds are not contained within an OVARY and instead lie exposed. This is in contrast to ANGIOSPERMS (flowering plants), where the seeds are contained in an ovary. The reproductive structures of gymnosperms are CONES, not flowers. Although still used, the term gymnosperm is no longer formally part of a classification scheme. Gymnosperms include conifers, cycads and ginkgoes. The cycads were mostly large, palm-like plants found in the tropics and subtropics that are now extinct. The maidenhair tree is the sole survivor of the ginkgoes.

gynoecium The female part of a flower, the collective name for the CARPELS.

H

habitat A localized area within a BIOME, for example a freshwater pond, a cave or a woodland. Habitats are varied and continually changing and may contain many species or only one. A microhabitat is one that has very specific living conditions, for example the area under a stone. The habitat largely provides for the needs of its members, which are POPULATIONS of individuals together forming a COMMUNITY.

habituation A form of learned behaviour in which an animal learns to ignore a repetitive stimulus that is neither harmful nor beneficial. *See also* CONDITIONING, IMPRINTING, LEARNING.

haem A complex organic molecule containing iron. Haem combines with the protein GLOBULIN to form HAEMOGLOBIN.

haemocoel The body cavity of some invertebrates, such as ARTHROPODS and MOLLUSCS, through which blood flows. *See* CIRCULATORY SYSTEM.

haemocyanin A blue RESPIRATORY PIGMENT found in MOLLUSCS and CRUSTACEANS.

haemodialysis A technique for removing the circulating waste products from the blood of patients with kidney failure. It uses the principle of DIALYSIS. The blood is passed through a kidney machine (dialyser) and toxic products are filtered out through a semipermeable membrane.

haemoerythrin A red/brown RESPIRATORY PIGMENT found in ANNELIDS.

haemoglobin In vertebrates and some invertebrates, a RESPIRATORY PIGMENT found in the RED BLOOD CELLS that is used to transport oxygen from the lungs to the body tissues. It is made of a haem group, which contains iron, and GLOBULIN (a protein). In humans, haemoglobin has a RELATIVE MOLECULAR MASS of 68,000. In other species, its relative molecular mass ranges from 16,000 to 3 million. The haem group is always the same, but the globulin and number of haem groups varies from species to species. In humans, there are four haem groups, each able to carry one oxygen molecule, so the amount of oxygen that can be carried in the blood is increased compared to species with fewer haem groups.

Haemoglobin combines easily with oxygen in regions of high oxygen concentration (e.g. in the lungs or gills) to form oxyhaemoglobin (which is bright red compared with the red colour of haemoglobin alone) and is transported in the blood to the body tissues. This is easily released where oxygen is at low concentration.

Foetal haemoglobin differs slightly from that of an adult and combines with oxygen even more easily, which allows it to obtain oxygen from the mother's haemoglobin in the placenta.

The carbon dioxide concentration affects the release and uptake of oxygen by haemoglobin (*see* BOHR EFFECT). Some carbon dioxide is carried by haemoglobin as carbaminohaemoglobin. Haemoglobin can also combine with carbon monoxide to form carboxyhaemoglobin, but this reaction is irreversible and results in death if too much carbon monoxide is inhaled.

See also MYOGLOBIN, THALASSAEMIA.

haemoglobinic acid A weak acid formed inside RED BLOOD CELLS when hydrogen ions combine with HAEMOGLOBIN. *See* BOHR EFFECT.

haemolymph The watery fluid that forms the blood of certain invertebrates. *See* CIRCULATORY SYSTEM.

haemolytic disease of the newborn *See* RHESUS DISEASE.

haemophilia An inherited disease in which the normal BLOOD CLOTTING CASCADE does not function. It is a sex-linked condition carried by females but only expressed in males. Sufferers lack factor VIII, one of the substances needed to form THROMBOKINASE, and therefore the insoluble blood protein FIBRIN does not form and bleeding does not stop. Treatment is to administer factor VIII, which can now be made by RECOMBINANT DNA technology. *See also* SEX LINKAGE.

hair A fine, long outgrowth from mammalian skin that consists of dead cells impregnated with KERATIN and MELANIN, which give the hair colour. Hair grows from follicles within the DERMIS, which are produced by inpushings of the MALPIGHIAN LAYER. Cells at the base of the follicle divide to produce hair. Muscles at the base of the hair can cause it to become erect, trapping air and providing insulation for some mammals. Hair can also have a sensory role.

hammer In the middle EAR. *See* MALLEUS.

haploid (*adj.*) Describing the presence of one set of CHROMOSOMES in the nucleus of a cell. GAMETES of higher organisms are haploid (as a result of MEIOSIS) while the other body cells are DIPLOID. Some plants are haploid, for example mosses and liverworts. *See also* POLYPLOID.

haptonasty *See* THIGMONASTY.

haptotropism *See* THIGMOTROPISM.

Hardy–Weinberg Equilibrium *See* HARDY–WEINBERG PRINCIPLE.

Hardy–Weinberg Principle, *Hardy–Weinberg Equilibrium* A mathematical representation of the theoretical relative frequencies of ALLELES within a given POPULATION. The principle assumes an isolated population with no NATURAL SELECTION, MUTATION or GENETIC DRIFT (so there is no gene flow into or out of the population).

These conditions are never met in a real population, but the principle is used in the study of gene frequencies. For example, there may be a DOMINANT allele 'A' and a RECESSIVE allele 'a'. In diploid individuals the combinations 'AA', 'Aa', 'aA' and 'aa' are possible. It might be assumed that the HOMOZYGOUS recessive 'aa' would in time be eliminated by natural selection, but the proportion of dominant to recessive alleles remains the same because the HETEROZYGOUS state 'Aa' or 'aA' acts as a reservoir for the recessive allele 'a'.

If p and q are the frequencies of the dominant 'A' allele and recessive 'a' allele' respectively', then the Hardy–Weinberg principle is expressed as:

$$p^2 + 2pq + q^2 = 1.0$$

This formula is useful in determining the frequency of any allele in a population.

Haversian canal Any of the tubes in the BONE that contain the blood and LYMPH VESSELS and NERVES.

hayfever An ALLERGIC REACTION of the upper respiratory tract and eyes caused by an abnormal sensitivity to pollen or other airborne particles. It is characterized by running nose, watering eyes and sneezing, caused by the HISTAMINE released during an allergic reaction. The histamine irritates the membranes of the nose and eyes causing them to swell. Antihistamines are used to treat hayfever and are most effective if taken before an attack.

HCG *See* HUMAN CHORIONIC GONADOTROPHIN.

heart *(See also diagram on following page)* A muscular organ that, by rhythmic contractions, pumps blood around the body of an animal with a CIRCULATORY SYSTEM.

In mammals, the heart consists of four chambers: two atria (*see* ATRIUM) and two VENTRICLES. Oxygenated blood enters the left atrium from the lungs (via the PULMONARY vein) and passes then through the BICUSPID VALVE into the left ventricle, where it is pumped out through the main ARTERY, the AORTA, to the general circulation. The right atrium receives deoxygenated blood from the circulation from the main vein, the VENA CAVA, passes through the tricuspid valve into the right ventricle, and then to the lungs via the pulmonary artery. The aorta and pulmonary artery also have non-return valves called semi-lunar valves. Oxygenated and deoxygenated blood never mix because the right and left sides of the heart are completely separate.

The muscle of the mammalian heart is a specialized CARDIAC MUSCLE that generates the heartbeat from within (myogenic) rather than via an external nerve impulse (neurogenic, as in insects).

A blood clot in the coronary arteries (which supply blood to the heart muscle itself) can lead to a heart attack, where the heart is deprived of blood. The severity of a heart attack depends on the position of the clot, whether it is in the main artery or a branch. Susceptibility to heart attacks is affected by factors such as smoking, diet (animal fats and high salt cause ATHEROSCLEROSIS), stress and lack of exercise.

Fish have a single circulation (*see* CIRCULATORY SYSTEM) with only two chambers in the heart, an atrium and a ventricle. Amphibians and reptiles have a double circulation, with two atria and one ventricle, so blood enters separately from the lungs and body but

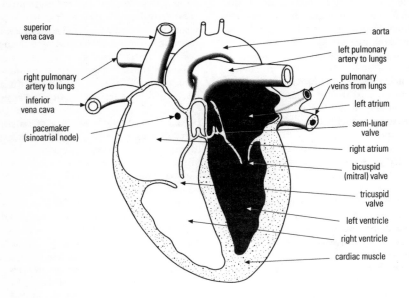

superior vena cava

aorta

left pulmonary artery to lungs

right pulmonary artery to lungs

pulmonary veins from lungs

inferior vena cava

left atrium

semi-lunar valve

pacemaker (sinoatrial node)

right atrium

bicuspid (mitral) valve

tricuspid valve

left ventricle

right ventricle

cardiac muscle

Structure of the human heart.

oxygenated and deoxygenated blood then mixes in a single ventricle.

See also PACEMAKER, PURKINJE FIBRES.

heart attack An acute medical condition characterized by a sudden severe pain in the chest and sometimes extending to the arms and throat. A heart attack is most commonly due to coronary thrombosis, a blood clot in an artery supplying the heart, and results in the heart being deprived of blood. The heart muscle deprived of blood is destroyed. This is also known as myocardial infarction. Susceptibility to this is affected by factors such as smoking, diet (animal fats and high salt cause ATHEROSCLEROSIS), obesity, high blood pressure, stress and lack of exercise. Although the term 'heart attack' is usually used to indicate thrombosis, an acute attack can be due to other conditions of the heart.

heartbeat The regular contractions and relaxation of the HEART and the accompanying sounds made by the opening and closing of the valves. *See also* PACEMAKER, PULSE.

helix A spiral in three dimensions.

helper T cell (T_H) A subset of T CELLS that help B CELLS in their ANTIBODY production. Helper T cells recognize foreign ANTIGEN when it is presented in combination with host cell antigens of the MAJOR HISTOCOMPATIBILITY COMPLEX. They are triggered to produce INTERLEUKIN-2, which in turn stimulates production of other CYTOKINES, including those which cause B-cell proliferation.

heparin A natural ANTICOAGULANT in the body that inhibits the action of THROMBIN.

hepatic (*adj.*) Relating to the LIVER.

Hepaticae A class of the phylum BRYOPHYTA, consisting of the liverworts. Liverworts are small flat or leafy plants similar to mosses and are found in damp conditions. They show ALTERNATION OF GENERATIONS.

hepatic portal vein The vein that transports blood rich in soluble digested food from the INTESTINE to the LIVER.

hepatocyte One of many cuboidal cells in the LIVER, with a large nucleus and containing many MITOCHONDRIA and GLYCOGEN granules in its CYTOPLASM. Hepatocytes also have MICROVILLI on their surface to facilitate exchange of materials between the blood and liver.

herbaceous perennial *See* PERENNIAL PLANT.

herbicide A PESTICIDE that kills weeds. Herbicides make up 40 per cent of all pesticides used. They act by stimulating AUXIN production and disrupting plant growth (*see* PLANT GROWTH SUBSTANCES).

herbivore An animal whose diet consists of green plants, seeds or fruit. The largest group of herbivores is the ZOOPLANKTON on the surface of the ocean, which feed on small algae. RUMINANTS such as cows, sheep and deer are herbivorous mammals and digest cellulose as the major part of their diet. They have specialized bacteria in their digestive system that are able to make the enzyme cellulase, which breaks down the cellulose in plant material. The teeth of ruminants are adapted to break down plant food more thoroughly, and their stomachs consist of four chambers to allow the bacteria breaking down cellulose to operate at the correct pH, isolated from the mammal's own GASTRIC JUICES. *See also* CARNIVORE, OMNIVORE.

hermaphrodite An organism that contains both female and male sex organs. In plants, this means that STAMENS and CARPELS are in the same flower. In animals, for example, earthworms, an individual can produce both sperm and ova. Despite this, hermaphroditic species usually cross-fertilize.

hesperidium A type of berry with a thick leathery outer layer and fluid-containing segments inside. Examples include the citrus fruits.

heterogametic (*adj.*) Describing the occurrence of a dissimilar pair of SEX CHROMOSOMES, as in human males who possess one X- and one Y-chromosome. *See* SEX DETERMINATION. *Compare* HOMOGAMETIC.

heterophagosome *See* PHAGOSOME.

heterosome A CHROMOSOME that is not the same in males and females; in other words, the SEX CHROMOSOMES. *Compare* AUTOSOME. *See also* SEX DETERMINATION.

heterosporous (*adj.*) Describing plants that have two types of SPORES. Microspores give rise to male GAMETOPHYTE generations, and megaspores give rise to female gametophytes (*see* ALTERNATION OF GENERATIONS). All SEED PLANTS are heterosporous, as are some CLUB MOSSES and FERNS. *Compare* HOMOSPOROUS.

heterotroph An organism that feeds on other animals or plants to obtain energy. All animals and fungi are heterotrophs. Heterotrophs can be HERBIVORES, CARNIVORES, OMNIVORES, SAPROTROPHS, DETRITIVORES and PARASITES. *Compare* AUTOTROPH.

heterozygous (*adj.*) In genetics, describing the presence of two different ALLELES at a particular LOCUS on a CHROMOSOME in a DIPLOID cell or organism. This is in contrast to the HOMOZYGOUS condition, where the alleles at a given locus are identical. RECESSIVE alleles are not expressed in the heterozygous state. Individuals in an outbreeding population will be heterozygous for some traits and homozygous for others.

hexose A MONOSACCHARIDE containing six carbon atoms in the molecule.

hibernation A state characterized by greatly reduced metabolic processes (reduced breathing and heart rate, fallen body temperature) that some animals enter to survive the winter. *See also* BIORHYTHM.

hindbrain One of three regions of the human brain. The hindbrain contains the MEDULLA OBLONGATA that controls activities such as RESPIRATION, HEARTBEAT and blood pressure. Overlying this is the CEREBELLUM, which controls the muscle movement needed for posture and locomotion. *See also* FOREBRAIN, MIDBRAIN.

hip girdle *See* PELVIC GIRDLE.

Hirudinea A class of the phylum ANNELIDA, consisting of leeches (e.g. *Hirudo*). Leeches possess suckers by which they attach themselves to a host to feed (they are therefore temporary PARASITES). They have no chaetae (bristles) and no distinct head and they are HERMAPHRODITES. Movement is by means of the suckers. Leeches are important medically. *Compare* POLYCHAETA, OLIGOCHAETA.

histamine A derivative of the amino acid histidine, released by MAST CELLS in response to the appropriate ANTIGEN during an inflammatory or allergic response. Histamine causes dilation of local blood vessels and an increase in their permeability to allow through, for example, ANTIBODIES, FIBRINOGEN and NEUTROPHILS, which are needed for repair or recovery of the site. During an allergic response, such as hayfever, the release of histamine causes some inflammation and this is responsible for the characteristic itching and sneezing. *See also* ALLERGIC REACTION, INFLAMMATION.

histidine A basic AMINO ACID, the precursor of HISTAMINE. It is one of the ESSENTIAL AMINO ACIDS.

histiocyte A MACROPHAGE found within TISSUE.

histochemistry The study of the distribution of molecules within the cells and matrices of TISSUES using staining techniques and MICROSCOPY.

histogram A BAR CHART that displays the frequency with which particular values or ranges of values are found in a set of data. The areas of the bars are proportional to the frequency, and successive ranges are placed side by side.

histology The study of TISSUE structure, mainly by means of staining and MICROSCOPY.

histone One of a group of proteins that is important in the packaging of EUKARYOTIC DNA. Histones, together with DNA, form CHROMATIN, which is the major component of CHROMOSOMES.

HIV (human immunodeficiency virus) The virus that causes AIDS. It is a RETROVIRUS that originated in Africa and was first identified in 1983. The virus infects HELPER T CELLS, which are needed for the IMMUNE SYSTEM to function. An infected (HIV-positive) person is unable to fight the HIV and is also susceptible to other diseases because the immune system is deficient. The virus is transmitted by blood or sexual secretions and therefore intravenous drug users, haemophiliacs, surgical patients receiving blood and those taking part in unprotected sex are all at high risk. The virus can remain dormant for about 8 years but the HIV-positive person is then a carrier of the virus. Pregnant women can pass the virus onto their children through the placenta or breast-milk. Transmission is prevented by the use of CONDOMS during sexual intercourse, and by screening donated blood and blood products for the virus.

holocrine gland An EXOCRINE GLAND in which cells are destroyed with the discharge of their secretion, for example SEBACEOUS GLANDS and SWEAT GLANDS. See also APOCRINE GLAND, MERO-CRINE GLAND.

homeostasis (*homeo* = same, *stasis* = standing) The maintenance of a constant internal environment within an organism. Homeostasis is important for efficient functioning of ENZYMES, which affects the entire organism.

There are many control systems to maintain homeostasis. Control of blood sugar levels is by the liver and involves interconversion of GLUCOSE and GLYCOGEN. This is regulated by the hormones INSULIN and glycogen from the PAN-CREAS. Respiratory gases (carbon dioxide and oxygen) are controlled by respiratory centres in the brain. Blood pressure is controlled by the rate of the HEARTBEAT. The HEART pumps blood and ensures the distribution of blood with its supply of essential materials to cells. OSMOREGULATION controls SOLUTE concentration and total body volume in an organism. Salt and pH levels must also be maintained.

Control of body temperature is called HOMEOTHERMY. Mammals and birds maintain a constant body temperature regardless of the external temperature and are called homeo-therms. Invertebrates, fish, amphibians and reptiles have a fluctuating body temperature, and obtain most of their heat from outside the body; they are called POIKILOTHERMS.

homeotherm, *endotherm* An animal exhibiting HOMEOTHERMY. *Compare* POIKILOTHERM.

homeothermy (*homeo* = same, *thermo* = heat) The maintenance of a constant body temperature regardless of the external temperature and usually higher than it. Homeothermy is characteristic of mammals and birds, which are called homeotherms or endotherms (warm-blooded). Most homeotherms have an insulating layer of fat or fur that helps prevent heat loss and heat gain. The HYPOTHALAMUS controls other ways of maintaining a constant body temperature (in response to messages received from skin sensors), such as reducing the metabolic rate, sweating, panting and vasodilation (widening of blood vessels) to increase heat loss, and increasing the metabolic rate, shivering and vasoconstriction (narrowing of blood vessels) to reduce heat loss. Some means of homeothermy are behavioural, such as HIBERNATION or being nocturnal to avoid daytime heat. Homeothermy enables a greater efficiency of metabolic functions in contrast to POIKILOTHERMY, in which the body temperature fluctuates.

homogametic (*adj.*) Describing the occurrence of a similar pair of SEX CHROMOSOMES, as in human females who possess two X-chromosomes. *See* SEX DETERMINATION. *Compare* HET-EROGAMETIC.

homoiotherm *See* HOMEOTHERM.

homologous (*adj.*) Describing a similarity in some aspect, such as structure, position or functional properties. Chromosomes that are similar in appearance are called homologous and associate in pairs during MEIOSIS. In CLASSI-FICATION, structures similar in appearance or origin are said to be homologous.

homosporous (*adj.*) Describing those plants that have one type of SPORE that gives rise to GAMETOPHYTES bearing both male and female reproductive organs. Examples include ferns and horsetails. *Compare* HETEROSPOROUS.

homozygous (*adj.*) In genetics, referring to the presence of two identical ALLELES at a particular LOCUS on a CHROMOSOME in a DIPLOID cell or organism. An individual homozygous for a particular trait will always pass this trait onto their offspring. RECESSIVE alleles are only expressed in the homozygous state. *See also* HETEROZYGOUS.

hormone Any molecule (usually of small molecular mass) secreted directly into the blood by ENDOCRINE GLANDS and transported to a target cell or organ, causing a specific response. A hormone is sometimes referred to as a chemical messenger and, unlike PROSTAGLANDINS, its effects are not just local. The PITUITARY GLAND controls the co-ordination of hormone action, and the HYPOTHALAMUS controls the pituitary gland.

Hormones regulate a number of body functions, for example general metabolism and growth (hormones from the THYROID GLAND), responses to stress or danger (hormones from the ADRENAL GLAND), blood sugar levels (hormones from the PANCREAS) and reproductive functions (hormones from the TESTIS, OVARY and PLACENTA).

Some hormones are amines, for example ADRENALINE, NORADRENALINE and THYROXINE, and may consist of only a few AMINO ACIDS. Others are POLYPEPTIDES of less than 100 amino acids, for example OXYTOCIN, ANTI-DIURETIC HORMONE, INSULIN and GLUCAGON. Others are larger (more than 300 amino acids) and constitute PROTEINS, for example PROLACTIN, FOLLICLE-STIMULATING HORMONE, LUTEINIZING HORMONE, THYROID-STIMULATING HORMONE, ADRENOCORTICOTROPHIC HORMONE and GROWTH HORMONE.

Some hormones are derived from LIPIDS and are called STEROID HORMONES, for example OESTROGEN, PROGESTERONE, TESTOSTERONE, CORTISONE and ALDOSTERONE. These are lipid-soluble and can pass through the cell membrane, so that when the hormone reaches its target cell it forms a complex with RECEPTOR molecule, passes through the cell membrane and influences some activity within the cell. Other hormones are proteins that are water-soluble and cannot pass through the cell membrane. They act by binding to a receptor molecule on the target cell membrane, which then activates a second messenger (a molecular signal) that initiates a specific chemical change within the cell (but is not exported by it). The water-soluble hormones usually mediate short-term effects, while lipid-soluble steroids mediate longer-term effects.

Hormones can act in a number of other ways (*see* individual entries). Hormones do not differ much from one species to another but their effects may differ. The use of the term hormone in the botanical context has now been replaced by the term PLANT GROWTH SUBSTANCE. *See also* ENCEPHALIN, ENDORPHIN, PHEROMONE.

hormone-replacement therapy (HRT) The oral administration of the hormones OESTROGEN and PROGESTERONE to women to help them overcome the effects of the reduction of these hormones during MENOPAUSE. Symptoms of the menopause, such as anxiety, hot flushes, irregular bleeding and osteoporosis (thinning of bone leading to increased fractures), may be controlled by HRT but the value of HRT is controversial and there are some side-effects.

hornwort A member of the phylum ANTHOCEROTAE.

horsetail A member of the phylum SPHENOPHYTA.

HRT *See* HORMONE-REPLACEMENT THERAPY.

human chorionic gonadotrophin (hCG) A PEPTIDE hormone secreted by the TROPHOBLAST and that is detected in the urine of pregnant women and forms the basis of pregnancy tests. Its role is to ensure the continued production of OESTROGEN and PROGESTERONE by the CORPUS LUTEUM until the PLACENTA fully develops.

Human Genome Project *See* GENOME.

human immunodeficiency virus *See* HIV.

human placental lactogen (hPL) A hormone produced by the human PLACENTA after about 5 weeks of pregnancy. It causes an increase in the maternal metabolism of fat instead of carbohydrate and is an INSULIN antagonist.

humoral immunity The response of an organism to invasion by a foreign object in which an ANTIBODY is produced. *Compare* CELL-MEDIATED IMMUNITY. *See also* IMMUNE RESPONSE, IMMUNITY.

humus An organic mixture in SOIL, consisting of dead plant and animal material and animal

waste products that are broken down by DECOM-
POSERS and DETRITIVORES. Humus has a high carbon content, is rich in minerals and is often
acidic. It is important in helping water retention
in sandy soil, and allows better aeration and
drainage in clay soils. In water-logged areas,
humus cannot be broken down fully by the
aerobic decomposers and accumulates as peat.

hyaloplasm *See* CYTOSOL.

hybrid The offspring resulting from a cross
between individuals of different SPECIES. In
animals, hybrids are usually infertile and
therefore cannot reproduce. An example that
occurs naturally is the mule, which is a cross
between a male horse and a female donkey. In
plants, hybrids are often produced to combine
desirable characteristics.

hybridization 1. The production of a HYBRID
individual or of hybrid cells by cell fusion. *See
also* MONOCLONAL ANTIBODY.

 2. In GENETIC ENGINEERING, the production
of a DNA hybrid. *See* RECOMBINANT DNA.

hybridoma *See* MONOCLONAL ANTIBODY.

hydration The combination of a substance with
water to produce a single product. It is the
opposite of dehydration.

hydrocarbon Any chemical compound consisting of hydrogen and carbon only. ETHENE is an
example.

hydrocortisone *See* CORTISOL.

hydrogen (H) A colourless, odourless gas; the
lightest of all the elements. Hydrogen is by far
the most abundant element in the universe,
making up about 75 per cent of the atoms.
Hydrogen is also widespread on Earth, forming many compounds, including water.

 Hydrogen is present in all ACIDS and these
will react with the more reactive metals to
release hydrogen gas. The presence of hydrogen in the laboratory is confirmed by lighting
a sample of the gas: it burns with a characteristic squeaky pop, caused by the high speed of
sound in the gas.

hydrogen bond A weak chemical bond formed
between electronegative atoms (such as oxygen, fluorine, nitrogen) in one molecule and
hydrogen atoms in another molecule. Hydrogen bonds are weak, non-covalent bonds
(about one tenth of the strength of normal
COVALENT BONDS) but they have considerable
effects on physical properties and are important biologically. Hydrogen bonding occurs in
water, where it is responsible for the anoma-

lous properties of water and its high boiling
point. Hydrogen bonding is important in living organisms, since it is responsible for the
secondary, tertiary and quarternary structure
in proteins. It also occurs between bases in the
chains of DNA (*see* BASE PAIR).

hydrolase One of a group of ENZYMES that is
involved in HYDROLYSIS reactions, such as
LIPASES.

hydrolysis Any chemical process whereby a substance is broken down into smaller molecules
with the addition of a water molecule.
Examples include ATP breakdown and digestion.

hydrophilic (*adj.*) (Greek = 'water-loving')
Describing a molecule that has a strong affinity for water. Such materials are generally soluble in water. *Compare* HYDROPHOBIC.

hydrophobic (*adj.*) (Greek = 'water-hating')
Describing a molecule that does not have a
strong affinity for water. *Compare* HYDRO-
PHILIC.

hydrophyte A plant that lives wholly or partially
under water, or in waterlogged soil. Examples
include swamp, marsh and aquatic plants,
such as rushes, reeds and water lilies. Since
there is a shortage of oxygen in such conditions, hydrophytes possess aeration tissue that
contains large spaces between the cells of the
stems and leaves. This gives the plants buoyancy, causing the leaves to rise to the surface to
get maximum light. Oxygen can be stored in
the spaces from where it can diffuse to the
roots for use in RESPIRATION. Stomata (*see*
STOMA) are reduced or absent in submerged
leaves (as water loss is not a problem), XYLEM is
poorly developed (as water conducting vessels
are not needed) and supporting SCLERENCHYMA
tissue is reduced (the water provides support
and the plant needs to be flexible to move with
the current). *See also* XEROPHYTE.

hydrosere A series of plant SUCCESSIONS originating in fresh water.

hydrostatic pressure Pressure produced by
fluids.

hydrostatic skeleton A type of SKELETON found
in certain invertebrates, such as earthworms,
where the body cavity is full of fluid. Because
liquid cannot be compressed, the structure is
resilient. Contraction of circular and longitudinal muscles alternately in one region that
then passes along the body causes peristaltic
movement (*see* PERISTALSIS).

hydrotropism The directional growth of a plant (or part of it) in response to water. Roots are positively hydrotropic, growing towards water, while stems and leaves show no response. *See also* TROPISM.

hydroxide Any compound containing a metal and the hydroxide ion, OH⁻.

Hydrozoa A class of the phylum CNIDARIA, including *Hydra* and the Portuguese man-of-war.

hyper- (*prefix*) Indicates above, higher, over, excessive. For example, hyperactive is extremely active, HYPERGLYCAEMIA is high levels of glucose. *Compare* HYPO-.

hyperglycaemia An abnormally high level of sugar in the blood.

hyperthyroidism A medical condition resulting from an overactive THYROID GLAND. This leads to thyrotoxicosis, which is an increase in metabolic rate. Treatment for this is destruction of part of the thyroid (by surgery or by administration of radioactive iodine). *See also* HYPOTHYROIDISM.

hypertonic solution A solution with a higher OSMOTIC PRESSURE compared to some other solution. In a pair of solutions, the hypertonic solution is the one with the greater osmotic pressure. *Compare* HYPOTONIC SOLUTION.

hypha (*pl.* **hyphae**) One of many thread-like filaments that together form the main body (or MYCELIUM) of a FUNGUS. Hyphae grow in length and by having side-branches, which form a network of tubes containing CYTOPLASM and nuclei. Substances, including food, pass along the hyphae by movement of the cytoplasm in a process called cytoplasmic streaming.

In some higher fungi (ASCOMYCETES and BASIDIOMYCETES), the hyphae are divided at intervals by septa (cross-walls) but there is still a central pore through which cytoplasm and nuclei can pass. In lower fungi (e.g. bread mould) no such divisions exist. At the ends of hyphae a structure called a SPORANGIUM grows that contains the SPORES by which the fungus reproduces.

hypo- (*prefix*) Indicates below, beneath, lower, incomplete. For example, HYPOGLYCAEMIA is

low levels of GLUCOSE, hypothermia is low body temperature. *Compare* HYPER-.

hypogeal (*adj.*) Describing seed GERMINATION where the COTYLEDONS remain below the ground.

hypoglycaemia An abnormally low level of sugar in the blood. This occurs in DIABETES.

hypophysis *See* PITUITARY GLAND.

hypothalamus A small region of the brain that plays a vital role in linking the NERVOUS SYSTEM and the ENDOCRINE SYSTEM and controls the activities of the latter. In the nervous system, the hypothalamus regulates the autonomic (involuntary) responses and controls behavioural patterns (such as feeding, thirst, sleeping and aggression) and physiological stability (such as water balance and temperature). It has a rich supply of blood vessels and monitors blood composition so it can adjust it when necessary.

In the endocrine system, the hypothalamus regulates the action of the PITUITARY GLAND, which in turn regulates many other ENDOCRINE GLANDS. It produces ANTI-DIURETIC HORMONE and OXYTOCIN, which are passed to the posterior pituitary for storage and secretion. Another example of hypothalamus control within the endocrine system is in its production of thyrotrophin-releasing factor (TRF), which passes to the anterior pituitary where it regulates production of thyroxine.

See also HOMEOTHERMY.

hypothyroidism A medical condition resulting from an underactive THYROID GLAND. This causes a reduction in an individual's metabolic rate, resulting in a reduced heart and ventilation rate, a lowered body temperature and obesity as well as mental sluggishness. A goitre may be seen (swelling in the throat). In infants the mental retardation is serious and leads to a condition called cretinism. Oral administration of the thyroid hormone thyroxine eliminates symptoms. *See also* HYPERTHYROIDISM.

hypotonic solution A solution with a lower OSMOTIC PRESSURE compared to some other solution. In a pair of solutions the hypotonic solution is the one with the lower osmotic pressure. *Compare* HYPERTONIC SOLUTION.

I

ICSH *See* INTERSTITIAL CELL-STIMULATING HORMONE.

IgA A class of IMMUNOGLOBULIN found mostly in MUCUS secretions, for example milk and saliva.

IgD A class of IMMUNOGLOBULIN, the function of which is uncertain.

IgE A class of IMMUNOGLOBULIN that is found on BASOPHILS or MAST CELL surfaces. It is associated with ALLERGIC REACTIONS, for example hayfever, asthma and parasitic infections.

IgG A class of IMMUNOGLOBULIN that represents 70 per cent of human immunoglobulin and is the main immunoglobulin of the secondary IMMUNE RESPONSE. IgG is the only immunoglobulin to cross the PLACENTA.

IgM A class of IMMUNOGLOBULIN that is produced early in response to an infection.

ileo-caecal valve The valve that controls the passage of food from the SMALL INTESTINE to the COLON.

ileum The region of the vertebrate SMALL INTESTINE between the DUODENUM and COLON that is mostly concerned with the absorption of food. The wall of the ileum is specialized for this by being extensively folded and having numerous projections facing inwards called VILLI, which themselves have minute projections called MICROVILLI that together form a BRUSH BORDER. This provides the large surface area over which absorption can take place. Most absorption is by ACTIVE TRANSPORT rather than DIFFUSION, which would also allow molecules to leave the body. The ileum has a good blood supply, which carries water-soluble materials, such as sugars, amino acids and minerals, to the LIVER, where their release to other parts of the body is regulated. Fats are absorbed by the lacteal (a small vessel in the villi) instead of blood capillaries and are transported in LYMPH VESSELS, entering the blood nearer to the heart.

Most of the water that is drunk is absorbed by the stomach and most of the water in digestive juices is absorbed by the ileum. The rest is reabsorbed by the COLON.

See also DIGESTIVE SYSTEM.

immiscible (*adj.*) Describing two liquids that will not mix together, such as hexane and water. When stirred, an EMULSION is formed, which will separate if left to stand, with the less dense material floating to the surface. *See also* MISCIBLE.

immune response The body's reaction to invasion by a foreign object, such as an organism, pollen grains, transplanted tissues or cancerous cells. It involves the recognition of ANTIGENS on the surface of the foreign material and the production of ANTIBODIES to combat the invasion. This antibody response is called HUMORAL IMMUNITY, which involves B CELLS. The immune response also involves CELL-MEDIATED IMMUNITY, in which T CELLS play a major role.

The first exposure to an antigen results in a primary immune response, in which the antigen is processed by a phagocytic cell (*see* PHAGOCYTE) such as a MACROPHAGE, so that it can be recognized by HELPER T CELLS, which then activate B cells. These then differentiate into either PLASMA cells, which secrete antibodies, or into memory cells, which are needed for future invasion by the same antigen. The primary immune response is characterized by the production of antibodies, usually of the IGM class (*see* IMMUNOGLOBULIN). However, the primary response is slow and low compared to the secondary immune response, when the memory B cells are stimulated by a second exposure to an antigen and rapidly secrete more antibody (usually IGG). The antibodies bind to the antigen, forming complexes that are destroyed by macrophages or neutralized in a variety of ways.

See also IMMUNITY.

immune system The components involved in the IMMUNE RESPONSE.

immunity The ability of an organism to resist an infection by harmful micro-organisms, and also to fight against cancerous cells. There is some innate immunity in animals that occurs naturally, provided, for example, by the skin barrier and stomach acids. Acquired immunity occurs as a result of exposure to ANTIGENS throughout life. Acquired immunity involves the production of antibodies (*see* ANTIBODY) in HUMORAL IMMUNITY and also CELL-MEDIATED IMMUNITY (CMI), which involves cells other than antibodies. The two responses are not exclusive, as cells produced in the CMI response are involved in the initiation of the antibody response and antibodies are usually present in CMI responses.

There are four main categories of acquired immunity. (i) Active natural immunity, in which the body's memory B CELLS (*see* LYMPHO-CYTES) respond to repeated exposure of an antigen by multiplying and releasing specific antibodies. (ii) Active induced immunity, in which an antigen is given by VACCINATION to induce an IMMUNE RESPONSE. (iii) Passive natural immunity, where antibodies are acquired by a foetus from its mother across the PLA-CENTA or in the COLOSTRUM. (iv) Passive induced immunity, in which specific antibodies are given by intravenous inoculation to a person being treated for a particular disease, for example tetanus and diphtheria in humans. Passive immunity is only temporary.

Some diseases, such as AIDS, attack the immune system and are therefore difficult to fight. The study of immunity is called immunology.

immunization *See* VACCINATION.

immunoassay Any of a number of techniques for quantitative analysis of a protein or other ANTI-GEN by exploiting its ability to bind to a specific ANTIBODY. An example of the use of such techniques is in measuring HUMAN CHORIONIC GONADOTROPHIN in urine as a means of determining pregnancy. Examples of immunoassays include ELISA and WESTERN BLOTTING.

immunocompromisation The lack of a fully functional IMMUNE SYSTEM due to some impairment acquired later in life. This is in contrast to IMMUNODEFICIENCY, which is inborn. A person could be immunocompromised as a result of an infection such as AIDS, or as a result of pregnancy, DIABETES or old age. People on immuno-suppressive drugs (*see* TRANSPLANTATION) would be immunocompromised.

immunodeficiency The impairment of one or more aspects of the IMMUNE SYSTEM, for example as a result of severe combined immune deficiency (SCID), which people are born

with. With SCID, a BONE MARROW transplant to replace the defective immune cells is the only chance of survival. Patients with AIDS are often referred to as immunodeficient but because AIDS is an acquired condition they are more correctly described as immunocompromised (*see* IMMUNOCOMPROMISATION).

immunoelectrophoresis A type of ELECTRO-PHORESIS in which a mixture of ANTIGENS is separated by electrophoresis, usually in an AGAR or AGAROSE gel, and the antigen of interest is identified by use of a specific ANTIBODY.

The antibody can be added in several ways. It can be added to a trough in the gel, close to the separated antigen. The antigen diffuses into the gel, forming a precipitate on contact with the corresponding antigen. In crossed immunoelectrophoresis, the separated antigen is subjected to further electrophoresis into a second gel containing the antibody. Precipitates are seen as arcs wherever corresponding antigen and antibody meet. This technique is useful for the comparison of a complex antigen mixture and can be used in the purification of an antigen.

immunofluorescence A technique used to detect a specific ANTIGEN by use of an ANTIBODY to which a fluorescent marker has been added. Antigen-antibody complexes form and can be detected using the appropriate illumination, for example in FLUORESCENCE MICROSCOPY.

immunogenic (*adj.*) Describing a substance that triggers an IMMUNE RESPONSE. *See also* ANTIGEN.

immunoglobulin A human GLYCOPROTEIN that is the ANTIBODY in an immune reaction to combat a foreign substance. There are five classes of immunoglobulin (IGA, IGD, IGG, IGE, IGM), which differ in their structure, the degree to which they polymerize and when they are produced.

The basic immunoglobulin unit consists of two heavy and two light POLYPEPTIDE chains, which form two 'Fab' regions containing the ANTIGEN-binding site, and an 'Fc' region that determines the biological properties of the immunoglobulin, for example where the antibody binds. Specificity of the antibody is determined by a sequence of AMINO ACIDS at one end of the Fab region.

Immunoglobulin can be obtained from blood PLASMA for administration to patients at high risk of contracting a disease.

See also IMMUNITY.

immunology The study of IMMUNITY.

immunosuppressive drug Any drug designed to suppress an IMMUNE RESPONSE. *See also* TRANS-PLANTATION.

implantation In mammals, the process by which a developing EMBRYO attaches to the wall of the UTERUS and stimulates the development of the PLACENTA. Following FERTILIZA-TION, the fertilized OVUM (ZYGOTE) divides by MEIOSIS until a BLASTOCYST is produced (the 64-cell stage). This takes about 7 days to travel down the FALLOPIAN TUBE, while dividing, before it implants in the ENDOMETRIUM lining the uterus. The outer layer of cells of the blastocyst develop into the TROPHOBLAST and later into the EXTRAEMBRYONIC MEM-BRANES. It is from one of these membranes, the CHORION, that VILLI develop, which invade the surrounding maternal tissue and eventually form the placenta. Delayed implantation can occur in some species, for example, badgers and polar bears, where the blastocyst forms but may not implant for many months until food supplies or the weather have improved.

imprinting A form of LEARNING that is an involuntary response to a specific stimulus at a sensitive time in an animal's development.

One example is the specific recognition by a young animal of its mother and of its own species. This imprinting can occur very soon after birth. Goslings learn to recognize the first moving object they see, so they can easily become imprinted on individuals (even of another species) other than their mother. Imprinted behaviour is fixed and not easily changed. *See also* CONDITIONING, HABITUATION.

impulse *See* NERVE IMPULSE.

inbreeding The mating of an animal with its close relatives. An inbreeding population shows less genetic variability than an OUTBREEDING population and continued inbreeding can result in harmful GENES being expressed, which in an outbreeding population would be RECES-SIVE. *Compare* OUTBREEDING.

incisor In mammals, a sharp TOOTH at the front of the mouth that is used for biting. Rodents such as rats have large incisors that continually grow and are adapted for gnawing. There are eight incisors in humans.

incus, anvil The middle of the three EAR OSSI-CLES in the mammalian middle EAR.

indehiscent (*adj.*) Describing a fruit that does not open spontaneously to shed its seeds, for example nuts.

indicator species A plant or animal with known ecological requirements whose presence in an area provides some information about the environmental conditions in that area. For example, some plants prefer acidic soil and some alkaline soil, so their presence or not indicates the soil type. Some LICHENS are sensitive to levels of sulphur dioxide in the air, so their absence could be indicative of air pollution.

induced fit *See* LOCK-AND-KEY MECHANISM.

inducible An ENZYME that is synthesized only when the SUBSTRATE is present. *Compare* CON-STITUTIVE.

infertility The inability of an organism to reproduce. Infertility can be due to the inability to produce GAMETES (no ova produced, or none or too few SPERM produced); blocked FALLOPIAN TUBES, so that sperm and ova cannot meet; or problems in maintaining the pregnancy, such as low hormone levels, recurrent spontaneous abortion, repeated unexplained early loss of embryos with a possible underlying immunological problem.

In humans it is now rare that problems of infertility cannot be overcome. The use of IN VITRO FERTILIZATION (IVF) has overcome the problem for many couples, particularly in cases of blocked fallopian tubes. Artificial insemination is another treatment that involves the artificial administration of sperm to a woman, either from her partner or from a donor (AID). Some of these treatments, especially AID and surrogacy, and the spare 'embryos' produced by IVF, raise difficult moral and ethical problems.

inflammation In immunology, the local response to injury or infection, outwardly involving redness and swelling of the surrounding tissue. HISTAMINE is released by MAST CELLS and causes dilation of local blood vessels and an increase in permeability to allow through antibodies, FIBRINOGEN and NEU-TROPHILS, which are needed for repair or recovery of the site. The presence of fibrinogen at the site of inflammation leads to FIBRIN forma-tion and clotting, to prevent haemorrhage and to trap any PATHOGENS to restrict the spread of infection. Neutrophils and later MONOCYTES (transforming into MACROPHAGES) pass through the blood capillaries and engulf pathogens and dead cells by PHAGOCYTOSIS, resulting in the accumulation of a yellowy liquid called pus (consisting of dead cells and bacteria) at the site of inflammation. *See also* ALLERGIC REACTION.

inflorescence In plants, a term used for a specific arrangement of flowers on a single main stalk. Inflorescences can be divided into cymose (definite) or racemose (indefinite) according to their method of branching. In a cymose inflorescence, the tip of the main stalk produces a single flower and subsequent flowers arise on lower side branches. In a racemose inflorescence, the main stalk increases in length by growing at its tip and flowers are borne along the whole length of the stalk, opening from below upwards. The oldest flowers, in a racemose inflorescence, are therefore found near the base. *See also* PEDICEL, PEDUNCLE.

infrared (*adj.*) Describing electromagnetic radiation with wavelengths greater than those of visible light and shorter than those of microwaves; that is, between 0.75 micrometres and 1 millimetre. Infrared radiation is emitted by all warm objects and can be detected by modified versions of photographic film or electronic devices. This is important for military purposes and for detecting live bodies under rubble. Many substances absorb infrared, which is a useful method of applying heat. Infrared radiation is also used in medical photography and treatment.

infrared radiation See INFRARED.

inhibitor A substance that slows down the rate of a chemical reaction or blocks it. *See also* ENZYME INHIBITION.

insect A member of the class INSECTA of the phylum ARTHROPODA.

Insecta A class of the phylum ARTHROPODA, consisting of the insects. It is a very diverse class of more than a million species. An insect's body is divided into three parts, the head, THORAX and ABDOMEN, with a pair of antennae on the head, three pairs of legs attached to the thorax and usually two pairs of wings (fore and hind), although not all insects are winged. The mouth parts can be adapted for biting and chewing, for example in bees and locusts, or for piercing and sucking, for example in butterflies and house flies. Some insects, for example lice and fleas, are PARASITES. An insect SENSORY SYSTEM is well developed: the antennae

act as feelers and detect odours, there is one pair of compound eyes and many insects can detect sound.

The skeleton of insects is external and hard (except at the joints) and contains CHITIN. The thorax is responsible for locomotion, the legs being adapted to walk on land, and in some, for example locusts, the hind legs are well developed to allow jumping. The wings are composed of two membranes with a strengthening framework of 'veins' between them. Differences in the wings form part of the classification of insects, for example the wings may be joined to operate as one pair, or hard, as in beetles, or adapted to assist in balance. The abdomen is responsible for metabolism and reproduction. Respiratory gases are carried to and from the body muscles by TRACHEAE, which are fine tubes opening to the outside at pores called SPIRACLES. The sexes in insects are separate; the female produces eggs and fertilization is internal. Many insects mate once and then die. The life cycle of an insect often involves several changes in appearance and way of life, from young to adult, and is called METAMORPHOSIS.

Economically, insects are very important to humans; they can be serious pests that need to be controlled with insecticides, they carry diseases such as malaria, sleeping sickness, typhoid and dysentery, and they can cause damage to stored food, wood and clothing. Insects are also useful as pollinators of fruit and crops, as scavengers, in the study of genetics (e.g. the fruitfly *Drosophila*) and as a source of products such as honey and silk. The study of insects is called entomology.

insecticide A PESTICIDE that kills hazardous or unwanted insects.

insoluble (*adj.*) Describing a compound that will not dissolve (normally in water).

inspiration Breathing in. *See* BREATHING.

insulin A protein HORMONE that is responsible for the regulation of blood sugar levels. Insulin is a 51 amino acid POLYPEPTIDE, made by the β-cells of the ISLETS OF LANGERHANS in the PANCREAS. Insulin is produced when blood sugar levels are high, for example after a meal, and has the effect of reducing the blood sugar levels. Insulin promotes uptake of free GLUCOSE by body cells (e.g. muscle and adipose cells) by altering the permeability of the CELL MEMBRANE to glucose. It also promotes uptake of amino acids by muscles and converts excess glucose (formed as a result of carbohydrate breakdown following a meal) into GLYCOGEN for storage in the liver. This is called GLYCOGENESIS. Deficiency in the production of insulin leads to a condition called DIABETES. Human insulin can now be produced from bacteria by GENETIC ENGINEERING techniques. *See also* GLUCAGON.

insulin-like growth factor (IGF) One of a group of GROWTH FACTORS that stimulate fat and other CONNECTIVE TISSUE cells.

integrated pest management Combined strategies for the control of pests. There are five major strategies employed in integrated pest management: PESTICIDES, BIOLOGICAL CONTROL, cultural control (such as using agricultural practices to change the pest's habitat), breeding pest-resistant species and sterile-mating control.

integrins A family of proteins involved in CELL ADHESION. They are found on the surface and across the membrane of cells, and are able to bind components of the EXTRACELLULAR MATRIX and also proteins on other cell surfaces, thereby performing adhesion functions.

integument In SEED PLANTS, the protective layer surrounding the OVULE. Most flowering plants have two integuments, while GYMNOSPERMS have only one. The integuments later form the TESTA or seed coat.

inter- (*prefix*) Indicates between or among, for example intercellular means between cells. *Compare* INTRA-.

intercostal muscle One of several muscles in TETRAPODS that allow the ribs to move up and down during breathing.

interferon Any one of a number of GLYCOPROTEINS that are made by cells infected with a virus and which confer on other, uninfected, cells immunity to the virus. This is achieved by unmasking genes that synthesize anti-viral proteins. Some viruses are still able to overcome this resistance.

Three groups of interferons (α, β, γ) are known in humans. α-interferons are produced by infected LEUCOCYTES, β-interferon by infected FIBROBLASTS and γ-interferon by LYMPHOCYTES in response to factors such as GROWTH FACTORS, rather than viral infection. Interferons are also produced by NATURAL KILLER CELLS and CYTOTOXIC T CELLS, which kill cancer cells and confer this ability to other normal lymphocytes. Thus interferons have great potential in the treatment of cancer as well as viral infections and attempts are underway to produce them on a large-scale by GENETIC ENGINEERING. α-interferon is already used clinically in the treatment of leukaemia.

interleukin A soluble substance involved in the communication between LYMPHOCYTES. Many interleukins are secreted by lymphocytes, but some are also produced by other cells. Interleukin-1 (IL-1) is secreted by macrophages following ANTIGEN stimulation and activates many aspects of INFLAMMATION, including the secretion of interleukin-2 by HELPER T CELLS. Interleukin-2 (IL-2) is required for the proliferation of T cells and stimulates the production of other CYTOKINES, such as B cell growth factor and γ-INTERFERON. IL-2 has been used clinically in the treatment of some cancers. Interleukin-3 (IL-3) stimulates the growth of MAST CELLS in some T cells. Interleukin-4 (IL-4) is thought to be a B cell growth factor. *See also* LYMPHOKINE.

interphase The interval in the CELL CYCLE between nuclear divisions (usually in MITOSIS but also in MEIOSIS). Interphase is often called the 'resting phase' because the chromosomes are not visible, but it is actually a period of intense activity when the cell ORGANELLES are duplicated and the DNA content is doubled.

interstitial cell Any one of several cells in the human TESTIS situated in the spaces between the SEMINIFEROUS TUBULES. The interstitial cells secrete TESTOSTERONE and are themselves stimulated by LUTEINIZING HORMONE.

interstitial cell-stimulating hormone (ICSH) *See* LUTEINIZING HORMONE.

interstitial fluid *See* TISSUE FLUID.

intestinal juice An alkaline secretion produced by the wall of the SMALL INTESTINE when stimulated by the presence of food. Intestinal juice contains mineral ions that neutralize the acid CHYME from the stomach, as well as enzymes that aid digestion. These enzymes include PROTEASES to break down PEPTIDES into smaller peptides and amino acids. Other enzymes break down carbohydrates, for example AMYLASE, MALTASE, LACTASE and SUCRASE.

intestine *See* SMALL INTESTINE, LARGE INTESTINE.

intra- (*prefix*) Indicates in, within or inside of. For example, intracellular refers to within a cell. *Compare* INTER-.

intra-uterine device (IUD), *coil* A plastic or copper coil inserted into the UTERUS as a method of CONTRACEPTION. A number of devices are available, all of which prevent IMPLANTATION of a fertilized egg by causing inflammation of the uterine lining. Once in place, an IUD provides continued protection that is 98 per cent effective, but there is a risk of infection that could lead to INFERTILITY. The IUD is not usually given to women who have not yet had a child.

intron A region of DNA in a GENE that does not code for a gene product. Introns are transcribed into MESSENGER RNA (*see* TRANSCRIPTION) but are removed (spliced) from the transcript before TRANSLATION. Introns are usually found in EUKARYOTIC genes, rarely in PROKARYOTES. Their functional significance is unknown. *Compare* EXON.

inulin A POLYSACCHARIDE that consists of a chain of FRUCTOSE molecules. Inulin is a storage CARBOHYDRATE in many plants, such as, the Dahlia. It is not found in animals.

inversion In genetics, a type of MUTATION. Inversion can be a POINT MUTATION in which two or more NUCLEOTIDE bases in a DNA sequence within a GENE are reversed. Inversion can also be the reversal of a section of a CHROMOSOME causing the genes within this section to be in the inverse order.

invertebrate An animal without a backbone, i.e. not a vertebrate. The majority of existing animal species are invertebrates, including SPONGES, FLATWORMS, ANNELIDS, ARTHROPODS, MOLLUSCS, ECHINODERMS and some invertebrate CHORDATES.

in vitro (= in glass) A process or experiment that takes place outside the body of a living organism, for example in the laboratory. *Compare* IN VIVO.

in vitro* fertilization (IVF), *embryo transfer A technique where FERTILIZATION of an OVUM and SPERM is performed in a test-tube, to overcome problems of INFERTILITY. The procedure was first carried out by Edwards (1925–) and Steptoe (1913–88) in Cambridge, UK, and the first IVF baby was born in 1978. The technique involves removing an ovum (after administration of a fertility drug to increase production) and sperm and mixing them in suitable culture conditions outside the body and transferring fertilized eggs back to the mother.

IVF is most suitable in cases of blocked FALLOPIAN TUBES but can also be used for women who have problems maintaining the pregnancy, when a surrogate (substitute) mother can be used.

in vivo A process or experiment that takes place within a living cell or organism. *Compare* IN VITRO.

involuntary muscle, *unstriated muscle, smooth muscle* MUSCLE under the control of the AUTONOMIC NERVOUS SYSTEM.

iodine (I) A dark violet, non-metallic element that is required as a MICRONUTRIENT by animals. It is a constituent of the thyroid hormone thyroxine, which controls metabolism. Deficiency causes cretinism in humans (*see* THYROID GLAND). In some other vertebrates, iodine is needed for METAMORPHOSIS. Iodine is not needed by higher plants.

ion An atom or molecule that is not electrically neutral, having gained or lost ELECTRONS. An ion that has lost one or more electrons (cation) is positively charged, whereas one that has gained electrons (anion) is negatively charged.

ion exchange chromatography A CHROMATOGRAPHY technique separates molecules according to their charge. The stationary phase, usually packed in a column, is electrically charged. When a mixture of liquids is passed down the column those with an affinity to the stationary phase (i.e. with an opposite charge) bind to it, while other groups are washed through the column. Bound molecules are then eluted from the column, using solutions of specific pH and ionic strength to obtain the separation required. Ion exchange is a very powerful separation technique and can be used to separate for example two proteins differing in a single AMINO ACID only.

ionic bond A chemical bond in which one or more electrons are transferred from one atom to another, creating a positive and negative ION. The attractive forces between these ions then holds the material together as a crystalline solid. These solids are generally hard, brittle materials with high melting points. They are usually soluble in water and insoluble in organic compounds. They always conduct electricity when molten or in solution.

ionizing radiation Any RADIATION that creates IONS in any matter through which it passes. This may be a stream of high-energy particles (such as electrons, protons, alpha-particles) or short-wavelength electromagnetic radiation (ultraviolet, X-rays, gamma-rays). Ionizing radiation causes extensive damage to the molecular structure of any substance. In biological tissue, the effects can be very serious, causing radiation burns and sickness, and sometimes causing cancer (particularly leukaemia).

iris A coloured diaphragm of circular and radial muscles in the EYE that can alter the size of the PUPIL to control the amount of light entering the eye.

iron (Fe) A hard, silver-grey metallic element, the fourth most abundant element in the Earth's crust, and the second most abundant metal. It is essential to all living organisms.

Iron is a component of electron carriers such as CYTOCHROMES, which are needed for respiration and photosynthesis. It is also a constituent of HAEMOGLOBIN and other respiratory pigments and some enzymes. It is required in the synthesis of CHLOROPHYLL. Iron deficiency in animals causes ANAEMIA and in plants chlorosis (yellowing of leaves).

irradiation The process of bombarding a surface with RADIATION, particularly IONIZING RADIATION. The term is especially used in the context of using ionizing radiation to kill bacteria in foodstuffs.

isidium (*pl. isidia*) A very small outgrowth from the surface of a LICHEN that comprises cells of both the ALGA and FUNGUS partners of the lichen. Isidia serve as structures of non-sexual reproduction by detaching from the lichen.

islets of Langerhans Groups of specialized cells in the PANCREAS that are concerned with regulation of blood sugar levels. There are two cell types: α-cells that secrete GLUCAGON and β-cells that secrete INSULIN.

isoleucine An ESSENTIAL AMINO ACID, present in CASEIN and body tissue. It is an ISOMER of LEUCINE.

isomer A chemical compound that possesses the same composition and molecular mass as another but differs in its physical or chemical properties. The differences are due to the structural arrangement of the constituent atoms. Structural isomers differ in the order in which the atoms are joined together. Sometimes these differences occur in the three-dimensional structure of the isomers and are seen only when models are constructed. Optical isomers are mirror images of each other, while geometric isomers arise due to the different spatial arrangement of atoms around a central plane of symmetry.

isomerase One of a group of enzymes that causes the rearrangement of groups within a molecule.

isometric growth Where the growth of a given feature is the same as the growth of the entire organism. For example, the leaves of most plants exhibit isometric growth. *Compare* ALLOMETRIC GROWTH. *See also* GROWTH.

isotope Any one of a series of atomic nuclei of the same element, each with the same number of protons in their nucleus but with a different number of neutrons, hence different masses. Most elements in nature consist of a mixture of isotopes, for example hydrogen, deuterium and tritium are isotopes of hydrogen.

IUD *See* INTRA-UTERINE DEVICE.

IVF *See* IN VITRO FERTILIZATION.

JK

Jacob–Monod theory A theory of GENE EXPRESSION put forward in 1961 by F. Jacob (1920–) and J. Monod (1910–1976). The theory explains PROKARYOTIC gene expression and is of some value in the explanation of EUKARYOTIC gene expression, although the latter is known to be more complicated. Jacob and Monod investigated the expression of the gene that codes for β-galactosidase, an enzyme that breaks down lactose. The theory explains the expression in terms of an OPERON, a group of genes that act together to allow certain enzymes to be switched on and off when needed.

jejunum The part of the SMALL INTESTINE between the DUODENUM and the ILEUM. Its main function is the absorption of digested food.

joint In animals with skeletons, the point at which bones meet. Joints usually allow movement. Synovial joints allow free movement and consist of ball-and-socket joints (e.g. the hip and shoulder) or hinge joints (e.g. the elbow, knee, toes and fingers). In ball-and-socket joints, the round end of one bone fits into the hollow end of another bone and can turn in any direction. In this type of joint, there is a layer of CARTILAGE for protection at the ends of the bones (called articular) and a fibrous covering called the synovial capsule around the joint. The inner layer of this is the synovial membrane, which secretes SYNOVIAL FLUID.

In hinge joints, the round end of one bone turns on the flat surface of another bone in one direction only. Other joints include gliding joints (e.g. between adjacent vertebrae and at the wrist and ankle), which allow one bone to move over the surface of another; pivot joints (e.g. between the first two vertebrae at the top of the VERTEBRAL COLUMN, allowing rotation of the head), which allow partial movement; and suture joints (such as those between the bones of the skull), which allow no movement. The bones of a joint are held together by LIGAMENTS.

joule (J) The SI UNIT of work and energy. One joule of work is done or energy transferred when a force of one NEWTON moves through a distance of one metre in the direction of the force. It replaces the CALORIE (1 joule is equal to 4.2 calories).

karyotype The characteristics of the set of CHROMOSOMES of a given species. Karyotype describes the number, size and shape of the chromosomes. For example, the human karyotype consists of 46 chromosomes, arranged in 23 pairs.

keratin In vertebrates, a fibrous sulphur-rich protein found in the skin that toughens the outer protective layer. Keratin is also found in hair, nails, hooves, feathers and horns.

ketone Any of a group of organic compounds containing a carbonyl group (C=O) attached to two carbon atoms. Ketones are named after the HYDROCARBON from which they are derived, with the ending -anone. An example is propanone (acetone, CH_3COCH_3). Ketones are liquids or low-melting point solids.

Ketones are formed by the oxidation of secondary ALCOHOLS and can be reduced back to secondary alcohols but are not themselves easily oxidized, unlike aldehydes. This feature is used to distinguish them from aldehydes in FEHLING'S TEST.

kidney In vertebrates, either one of a pair of major organs responsible for excretion of waste products and maintaining the balance of water and solutes in the body tissues and blood (by OSMOREGULATION). The kidneys are located on the back wall of the ABDOMEN and are 7–10 cm long and 2.5–4 cm wide. Blood enters from the RENAL artery and leaves via the renal vein.

The functional unit of the kidney is the NEPHRON, a filtering unit used to form URINE. There are over a million nephrons in a human kidney. The nephron consists of a knot of blood capillaries, called the GLOMERULUS, and a long tubule with several clearly defined regions. Some substances pass out of the blood through the glomerulus and enter the tubule as a filtrate. As the filtrate passes along the tubule, some of the substances are selectively reabsorbed, including water, and eventually urine is formed, which enters the URETER for excretion.

Two hormones are important in controlling osmoregulation by the kidney. ANTI-DIURETIC HORMONE (ADH) controls the permeability of the walls of part of the tubule and therefore the degree to which water is drawn out from them. ALDOSTERONE is important in regulating the sodium ion concentration in the kidney. If one kidney is lost, the other usually becomes enlarged and takes over the function of both. If both kidneys are defective, DIALYSIS is needed to sustain life.

See also ADRENAL GLAND, BOWMAN'S CAPSULE, LOOP OF HENLE.

kilo- (k) A prefix before a unit indicating that the size of the unit is to be multiplied by 10^3, for example kilojoule (kJ) is equivalent to one thousand joules.

kilogram The SI UNIT of mass. One kilogram is defined as the mass of the International prototype kilogram, a cylinder of platinum-iridium alloy kept at Sèvres, near Paris, France.

kingdom The major category in the CLASSIFICATION of organisms. The most common classification system today has five kingdoms, consisting of ANIMALIA (all ANIMALS), PLANTAE (all PLANTS), FUNGI (all fungi), PROTOCTISTA (including all ALGAE and PROTOZOA) (these four kingdoms consist of EUKARYOTES) and prokaryotae (all PROKARYOTES). Kingdoms are divided into phyla (see PHYLUM).

Klinefelter's syndrome A genetic disorder affecting men in which individuals gain an X-chromosome and are genetically XXY, XXXY or XXXXY instead of the normal XY. They are phenotypically males but with underdeveloped testes, infertility and with some female development, notably breast enlargement.

Krebs cycle, *TCA cycle, tricarboxylic acid cycle, citric acid cycle* The second stage of cellular RESPIRATION. In the Krebs cycle, the three-carbon compound pyruvate, formed from GLUCOSE during the first stage of GLYCOLYSIS, is converted, in the presence of oxygen, to

carbon dioxide and hydrogen atoms. The Krebs cycle is named after its discoverer Hans Krebs (1900–81).

Before entering the cycle, pyruvate combines with a compound called COENZYME A to form acetyl coenzyme A, which then enters the Krebs cycle to combine with OXALOACETIC ACID. The cycle itself generates very little energy for the cell, but the hydrogen atoms enter the ELECTRON TRANSPORT SYSTEM, where they later generate energy in the form of ATP. The carbon dioxide is removed as a waste product. The process is cyclic with the end-product, oxaloacetic acid, being the substrate for the next round.

Acetyl coenzyme A can be derived not only from the breakdown of sugars but also from the breakdown of fats or proteins. Fats are broken down to GLYCEROL (which is incorporated into glycolysis) and FATTY ACIDS (which are converted to acetyl coenzyme A). Protein is broken down into its constituent AMINO ACIDS, some of which are converted to acetyl coenzyme A, while others enter the Krebs cycle at various points, depending on their carbon content.

In cells with MITOCHONDRIA, the enzymes needed for the Krebs cycle are found mostly in the mitochondrial matrices and so the reactions occur there. In PROKARYOTES, the enzymes are free in the CYTOPLASM. The hydrogen atoms released are carried to the electron transport system by the carriers NAD or FAD – FAD yields less ATP than NAD.

In addition to degrading pyruvate to carbon dioxide and providing hydrogen atoms for the electron transport system, the intermediate compounds of the Krebs cycle are needed to make other substances, including fatty acids, amino acids and CHLOROPHYLL. This cycle is therefore considered to play a key role in a cell's biochemistry.

Kupffer cell A MACROPHAGE found in the liver, forming part of the RETICULOENDOTHELIAL SYSTEM.

kwashiorkor A type of malnutrition that occurs when there is a lack of protein, and therefore ESSENTIAL AMINO ACIDS, in the diet of young children. The body is swollen, hair is soft and changes colour and growth is retarded.

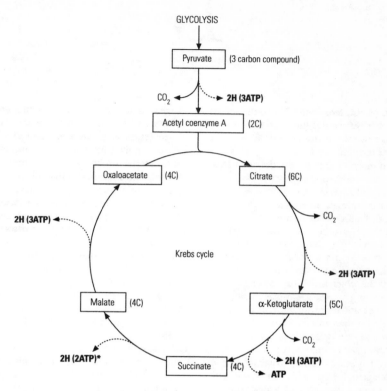

GLYCOLYSIS

Pyruvate (3 carbon compound)

CO_2 ← 2H (3ATP)

Acetyl coenzyme A (2C)

Oxaloacetate (4C)

Citrate (6C)

2H (3ATP) ←

CO_2

Krebs cycle

2H (3ATP)

Malate (4C)

α-Ketoglutarate (5C)

CO_2

2H (2ATP)*

2H (3ATP)

Succinate (4C)

ATP

*2H carried by FAD rather than NAD therefore yields less ATP

Summary of the Krebs cycle.

L

lachrymal gland, *tear gland* The gland in the eye of some vertebrates that secretes a watery substance, which lubricates and cleanses the surface of the eye.

lactase An enzyme that breaks down LACTOSE into GLUCOSE and GALACTOSE.

lactate fermentation A type of FERMENTATION, common in animals, in which LACTIC ACID is produced. This type of fermentation occurs where there is a temporary lack of oxygen, thereby preventing AEROBIC RESPIRATION. The best example is in a muscle during vigorous exercise. Pyruvate from GLYCOLYSIS is converted to lactic acid, which builds up, causing cramps and preventing the muscle functioning properly. Tissues are more tolerant to this lactic acid build up than to a build up of pyruvate which would otherwise occur. During lactate fermentation, an OXYGEN DEBT is accumulated, which is paid back when oxygen becomes available again, by the oxidation of the lactic acid to carbon dioxide and water with the release of energy. Lactate fermentation thus allows animals to withstand short periods without oxygen. In some organisms, for example parasitic worms, the lactic acid is simply excreted, so there is no oxygen debt to be paid.

Animals living in ponds or rivers with fluctuating levels of oxygen also make use of lactate fermentation. It also occurs in certain bacteria and fungi, for example LACTIC ACID BACTERIA, in the souring of milk.

lactation In mammals, the production of milk by the MAMMARY GLANDS in response to the birth of their offspring. After the birth, milk production is stimulated by the hormone PROLACTIN, secreted by the PITUITARY GLAND. The sucking of the nipples of the mammary gland stimulates the release of OXYTOCIN from the pituitary, which causes contraction of the breast alveoli (*see* ALVEOLUS) resulting in the expression of milk (the 'let down' reflex).

Milk produced for the first few days following the birth is COLOSTRUM, which is an important source of antibodies (those that are too large to cross the placenta). Full milk produced after a few days contains fat, protein (CASEIN), sugar (LACTOSE), minerals and antibodies. Human milk (but not cows' milk) contains antibacterial antibodies and an iron- binding protein called lactoferrin that inhibits bacterial growth in the baby's intestine. Milk production can continue as long as the baby is suckling.

lacteal A small vessel in the VILLI of the SMALL INTESTINE through which fats are absorbed. *See also* ILEUM.

lactic acid, *2-hydroxypropanoic acid* ($CH_3CHOHCOOH$) A colourless, odourless organic liquid produced by LACTIC ACID BACTERIA during FERMENTATION. It is also produced by muscle cells when they are exercising vigorously and experience OXYGEN DEBT and LACTATE FERMENTATION.

lactic acid bacteria A group of bacteria used in the dairy industry in the manufacture of cheese and yoghurt. When added to pasteurized milk (*see* PASTEURIZATION), lactic acid bacteria convert LACTOSE to LACTIC ACID. The fall in pH causes the milk to separate into solid curd and liquid whey. It is the curd that is used to make cheeses.

Other micro-organisms can be added to give variety to the ripening of cheese, for example, *Penicillium* is added to blue cheese, and the ripening of soft cheeses is helped by fungi growing on their surfaces. Lactic acid bacteria are also used in the manufacture of yoghurt; the starting material is pasteurized milk with the fat removed.

lactoferrin A protein present in human milk that inhibits bacterial growth. *See* LACTATION.

lactose A DISACCHARIDE made from the combination of GLUCOSE and GALACTOSE. Lactose is found in mammalian milk (5 per cent in cows' milk) and is important to the suckling young.

lacuna A space or cavity, such as the spaces in BONE that contain the bone cells.

laevorotatory (*adj.*) Describing a compound that rotates the plane of polarization of plane-polarized light (light that oscillates in one direction only) to the left (anticlockwise). This used to be denoted by the prefix *l*- but (–) is now used. This is not to be confused with the prefix L- used to indicate the configuration of CARBOHYDRATES and AMINO ACIDS.

lag phase *See* BACTERIAL GROWTH CURVE.

lamella Any thin layer, plate or membrane, especially the thin layers of tissue of which bone is formed.

lamina The blade of a leaf. *See* LEAF.

lamina flow cabinet A cabinet used in the laboratory for TISSUE CULTURE. The cabinet, or hood, is kept sterile by an air-flow system, which not only provides a suitable environment in which to perform tissue culture but also protects the worker from contamination by the culture.

large intestine Part of the digestive tract between the SMALL INTESTINE and the ANUS. The large intestine is 1.5 m long and 6 cm in diameter. Like the small intestine, it consists of muscular tubes with a folded inner lining increasing the surface area for absorption of food. It is linked to the small intestine by the CAECUM, which leads to the APPENDIX. The ileocaecal valve controls passage of food from the small intestine to the COLON. Water, minerals and vitamins in digestive secretions that have not been absorbed by the ILEUM are reabsorbed in the colon, and anything remaining is formed into FAECES. Excess calcium and iron salts are passed from the blood through the large intestine for removal with the faeces. Movement of faeces towards the RECTUM (for temporary storage) is aided by MUCUS secreted by the wall of the large intestine. The faeces are then expelled through the anus. *See also* DIGESTIVE SYSTEM.

Lamarckism A theory of EVOLUTION, put forward by the French naturalist Jean-Baptiste Lamarck (1744–1829), that was never widely accepted. Lamarck suggested that useful characteristics acquired by an organism during its lifetime would be inherited by its offspring, while disuse of other characteristics would result in their eventual disappearance from the species. For example, he suggested that giraffes have long necks because they continually stretch to reach for high leaves. According to his theory, giraffes with longer necks through stretching would be able to pass this acquired characteristic onto their offspring. Although

now discredited, Lamarck did influence the thinking of Charles Darwin (1809–82), who later proposed the widely accepted theory of NATURAL SELECTION. *See also* DARWINISM.

laminins Proteins of the EXTRACELLULAR MATRIX thought to be concerned with stimulation of cell division.

larva (*pl. larvae*) A form of many animals as they hatch from the egg and before they develop into sexually mature adults. Larvae may be very different from the adult, for example, tadpoles and caterpillars differ significantly from their adult forms. *See also* METAMORPHOSIS.

larynx, *voicebox* A cavity at the opening of the TRACHEA where it meets the PHARYNX. The larynx contains a series of LIGAMENTS called the vocal cords that vibrate, producing sound waves. It is the Adam's apple in humans. The complexity of the vocal cords varies; reptiles have none but in birds there is a structure called a syrinx with well-developed vocal cords.

lasso cell A cell type characteristic of CTENOPHORA, for example comb-jellies. Lasso cells are found in the tentacles of these organisms and are sticky for capturing prey. They do not penetrate the prey.

leaching The process in which the passage of water or some other solvent through a porous solid material causes some part of that solid to be dissolved in the solvent. In particular, leaching is the washing away of substances out of the soil. This can be a result of DEFORESTATION, in which case useful nutrients are lost and soil fertility is reduced. Where fertilizers leach out of the soil, this can lead to POLLUTION of rivers, etc.

leaf A structure growing out from the stem of a plant, usually the main site of PHOTOSYNTHESIS. A leaf is made up of three parts: the sheath (or leaf base), the petiole (or stalk), and the lamina (or blade). A leaf blade is made up of MESOPHYLL cells, between the upper and lower EPIDERMIS, and a waxy CUTICLE that prevents water loss. The mesophyll cells contain numerous CHLOROPLASTS in which photosynthesis takes place.

Leaves have evolved many ways of ensuring they obtain sufficient light for photosynthesis: they have a large surface area to capture sunlight, they are thin and have a staggered arrangement on the plant to allow as much light through as possible, and chloroplasts are numerous and can move within the mesophyll

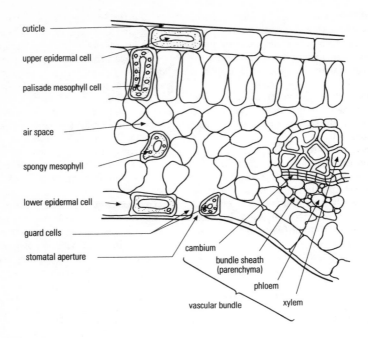

cuticle

upper epidermal cell

palisade mesophyll cell

air space

spongy mesophyll

lower epidermal cell

guard cells

stomatal aperture

cambium

bundle sheath (parenchyma)

phloem

vascular bundle　　xylem

Structure of a dicotyledonous leaf (cross-section through the blade).

to be in the best position for absorption of light. Leaves are also phototropic, moving towards the source of light (*see* TROPISM).

Evergreen leaves are called persistent, whereas those that fall off in the autumn are deciduous. A simple leaf, for example that of a beech tree, is undivided, whereas a compound leaf, for example that of a horse chestnut tree, is divided into several leaflets. Pinnate leaves are a form of compound leaves where the leaflets are arranged in rows either side of the midrib, for example in the ash.

learning Behaviour in animals that is acquired and modified in response to experience and changes in the environment. Learning contrasts with innate or instinctive behaviour, which is inborn and usually cannot be modified. Forms of learning include HABITUATION, CONDITIONING and IMPRINTING. The highest form of learning is insight or intelligent behaviour, which involves the recall and adaptation of previous experiences to a new situation.

lecithin A PHOSPHOLIPID that is an important component of CELL MEMBRANES of plants and animals.

lectin Any one of a group of proteins and GLY-COPROTEINS that can cause AGGLUTINATION of cells by cross-linking cell-surface carbohydrates and other ANTIGENS.

legume A fruit plant with a pod that opens on both sides to release its dry seeds, for example in peas, beans and clovers. Legumes are members of the family Leguminosae. Their roots have ROOT NODULES that contain nitrogen-fixing bacteria, which improve soil fertility by fixing atmospheric nitrogen (*see* NITROGEN FIXATION). Legumes are therefore important in agriculture. The edible seeds are called pulses.

Leguminosae A family of flowering plants having LEGUMES as fruit and ROOT NODULES.

lens In general, any curved piece of material that focuses light to form an image. In the EYE, the lens is a transparent, biconvex (curving outwards) structure behind the IRIS, which focuses light entering the eye onto the RETINA. The lens is held in place by the suspensory ligaments which attach it to the CILIARY MUSCLES. The lens is able to change its shape to focus on near or far objects as a result of the action of the ciliary muscles. *See also* ACCOMMODATION.

lenticel One of several small pores found in the trunk of trees or outer layer of woody stems through which gas exchange occurs. Lenticels consist of regions of loosely packed cells with air spaces between them, and are a site of some water loss (*see* TRANSPIRATION).

leucine One of the most common ESSENTIAL AMINO ACIDS. It is produced in several ways, in particular by the digestion of proteins by ENZYMES in the PANCREAS.

leucocyte *See* WHITE BLOOD CELL.

leucocytosis An abnormal increase in the number of WHITE BLOOD CELLS in the blood.

leucopenia An abnormal reduction in the number of WHITE BLOOD CELLS in the blood.

leucoplast A PLASTID that is a colourless food storage body found in the cells of plant tissue not normally exposed to light. Leucoplasts include the STARCH-storing amyloplasts found in roots of many plants, oil-storing elaioplasts and protein-storing aleuroplasts.

LH *See* LUTEINIZING HORMONE.

lichen A member of the kingdom FUNGI which is a symbiotic association (*see* SYMBIOSIS) between an ALGA and a fungus. The fungal partner (mycobiont) is usually a member of the ASCOMYCOTA and the algal partner (phycobiont) is either a CHLOROPHYTA (green alga) or a CYANOBACTERIA. The phycobiont provides food for the mycobiont through its photosynthesis and the mycobiont provides protection for its partner. Lichens are morphologically distinct from the form of either partner. They may form a thin, flat crust or be erect and branching. They are often seen on tree trunks. Lichens reproduce non-sexually by means of SOREDIA or ISIDIA. The fungus can reproduce sexually (by the production of SPORES) but new lichens will only form if an algal partner happens to be present, otherwise the fungus will die. Lichens are very slow-growing but can survive in places where other plants cannot, for example mountain and arctic regions, bare soil, rocks, tree trunks and fences. Some lichens are used as INDICATOR SPECIES to monitor POLLUTION in an area, since they are sensitive to air pollutants such as sulphur dioxide. Some are useful in the production of dyes, such as LITMUS, and others are sources of medicines or cosmetics.

ligament A band of strong, flexible CONNECTIVE TISSUE attaching two bones at a joint and restricting movement of the joint to prevent dislocation. Yellow elastic ligaments consist mostly of elastic fibres and are more extensible than ligaments made of COLLAGEN fibres.

ligase One of a group of enzymes that forms bonds between two molecules using energy from the breakdown of ATP. DNA ligase in particular is very useful in GENETIC ENGINEERING because it joins DNA chains specifically cut by RESTRICTION ENZYMES.

lignin A chemical substance found in the CELL WALLS of plants, composed of rings of carbon atoms joined in a chain. Lignin confers strength to the plant, and is difficult to digest so provides protection from attack by many organisms. Lignin is found in all wood and is therefore of commercial importance. There are some regions of the cell wall where PLASMODESMATA are present and lignin is not deposited, so that water and dissolved minerals can pass between cells. *See also* SCLERENCHYMA.

line transect *See* TRANSECT.

linkage In genetics, the association of two or more GENES occurring on the same CHROMOSOME that therefore tend to be inherited together. Not all genes together on a chromosome will be inherited together because RECOMBINATION processes can occur, such as CROSSING-OVER of chromosome pairs, which shuffles the genetic material. Genes closer together on a chromosome are more likely to remain linked and be inherited together. SEX LINKAGE refers to genes carried on the SEX CHROMOSOME that may have nothing to do with the sex of an organism but are linked to it by their position. *See also* MENDEL'S LAWS.

linoleic acid An ESSENTIAL FATTY ACID present in many plant fats and oils, for example groundnut oil, linseed oil, soya-bean oil. Linoleic acid is a polyunsaturated fatty acid with two double COVALENT BONDS.

linolenic acid An ESSENTIAL FATTY ACID present in some plant oils, for example linseed and soya-bean oil and in algae. Linolenic acid is a polyunsaturated FATTY ACID with three double COVALENT BONDS.

lipase One of a group of enzymes responsible for breaking down fats into FATTY ACIDS and GLYCEROL during digestion. Lipase is made by the PANCREAS and requires a slightly alkaline environment to function.

lipid Any one of a large group of organic compounds that are the major constituents in plant and animal fats, waxes and oils. Lipids are all ESTERS OF FATTY ACIDS, and are soluble in alcohol but not water. The most common alcohol with which fatty acids react to form lipids is GLYCEROL. The lipids formed in this way are called GLYCERIDES and can be mono, di or tri, depending on how many hydroxyl (OH) groups from glycerol have combined with the fatty acids, for example TRIGLYCERIDES have three fatty acids attached to glycerol. The properties of a lipid largely depend on the fatty acids present, because the alcohol is usually the same (glycerol). PHOSPHOLIPIDS are lipids containing glycerol in which one of the fatty acids is replaced by phosphoric acid.

Lipids provide an energy store in plants and animals, and in the form of fat give some protection and insulation to animals and their internal organs. Lipids also provide waterproofing for plants and animals, as oily secretions and waxes formed by the combination of fatty acids with an alcohol other than glycerol. The EMULSION TEST (which gives a cloudy solution when lipid and alcohol mix) or the Sudan III test (a red dye that detects fats and oils) detect lipids in a solution.

lipoprotein A complex of LIPIDS and protein that is a structural component of all cell membranes. CHOLESTEROL is transported in the blood within such a complex, either free or esterified (*see* ESTER) to FATTY ACIDS.

litmus A naturally occurring vegetable dye that can be used as an indicator. A solution of litmus is red under acidic conditions but blue in alkaline solutions.

litmus paper Paper soaked with LITMUS, used to test for acidic or alkaline solutions.

litre (l) A unit of volume, correctly called the decimetre cubed (dm^3) in the SI system. One litre is equal to 10^{-3} m^3.

liver In vertebrates, a large organ derived from the ENDODERM that makes up 3–5 per cent of the body weight (2 kg). The liver is located in the upper ABDOMEN, below the DIAPHRAGM to which it is attached. It has many functions concerned with DIGESTION and maintaining a constant internal environment.

Oxygenated blood from the AORTA enters the liver via the HEPATIC artery, and blood leaves via the hepatic vein. The hepatic portal vein supplies blood from the INTESTINE that is rich in soluble digested food. Under the light microscope, the liver can be seen to be made up of lobules, the functional unit of which is called the acinus. The hepatic artery and hepatic portal vein combine to form channels called sinusoids, where the blood mixes and allows exchange of materials between the blood and liver cells lining the sinusoids. These liver cells are called HEPATOCYTES and they secrete the BILE fluid that assists in the breakdown and absorption of fats in the small intestine. Kupffer cells are also found lining the sinusoids; these are phagocytic cells (*see* PHAGOCYTOSIS) important in removing foreign bodies entering from the intestine and destroying worn-out or damaged blood cells.

The liver also converts excess GLUCOSE to GLYCOGEN for storage until needed, a process called GLYCOGENESIS. The liver plays a key role in the maintenance of blood glucose levels and can return glucose to the blood when needed by the breakdown of glycogen (GLYCOGENOLYSIS) or by converting protein and fats into glucose (GLUCONEOGENESIS). *See also* PANCREAS.

The liver also breaks down excess amino acids by DEAMINATION and can convert one amino acid to another by TRANSAMINATION to replace deficient non-essential amino acids. Other functions include the synthesis of many important PLASMA PROTEINS (such as ALBUMIN and FIBRINOGEN), storage of vitamins and minerals (e.g. iron, potassium, copper and zinc), breakdown of hormones, removal of toxins (such as nicotine and alcohol by converting to safer chemicals), storage of blood and production of heat (under control of the HYPOTHALAMUS).

liverwort A member of the class HEPATICAE.

loam soil A type of SOIL consisting of a mixture of sand, clay, silt and organic matter. Loam soil is the most fertile type of soil, able to support a variety of plant growth, depending on the precise ratio of its components.

lock-and-key mechanism A proposed mechanism of ENZYME action. The three-dimensional structure of the enzyme provides a specific ACTIVE SITE that can be compared to a lock, while the SUBSTRATE molecule it acts upon is the key. According to this theory, only the correct substrate will fit the active site of an enzyme. An enzyme–substrate complex forms that is a different shape to the substrate alone, and it therefore falls away from the enzyme, leaving it free to attach to another substrate molecule. Modern interpretations of this theory suggest that the AMINO ACIDS, forming the active site alter their relative three-dimensional positions as the substrate binds. This is termed an induced fit.

locus The position of a GENE within a DNA molecule. The position of a given locus can be determined by CHROMOSOME MAPPING. *See also* ALLELE.

log phase, *logarithmic phase See* BACTERIAL GROWTH CURVE.

loop of Henle The part of the KIDNEY that is responsible for reabsorbing water, sugars and other useful minerals from the glomerular filtrate (*see* GLOMERULUS). The loop forms part of a long tubule within the functional unit called the NEPHRON, of which there are over a million in the kidney.

The loops of Henle are located in the inner region of the kidney, the MEDULLA. A loop is connected at one side by the proximal convoluted tubule to the BOWMAN'S CAPSULE, and at the other side by the distal convoluted tubule to a collecting duct. Much reabsorption is carried out by the proximal convoluted tubule, and the loop of Henle concentrates the filtrate further. The degree to which the loop can concentrate the filtrate passing through it is related to the length of the loop, for example it is very long in the desert rat, where it is an advantage for URINE to be very concentrated.

A COUNTERCURRENT SYSTEM operates in the loop of Henle, in which sodium and chloride ions are removed by ACTIVE TRANSPORT from the wider, less permeable ascending limb into the interstitial spaces between the limbs of the loop and into the associated, parallel blood capillary, the vasa recta. The concentration of tissue fluid and blood is greater at the apex of the loop (in the medulla). This high concentration causes water to be drawn out of the more permeable descending limb of the loop by OSMOSIS and enter the vasa recta.

lumen An inner, open space within tubular structures, such as blood vessels and the digestive system. In plants, it is the space remaining within the CELL WALL of a cell that has lost its contents.

lung A large, delicate spongy organ in which gaseous exchange takes place. Humans and most other TETRAPODS have two lungs in the THORAX, which are protected by the ribcage. The bulk of the lung tissue is made up of millions of sac-like structures called alveoli (see ALVEOLUS) that consist of thin, moist, membranous sheets over which blood passes, so that

oxygen can enter the bloodstream and carbon dioxide waste can leave. BREATHING causes air to enter and leave the lungs. Each lung is made airtight by two pleural membranes (the PLEURA), which form the PLEURAL CAVITY between them and into which pleural fluid enters, which lubricates the lungs to allow for expansion. Pleurisy is an infection of the pleural cavity. *See also* RESPIRATORY SYSTEM.

luteinizing hormone (LH), *interstitial cell-stimulating hormone,* (ICSH) A GLYCOPROTEIN produced by the PITUITARY GLAND. In females, LH stimulates OVULATION (release of an OVUM from the OVARY), transforms the ruptured GRAAFIAN FOLLICLE into the CORPUS LUTEUM and controls the levels of OESTROGEN and PROGESTERONE if pregnancy follows. In males, LH stimulates the INTERSTITIAL CELLS of the testes (*see* TESTIS) to produce the hormone TESTOSTERONE. LH is under the control of the HYPOTHALAMUS, which releases gonadotrophic-releasing factor that stimulates LH but is itself inhibited by the testosterone it stimulates.

luteotrophic hormone (LTH) *See* PROLACTIN.

luteotrophin *See* PROLACTIN.

lyase Any one of a group of enzymes that causes the addition or removal of a chemical group other than by HYDROLYSIS.

Lycopodophyta A phylum of the PLANT kingdom consisting of the club mosses. Club mosses are small, evergreen plants with trailing or upright stems that have small spirally arranged leaves. They have sporangia (*see* SPORANGIUM) that are usually in cones. Some species are HOMOSPOROUS and others are HETEROSPOROUS.

lymph The fluid carried by the LYMPHATIC SYSTEM. It is a clear liquid made of tissue fluid and WHITE BLOOD CELLS (mostly LYMPHOCYTES).

lymphatic *See* LYMPH VESSEL.

lymphatic duct A major vessel of the LYMPHATIC SYSTEM. It is formed from the convergence of LYMPH VESSELS from the right side of the head, THORAX and arm. Other lymph vessels drain into the THORACIC DUCT. These two ducts carry LYMPH up to the main veins near the heart.

lymphatic system, *lymphoid system* In vertebrates, a series of vessels carrying LYMPH throughout the body, and the organs associated with the production and storage of LYMPHOCYTES. The lymphatic system forms a second CIRCULATORY SYSTEM (the other being blood) that provides nutrients and WHITE

BLOOD CELLS to body tissues and drains waste products away from body tissues.

Lymph capillaries are the smallest vessels and are found in all body tissues except nervous tissue. They are similar to blood capillaries but their walls are more permeable and allow even bacteria through. Lymph capillaries merge to form larger tubes called LYMPH VESSELS that pass lymph into large veins near to the heart. Along the lymphatics are a number of LYMPH NODES that are responsible for removing foreign bodies from lymph and are important in the IMMUNE RESPONSE. Lymph nodes, along with the SPLEEN, parts of the digestive system and respiratory and urinogenital regions, form secondary lymphoid tissue that are the sites of migration of lymphocytes from their site of production. The THYMUS, embryonic liver and adult BONE MARROW form the primary lymphoid tissue, where lymphocytes are produced.

Some vertebrates, for example amphibians, have a lymph heart that pumps lymph around the body. Mammals rely on muscular contractions (of skeletal muscles pushing on lymph vessels), inspiratory movements (drawing lymph up to the THORAX) and the HYDROSTATIC PRESSURE of tissue fluid leaving the blood capillaries to push lymph along the lymphatic system.

lymph node In mammals, a small mass of tissue found at a number of sites along the major LYMPH VESSELS. Lymph nodes are a component of secondary lymphoid tissue (*see* LYMPHATIC SYSTEM). LYMPH passes through the nodes and is filtered by the engulfing action of MACROPHAGES contained in the lymph nodes, which remove foreign bodies such as bacteria. Lymph nodes contain many LYMPHOCYTES and are therefore important in the IMMUNE RESPONSE. Each node has its own blood supply. Larger lymph nodes include tonsils (in the throat of humans) and adenoids (at the back of the nose) and are often mistaken for glands. The SPLEEN, found near the stomach in vertebrates, is the largest mass of lymphoid tissue.

lymphocyte In vertebrates, a WHITE BLOOD CELL with a large nucleus that is found in blood and LYMPH and is involved in the IMMUNE RESPONSE. There are two types of lymphocytes, called B CELLS and T CELLS, both of which are formed in BONE MARROW and later settle in the SPLEEN or a LYMPH NODE. B cells are responsible for the

production of ANTIBODY to fight invasion by a foreign ANTIGEN, and also for the production of memory cells by which repeat infections are more quickly eliminated. T cells are involved in CELL-MEDIATED IMMUNITY. There is some overlap of their functions. *See also* LYMPHATIC SYSTEM.

lymphoid system *See* LYMPHATIC SYSTEM.

lymphoid tissue Tissue responsible for the production and maturation of LYMPHOCYTES. The THYMUS (bursa of fabricus in birds), embryonic liver and adult BONE MARROW form the primary lymphoid tissue where lymphocytes are produced. They migrate then to secondary lymphoid tissue, comprising the SPLEEN, LYMPH NODES, parts of the DIGESTIVE SYSTEM and respiratory and urinogenital regions. Lymphoid tissue forms part of the LYMPHATIC SYSTEM.

lymphokine Any soluble substance produced by a LYMPHOCYTE that is involved in communication between cells of the IMMUNE RESPONSE. Those produced by T CELLS are often called INTERLEUKINS (although other cells can make these too). B CELLS can produce lymphokines but this seems to be less important in CELL-MEDIATED IMMUNITY. Other examples include γ-INTERFERON, B-cell growth factor and macrophage migration inhibition factor. *See also* CYTOKINE

lymphoma A TUMOUR made of tissues of the LYMPHATIC SYSTEM, for example in Hodgkin's disease.

lymph vessel, *lymphatic* Any vessel of the LYMPHATIC SYSTEM that is like a vein with a non-return valve so that circulation is in one direction only (towards the heart). Lymph vessels from the right side of the head, THORAX and arm combine to form the right lymphatic duct, and the rest of the vessels drain into the THORACIC DUCT. Both sides then pass lymph into large veins near the heart. Along the lymphatics are a number of LYMPH NODES.

lysin An ANTIBODY that chemically destroys foreign substances.

lysine A basic AMINO ACID essential for growth. It is cannot be synthesized by the body and must be present in the diet. Lysine is found in particular in HISTONE proteins. Lysine is used to fortify foods and feeds and in culture media. It is produced by the HYDROLYSIS of certain proteins.

lysis The destruction of a cell by rupturing its CELL MEMBRANE or CELL WALL. *See also* LYSOSOME.

lysosome (*Lysis* = splitting, *soma* = body) A membrane-bound EUKARYOTIC cell ORGANELLE, formed by the GOLGI APPARATUS. Lysosomes are 0.2–0.5 μm in diameter and are similar to spherical MITOCHONDRIA but lack an internal structure. They contain a large number of enzymes (mostly HYDROLASES) in acid solution that are involved in autolysis, the complete cell breakdown after a cell's death, and in digesting worn-out organelles.

A lysosome that has not been involved in hydrolytic activity is referred to as a primary lysosome, and the fusion of primary lysosomes with each other or material for digestion results in secondary lysosomes that are larger (*see* PHAGOSOME).

Lysosomes are abundant in WHITE BLOOD CELLS, where the enzymes attack ingested bacteria. In these cases material is digested within the lysosome, but enzymes can also be released outside the cell to break down other cells in a controlled selective way, for example during METAMORPHOSIS. Similar structures are seen in plants and fungi.

M

macroevolution *See* EVOLUTION.

macromolecule A very large MOLECULE. Examples of substances with macromolecules are POLYMERS, PROTEINS and HAEMOGLOBIN.

macronutrient Any chemical substance required in relatively large amounts by plants and animals for their normal growth and development. Macronutrients include NITRATES, PHOSPHATES, SULPHATES, CALCIUM, SODIUM, IRON, CHLORINE, POTASSIUM and MAGNESIUM. Other substances are required in much smaller amounts and are called MICRONUTRIENTS. Substances usually fall into the same category for plants and animals but there are some exceptions, such as chlorine, which is a major element in animals but a TRACE ELEMENT in plants. These essential nutrients may fulfil one or more of a variety of metabolic roles.

macrophage In vertebrates, a type of WHITE BLOOD CELL with many functions, mostly concerned with the removal of foreign or dead cells or debris (after infection or injury). Like PHAGOCYTES, macrophages can ingest particles (foreign or food) from their surroundings, which are subsequently broken down or destroyed, but macrophages are larger than other phagocytes, with a longer life span, and are found all over the body. In the liver (as KUPFFER CELLS) and SPLEEN, macrophages ingest worn-out blood cells. In the lungs, they ingest dust and other inhaled particles, and in LYMPH NODES and the LYMPHATIC SYSTEM they are concerned with fighting infection and the IMMUNE RESPONSE. Macrophages are important in the immune response as ANTIGEN-PRESENTING CELLS. This system of circulating tissue macrophages forms the RETICULOENDOTHELIAL SYSTEM. *See also* MAJOR HISTOCOMPATIBILITY COMPLEX, OPSONIN, T CELL.

macrophage-activating cell A type of T CELL that produces LYMPHOKINES to activate MACROPHAGES.

macroscopic (*adj.*) Visible to the naked eye.

macula A region in the inner EAR that is concerned with the position of the body in relation to gravity. It consists of a gelatinous mass within which are embedded sensory hair cells, the hairs of which are embedded in calcium carbonate deposits called OTOLITHS. The otoliths detect movement of the head, which in turn moves the gelatinous mass and displaces the sensory hairs. This generates a NERVE IMPULSE that is transmitted to the brain.

mad cow disease *See* BOVINE SPONGIFORM ENCEPHALOPATHY.

magnesium (Mg) A silvery, lightweight, metallic element that is widespread in nature. It is an essential element for plants and animals. In plants, it is a constituent of CHLOROPHYLL and is therefore essential for their growth. In animals, magnesium is a COFACTOR for some enzymes and is involved in the transmission of NERVE IMPULSES. It is also a component of bones and teeth.

magnetic resonance imaging (MRI) An imaging technique increasingly used in medicine. The patient is placed in a magnetic field, which aligns the spins of the protons in hydrogen atoms throughout the patient. A pulsed electromagnetic field destroys this alignment, and the signals produced as the protons realign themselves have frequencies characteristic of the chemical environment of the individual protons. TOMOGRAPHY techniques are used to produce a picture of the body in a series of slices. Unlike X-RAYS, MRI is non-ionizing (*see* IONIZING RADIATION) and can produce clear images of soft tissues. This makes it an invaluable tool for locating tumours and tissue abnormalities. *See also* NUCLEAR MAGNETIC RESONANCE.

magnification The extent to which an image is larger than the original object, in a microscope or telescope for example.

magnification = image size/object size

major histocompatibility complex (MHC) A mammalian GENE complex encoding ANTIGENS on the surface of most body cells that are unique to the individual. The MHC was originally identified as the region encoding antigens important in graft rejection (*see* TRANSPLANTATION).

The MHC regulates T CELL activity in such a way that foreign antigens are only recognized when they are in association with MHC molecules. There are two classes of MHC molecules: class I molecules are found in low levels on most body cells, but particularly on T cells, and class II molecules are found mostly on B CELLS and MACROPHAGES.

Malacostraca The largest subclass of CRUSTACEA, which includes crayfish, prawns, shrimps, crabs, lobsters and woodlice.

malaria An infectious disease in the tropics caused by the protozoan PARASITE *Plasmodium* carried by the *Anopheles* mosquito. The disease is characterized by a periodic high fever and enlarged SPLEEN and affects about 200 million people a year, more than any other organism in the tropics. There are four protozoa of the genus *Plasmodium* that can cause malaria, the most dangerous being *P. falciparum.*

The life cycle of the parasite is complex and involves an asexual stage in the liver and RED BLOOD CELLS of humans and a sexual stage beginning in humans and continuing in the mosquito. When humans are bitten by an infected mosquito, the parasite enters the blood in the form of sporozoites (a circle-shaped form of parasite). These enter the liver, where they quickly divide to produce cells termed merozoites, which can reinfect other liver cells or enter red blood cells. They divide further and eventually cause the cell to rupture, releasing merozoites into the blood. The red blood cells rupture in synchrony every 2–3 days, causing the characteristic high fever. Some of the merozoites develop into GAMETES and remain dormant until a mosquito bites the human and takes the gamete into its stomach, where the gametes are fertilized and the ZYGOTE buries into the stomach wall. MEIOSIS occurs to produce large numbers of sporozoites, which are released into the mosquito's body cavity and eventually to the SALIVARY GLANDS, to pass back to another human victim. The mosquito is the secondary host or VECTOR.

No vaccine is available, mainly due to the ANTIGENIC VARIATION shown by the parasite. Mosquitoes are largely resistant to insecticides, but synthetic drugs are used in the prevention or treatment of the disease. Drugs are most effective on the sporozoite and gamete stages, but relapses can occur due to the reservoir of merozoites in the liver. Some protection from malaria is provided by SICKLE-CELL DISEASE, which alters the shape of red blood cells, making them harder to infect.

malic acid An organic acid found particularly in apples but also other fruit. Malic acid is present in all living cells and is an intermediate of the KREBS CYCLE.

malignant (*adj.*) Of a TUMOUR, uncontrollable, rapidly spreading, or resistant to treatment. *See* CANCER.

malleus, *hammer* One of the three EAR OSSICLES in the mammalian middle EAR.

Malpighian layer, *germinative layer* The deepest layer of skin EPIDERMIS that consists of actively dividing cells that move up through the layers of the epidermis. The Malpighian layer extends into the DERMIS, and gets its food and oxygen from capillaries in the dermis because there are no blood vessels in the epidermis. The brown pigment MELANIN is produced by the Malpighian layer. *See also* HAIR.

maltase An enzyme that breaks down MALTOSE into its constituent GLUCOSE molecules.

maltose A DISACCHARIDE made from the combination of two GLUCOSE molecules. Maltose is a major constituent of malt and is therefore important in the manufacture of beer and whisky.

mammal A member of the vertebrate class MAMMALIA in the phylum CHORDATA.

Mammalia A CLASS of vertebrates characterized by the possession of MAMMARY GLANDS in the female, in addition to the presence of body hair (although in variable amounts), lungs, a lower jaw consisting of two bones, a middle ear consisting of three bones, and no nucleus in the RED BLOOD CELLS. There are three groups: (i) the placental mammals (the most advanced), with young that develop inside their mother within a uterus nourished by a placenta; (ii) MARSUPIALS; and (iii) MONOTREMES (the least evolved). Placental mammals have dominated the globe and are found in all parts of the world.

Mammals are very varied in size and shape, from a shrew to a blue whale, and live in a variety of habitats, for example lions, monkeys and humans on land, and whales in the sea. All mammals are HOMOIOTHERMS (warm-blooded; able to maintain a constant body temperature), and all are heterodonts (have different types of teeth). There are 4,000 species of mammals, including the ORDERS of primates (humans, apes), rodents (rats, mice), carnivores (cats, dogs), lagomorphs (rabbits) and cetaceans (whales, dolphins).

mammary gland In female mammals, a milk-producing gland that is only active after the birth of their young. Mammary glands are formed on the chest or ABDOMEN, and the number varies from 2 to 20, according to species. Mammary glands develop during puberty from EPIDERMIS, and during PREGNANCY the hormones OESTROGEN and PROGESTERONE cause the development of milk glands that form a series of branching ducts ending in secretory alveoli (sacs; see ALVEOLUS). After birth of the foetus, milk production (LACTATION) is stimulated by the hormone PROLACTIN secreted by the PITUITARY GLAND. The sucking of the nipples of the mammary gland stimulates OXYTOCIN release from the pituitary, which results in the expression of milk. Milk production can continue as long as the baby suckles.

mandible 1. In ARTHROPODS, a part of the mouthparts that is involved in biting and crushing food.

2. In vertebrates, the lower jaw.

manganese (Mn) A grey metallic element that is a MICRONUTRIENT required by living organisms. Manganese is a COFACTOR for several enzymes and involved in bone development. Deficiency in animals causes bone deformations and in plants a mottling of the leaves.

maramus See DIET.

mark, release, recapture In biology, a technique used to estimate the size of a population of mobile animals that can be tagged or marked in some way. A known number of animals are captured, tagged or marked and released into the population again. At a later date, a given number of animals are collected at random and the percentage that are marked is recorded. Population size is calculated on the assumption that the percentage of marked animals in the second capture is the same as the percentage of marked animals in the population as a whole. The marked animals must have enough time to distribute themselves evenly throughout the population.

A problem with this method is that marked animals may be more likely to predation if the marking makes them more conspicuous. Also,

it does not take into account the death or migration of marked animals.

Examples of this type of sampling are the marking of arthropods by means of a dab of non-toxic paint on their backs, and the attachment of ear tags to mammals and rings to the legs of birds.

See also QUADRAT, TRANSECT.

marsupial A mammal, of the subclass Metatheria, whose young are born at an early stage of development and mature further within a pouch on the mother's body. The MAMMARY GLANDS are located within the pouch and provide nourishment for the young. Marsupials are restricted to Australasia and South America. Examples include kangaroos, koala bears and opossums.

mass flow, *pressure flow* A theory put forward in 1930 to try to explain the TRANSLOCATION of materials in the PHLOEM of VASCULAR PLANTS. It is known that the products of photosynthesis (soluble carbohydrates) are translocated in the phloem but the flow rate is too fast to be explained by DIFFUSION.

In the mass flow theory, photosynthesizing cells in the leaf accumulate soluble carbohydrates, which lowers their water potential and draws water into the cells from the XYLEM. This causes an increase in their pressure potential. In the roots, the reverse occurs: carbohydrates are used or stored and the water potential of these cells is increased, which lowers their pressure potential. There is therefore a gradient pressure potential between the source of carbohydrates and their point of utilization. Liquid flows along this gradient, from the leaves to other tissues, through the SIEVE TUBE ELEMENTS of the phloem. The process is passive, except for the ACTIVE TRANSPORT of sugars from the COMPANION CELLS to the sieve elements, which would be necessary as this is against a concentration gradient. Although the mass flow theory is widely accepted it still leaves some questions unanswered.

mast cell A cell similar to the blood BASOPHIL but containing different granules in its CYTOPLASM. Mast cells are often found in CONNECTIVE TISSUE. The numbers of mast cells often increase in certain allergies and they are involved in the IMMUNE RESPONSE to parasites. See also HISTAMINE.

matrix See EXTRACELLULAR MATRIX.

mean See AVERAGE.

mechanoreceptor A RECEPTOR cell that detects pressure changes, gravity and vibrations (sound). *See also* SENSE ORGAN, EAR.

median The value that is midway between the largest and the smallest values found in a set of data. *See also* AVERAGE.

medical imaging The name given to a range of techniques used to obtain images of the inside of a human body (usually to diagnose illness) without resorting to surgery. The techniques are often based on the transmission of X-rays, the reflection of ULTRASOUND and on NUCLEAR MAGNETIC RESONANCE. *See also* ENDOSCOPE, MAGNETIC RESONANCE IMAGING, RADIOGRAPH, TOMOGRAPHY, ULTRASOUND IMAGING.

medulla 1. In animals, the central part of some organs such as the ADRENAL GLAND, brain or kidney.

2. In plants, the central part of a stem or root, usually consisting of PARENCHYMA. *See also* CORTEX.

medulla oblongata Part of the vertebrate HINDBRAIN. The medulla oblongata is an enlarged region connected to the SPINAL CORD, and contains centres controlling respiration, blood pressure, the rate and strength of the heartbeat, in addition to other activities, such as swallowing, coughing, taste and touch. *See also* BRAIN.

medusa The free-swimming structural form in the life cycle of the CNIDARIA.

mega- (M) A prefix indicating that the size of a unit is to be multiplied by 10^6. For instance, one megawatt (MW) is equal to one million watts.

meiosis, *reductive division* A process of CELL DIVISION in which each daughter cell formed contains half the number of CHROMOSOMES compared with the parent cell. In animals, it occurs in the formation of GAMETES as part of SEXUAL REPRODUCTION, so that when female and male gametes fuse during FERTILIZATION the resulting cells regain a full set of chromosomes, instead of doubling their total number of chromosomes. In plants, meiosis is part of SPORE formation.

The stages of meiosis are similar to MITOSIS, but two divisions occur (I and II):

(i) In prophase I, the chromosomes become visible and, unlike in mitosis, associate in their homologous (like) pairs (*see* CHROMOSOME). CROSSING-OVER may occur. The nuclear membrane disintegrates and a system of protein tubules called the spindle forms.

(ii) In metaphase I, the pairs of chromosomes associate together at the equator of the spindle, but orientate towards opposite poles of the cell (this orientation is random relative to other pairs).

(iii) In anaphase I, the spindle contracts and pulls the chromosome pairs apart to opposite poles.

(iv) In telophase I, a nuclear membrane forms around each group, the spindle disappears and the cell may divide (to yield two cells each with a NUCLEUS and half the number of chromosomes) and enter INTERPHASE (but with no DNA replication; *see* MITOSIS). Some cells go straight to prophase II.

Stages of meiosis.
a) Prophase I

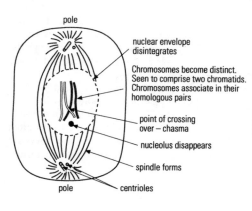

pole
nuclear envelope disintegrates
Chromosomes become distinct. Seen to comprise two chromatids. Chromosomes associate in their homologous pairs
point of crossing over – chasma
nucleolus disappears
spindle forms
pole   centrioles

b) Metaphase I

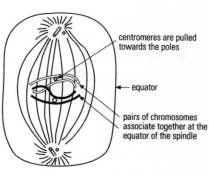

centromeres are pulled towards the poles
equator
pairs of chromosomes associate together at the equator of the spindle

Stages of meiosis (continued).

c) Anaphase I

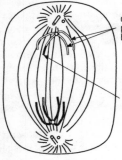

one of the homologous
pair of chromosomes
being pulled to a pole

piece of chromatid from
sister homologous
chromosome (exchanged
during cross-over in
prophase 1)

f) Metaphase II

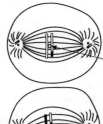

chromosome lying at
the equator of the
spindle – no association
of homologous pairs

d) Telophase I

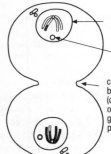

nuclear
envelope re-forms

nucleolus
re-forming

cell dividing
by constriction
(does not always
occur – some cells
go straight to
prophase II)

g) Anaphase II

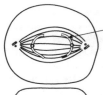

chromatids
separate and
move to opposite
poles

e) Prophase II

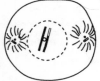

spindle fibres form
at right angles to
previous spindle axis

nuclear envelope
disintegrates

h) Telophase II

Cells divide to give
four daughter cells
each with the haploid
number of chromosomes.
None of the chromosomes
in this example are
the same as the originals

nuclear membrane
reforms

(v) In prophase II, the nuclear membrane disintegrates and the spindle forms at right angles to that of the first mitotic division.

(vi) In metaphase II, anaphase II and telophase II, the chromosomes arrange themselves (not in pairs) on the equator of the spindle, the CHROMATIDS are pulled apart to opposite poles, and nuclear membranes are re-formed as in a typical mitotic division. The end result is four cells each with a nucleus and half the number of chromosomes.

The main purpose of meiosis is to yield offspring that are different to their parent so that they can adapt to a changing or new environment. The genetic variability comes from the fusion of two gametes to give a different set of chromosomes to that of each parent. The crossing-over that occurs in prophase I and the random orientation of chromosome pairs during metaphase I provide further genetic variety.

melanin A brown pigment, produced by the MALPIGHIAN LAYER of EPIDERMIS, that determines skin, hair and eye colour and absorbs ultraviolet light, thereby protecting the tissues underneath from sunlight. The amount of melanin in skin depends on genetic and environmental factors. *See also* MELATONIN.

melatonin A hormone-like substance secreted by the PINEAL GLAND. Production of melatonin is inhibited by light, so its levels are greatest during the night. In this way, the pineal gland keeps track of changes in daylength. Melatonin helps regulate certain seasonal changes in animals, such as the reproductive cycle of seasonally breeding animals. It also controls skin colour changes in certain animals by triggering the aggregation of the pigment MELANIN.

membrane A layer surrounding cells or ORGANELLES that is made of LIPIDS and proteins. The membrane controls passage of molecules into and out of the cell or organelle. *See also* CELL MEMBRANE.

membrane potential In a living cell, the POTENTIAL DIFFERENCE across the CELL MEMBRANES. *See* NERVE IMPULSE.

memory cell A type of mature B CELL produced in response to an infection. Instead of producing large amounts of specific ANTIBODY, a memory cell remains in the circulation ready to be reactivated to produce antibody quickly if a second exposure to the initial ANTIGEN occurs. *See also* PLASMA CELL.

menarche *See* MENSTRUAL CYCLE.

Mendel's laws The original laws of inheritance set out by an Austrian monk, Gregor Mendel (1822–84), who studied the transmission of different features of the garden pea from parents to offspring. Although at that time GENES and DNA had not been discovered, much of what Mendel found forms the basis of our understanding of genetics.

Mendel's first law is segregation: an individual possesses a pair of ALLELES, one from each parent, that separate during MEIOSIS and go to different GAMETES so that they pass to different offspring, without blending. The offspring will be like one parent in a particular characteristic, for example eye colour, and not a mixture of both parental characteristics. This is also referred to as monohybrid inheritance: the inheritance of a single characteristic only. Which genetic feature is outwardly expressed depends on which allele is DOMINANT.

Mendel's second law is independent assortment: each member of an allelic pair can combine randomly with each of another pair. When considering the simultaneous inheritance of two characteristics (dihybrid inheritance), the presence of one does not affect the inheritance pattern of the other. This law does not hold true for linked genes (*see* LINKAGE), which tend to be inherited together.

Based on Mendel's laws, for simple genetics (excluding, for example, linkage and MUTATION) it is possible to predict the ratios of characteristics among offspring where the characteristics of the parents are known. This is called Mendelism.

See also DIHYBRID CROSS, MONOHYBRID CROSS.

Mendelism The theory of inheritance originally proposed by Gregor Mendel (1822–84). *See* MENDEL'S LAWS.

meninges In vertebrates, three membranous coverings enclosing and protecting the BRAIN and SPINAL CORD. The meninges consist of an innermost pia mater, a middle arachnoid and an outermost dura mater membrane. Between the arachnoid and the pia mater membranes is a space filled with CEREBROSPINAL FLUID; a clear, colourless solution of glucose and mineral ions and a few white blood cells but no protein. The fluid supplies nutrients and acts as a shock-absorber for the CENTRAL NERVOUS SYSTEM.

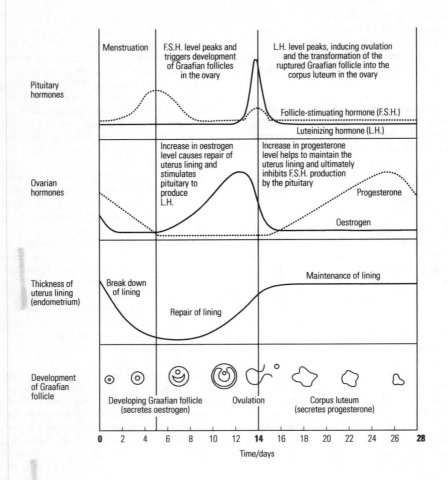

Summary of the menstrual cycle.

meniscus The name given to the curved shape of a liquid surface in a tube, or any similarly shaped object.

menopause The time in a woman's life when her reproductive capacity ends (usually between the age of 45 and 50 years). The changing levels of the hormones OESTROGEN and PROGESTERONE can cause some symptoms associated with the menopause, such as hot flushes and osteoporosis (thinning of the bones), although often the menopause passes uneventfully. In more severe cases of post-menopausal problems, HORMONE-REPLACEMENT THERAPY (HRT) is administered to replace some of the hormones and so relieve some of the symptoms.

menstrual cycle The cycle that occurs in female mammals of reproductive age to prepare the body for pregnancy. In humans, the start of the cycle is at puberty (at an average age of 12 years) and lasts until the MENOPAUSE (at an average age of 45–50 years). The first cycle is called menarche.

Each cycle lasts about 28 days, during which a GRAAFIAN FOLLICLE matures in the OVARY and releases its OVUM (ovulation) at about day 14. The CORPUS LUTEUM then develops and produces PROGESTERONE. This causes the lining of the UTERUS to thicken and fill with blood vessels in preparation for IMPLANTATION of a fertilized egg. If no FERTILIZATION occurs, the corpus luteum dies and the uterine lining is shed, causing the loss of blood for about 5 days which is menstruation. The cycle is controlled by hormones including OESTROGEN and progesterone. It is the changing levels of these hormones that can cause pre-menstrual tension (PMT) and also menopausal symptoms, such as hot flushes and OSTEOPOROSIS.

In many mammals, the reproductive cycle is called the OESTROUS CYCLE.

See also ENDOMETRIUM, HORMONE-REPLACE-MENT THERAPY, PILL.

menstruation A loss of blood from the UTERUS of women occurring about every 28 days, due to the breakdown of the lining of the uterus if fertilization of the egg does not occur. Menstruation is commonly called a 'period' and lasts for about 5 days. See MENSTRUAL CYCLE.

meristem In plant tissue, a region of actively dividing cells that produces new tissue. Apical meristems are found at the growing tips of stems or roots and cause an increase in their length. Lateral meristems, for example the CAMBIUM, cause an increase in girth. The rigid cell wall of plant cells restricts their ability to grow, which is why meristems, which are actively growing immature cells, are needed. Meristem cultures involve growing shoots in nutrient mediums into new plants, which is used in plant propagation.

merocrine gland Any EXOCRINE GLAND that remains intact after secreting its product, for example SALIVARY GLANDS. See also APOCRINE GLAND, HOLOCRINE GLAND.

merozoite A stage in the life cycle of protozoans from the phylum APICOMPLEXA, such as the *Plasmodium* parasite. See MALARIA.

mesocarp The thick fleshy layer in the wall of a fruit, underneath the outer EXOCARP.

mesoderm The middle GERM LAYER of a developing animal EMBRYO. The mesoderm forms a NOTOCHORD that later becomes the VERTEBRAL COLUMN in vertebrates, and also gives rise to muscle, kidneys, the heart, blood cells, the reproductive system, eyes and connective tissue. The mesoderm is absent from primitive animals.

mesophyll The layer of tissue in a LEAF blade between the upper and lower EPIDERMIS. The mesophyll comprises two layers. The palisade

mesophyll is just below the upper epidermis, and below this is the spongy mesophyll.

The palisade layer consists of closely packed columnar cells containing numerous CHLOROPLASTS, in which PHOTOSYNTHESIS takes place (during daylight). The palisade cells are adapted to receive the components of photosynthesis: water from the nearby XYLEM, carbon dioxide from the air through the stomata (see STOMA), and sunlight, the passage of which is maximized by the vertical arrangement of the cells (with fewer cross-walls).

The spongy mesophyll consists of loosely arranged cells of a regular shape with fewer chloroplasts but more spaces for rapid DIFFUSION of gases (entering with the air through the stomata).

mesosome An infolding of the CELL MEMBRANE of bacteria that contains the ELECTRON TRANSPORT SYSTEM. Mesosomes are thought to function similarly to MITOCHONDRIA.

mesothelium A single layer of flattened cells lining the abdominal cavity and THORAX and forming a part of SEROUS MEMBRANES. Mesothelium is similar to EPITHELIUM except that it is derived from MESODERM. *Compare* ENDOTHELIUM.

messenger RNA (mRNA) A type of RNA that acts as a template for PROTEIN SYNTHESIS in a cell. Messenger RNA is a single-stranded molecule thousands of NUCLEOTIDES long, formed into a helix, and represents less than 5 per cent of the total RNA in a cell. It is made in the nucleus and is a mirror copy of part of one of the DNA strands (see TRANSCRIPTION). From the nucleus, it enters the CYTOPLASM and associates with the RIBOSOMES, where it is subsequently involved in the production of a polypeptide (see TRANSLATION). Messenger RNA is very unstable. See also GENE SPLICING, TRANSFER RNA, START CODON, STOP CODON.

metabolic rate A measure of the energy used up by an organism in a given time period. It is affected by the level of activity of the organism and other factors, such as temperature. See also BASAL METABOLIC RATE.

metabolism The chemical processes occurring within a living organism. Metabolism is a continual process of building up of body tissue (anabolism) and breaking down of living tissue into energy and waste products (catabolism). The control of metabolism is complex, involving HORMONES and ENZYMES.

Metabolic rate refers loosely to the metabolic activity of an organism, measured by the respiratory rate (oxygen is used as a guide to metabolic activity). *See also* METABOLITE, THYROID GLAND.

metabolite A substance required for or produced by METABOLISM. Primary metabolites are involved in essential processes, whereas secondary metabolites are required for or produced by non-essential reactions. Secondary metabolites can be important in defence and are characteristic of a particular organism. For example, fungi produce toxic secondary metabolites and plants produce some that make them less palatable.

metameric segmentation The division of the body of an animal into similar segments. *See also* ANNELIDA.

metamorphosis A stage in the life cycle of most insects and amphibians and some fish, during which the body changes dramatically from one form to another. The changes involve major tissue reorganization, such as moulting (outer skin is shed, allowing the new soft body to alter its size and shape), and changes in GENE EXPRESSION. These changes are under hormonal control and occur relatively rapidly. The change from a tadpole to a frog is an example of metamorphosis.

Higher insects undergo complete metamorphosis, in which the LARVAE bear no resemblance to the adults. The larvae develop into PUPAE, which is a resting stage where no food is taken in and during which the organs and tissues of the young change into those of the adult. An example is the change of a caterpillar to a pupa, called a chrysalis, and then to a butterfly. Lower insects experience incomplete metamorphosis, in which the young develop through a series of stages resembling the adult, to the mature adult.

metaphase A stage in MITOSIS and MEIOSIS.

metaphloem In plants, primary PHLOEM tissue that develops from the PROTOPHLOEM when the shoot or root has completed its elongation. Like METAXYLEM, the cells of the metaphloem are larger than the protophloem and tend to crush the latter.

metastasis A secondary TUMOUR that develops from cells from the primary tumour that have been shed and have travelled through the blood or LYMPHATIC SYSTEM to another part of the body.

Metatheria A subclass of MAMMALIA comprising the MARSUPIALS. *Compare* EUTHERIA, PROTOTHERIA.

metaxylem In plants, primary XYLEM tissue that develops from the PROTOXYLEM when the shoot or root has completed its elongation. Metaxylem consists of larger cells than the protoxylem, which therefore tends to be crushed. Metaxylem walls have more LIGNIN than the protoxylem.

metazoa (*sing. metazoan*) A term used in earlier classification systems to describe what now constitutes the kingdom ANIMALIA.

methane (CH_4) A colourless, odourless gas, the simplest HYDROCARBON and the main component of natural gas. It is emitted by decaying vegetable matter and is therefore found in marshlands (as marsh gas) and is given off during SEWAGE DISPOSAL. Methane contributes to the GREENHOUSE EFFECT.

methionine An ESSENTIAL AMINO ACID containing sulphur, present in many proteins. *See also* START CODON.

metre The fundamental unit of length in the SI system (*see* SI UNIT). The metre was originally defined as the length of a standard metal bar, then in terms of a certain number of wavelengths of light, but now defined as the distance travelled by light in a vacuum in $1/(299,792,458)$ seconds.

metric system Any system of measurements based on the METRE as the unit of length, the GRAM as the unit of mass and the second as the unit of time, or on some multiple of these units. In science, the system almost universally used is the SI system of units, which defines units for all physical quantities, derived from seven base units. *See* SI UNITS.

MHC *See* MAJOR HISTOCOMPATIBILITY COMPLEX.

micelle A small group of molecules loosely clumped together, in a COLLOID for example.

micro- (μ) A prefix indicating that a unit is to be multiplied by 10^{-6}. For instance, the micrometre (μm) is equal to one millionth of a metre.

microbiology The study of MICRO-ORGANISMS. Microbiology has many applications, for example in medicine, industry and genetic engineering.

microbody *See* PEROXISOME.

microevolution *See* EVOLUTION.

microglial cell A type of small GLIAL CELL with an irregular shape. They occur more frequently in

GREY MATTER than WHITE MATTER. Microglial cells may be PHAGOCYTES.

microhabitat A HABITAT with very specific living conditions, such as a fallen log in a forest.

micrometer An instrument for the accurate measurement of small objects (typically up to 10 cm).

micronutrient Any chemical substance required in very small amounts by plants and animals for their normal growth and development. They are often found in COFACTORS and COENZYMES. The micronutrients include the elements manganese, copper, iodine, cobalt, zinc, molybdenum, boron, selenium, chromium, silicon and fluorine. These are also known as trace elements. VITAMINS can also be considered to be micronutients. These essential nutrients may be required for one or more metabolic roles. *Compare* MACRONUTRIENT.

micro-organism Any organism too small to be seen by the naked eye and visible only under a microscope. Micro-organisms include VIRUSES, BACTERIA, PROTOZOA, YEASTS and some ALGAE. 'Micro-organism' is a general term that is not significant for classification purposes.

micropyle In SEED PLANTS, a small hole at one end of the OVULE where the protective INTEGUMENT(S) is (are) absent. During FERTILIZATION, the POLLEN TUBE containing the male GAMETE passes through the micropyle and so gains access to the egg cell within the ovule. In a mature SEED, the micropyle is seen as a small pore in the seed coat through which water enters at the start of GERMINATION. *See also* DOUBLE FERTILIZATION.

microscope An optical device that uses a system of lenses to magnify objects too small to be seen in fine detail with the naked eye. In 1665, Robert Hooke (1635–1703) carried out the first microscopic examination of cells in cork, and Anton van Leeuwenhoek (1632–1723) recorded bacteria in 1683.

A simple microscope has a single lens but limited powers of magnification, whereas a compound microscope uses two lenses and light passes from an object through the first lens (objective) to produce a magnified image that is then magnified further by the second lens (eyepiece). The total magnification is the product of the magnification of each lens and is maximum of 1,500–2,000 in a light microscope. The RESOLUTION of a light microscope is limited to two points 0.2 μm apart. The thinner the material observed, the greater the clarity of image.

Preserved or fixed tissues are usually embedded in paraffin wax and thin sections (3–20 μm) are cut using an instrument called a microtome. Even thinner sections can be obtained by freezing the tissue in liquid nitrogen and cutting sections in a cryostat at –200°C. Various staining methods are used to enhance the image seen and to highlight specific cells or materials.

See also ELECTRON MICROSCOPE, FLUORESCENCE MICROSCOPY, PHASE CONTRAST MICROSCOPY.

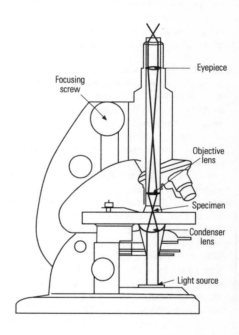

Light microscope.

microscopy The study or use of MICROSCOPES.

microtome A machine for cutting thin sections of tissue (frozen or embedded in paraffin wax) to be used in a light MICROSCOPE. The sections are 3–20 μm thick. A steel knife is used.

microtubule A hollow cylinder (about 25 nm in diameter), made of protein filaments (threadlike structures), that is an essential component of the CYTOSKELETON in almost all EUKARYOTIC cells. Microtubules are also vital components of CILIA, FLAGELLA, and the SPINDLES formed in MITOSIS and MEIOSIS.

microvilli (*sing.* ***microvillus***) Minute finger-like projections on the surface of many EUKARYOTIC cells, particularly EPITHELIA cells concerned with absorption, for example in the SMALL INTESTINE. Each microvillus is about 1 μm long and 0.1 μm in diameter. Microvilli can extend and retract and form a BRUSH BORDER, increasing the surface area over which absorption can take place.

midbrain The smallest of three regions of the human brain, linking the FOREBRAIN and HINDBRAIN. This region contains the visual and auditory centres, for example controlling movement of the head to fix on an object or sound. *See also* PINEAL GLAND.

middle lamella A thin layer of material between plant CELL WALLS that binds adjacent cells together.

migration The seasonal movement of certain animals, mostly birds and fish, to distant lands for breeding or feeding. For example, some birds fly south during the winter months and return to their breeding ground in the spring. In some species, for example locusts, the return journey is made by the next generation. Sometimes whole species migrate over a period of many years. The precise method of migration is unclear.

milk The fluid secreted by the MAMMARY GLANDS of the female mammals to provide nourishment for their offspring. *See* LACTATION.

milk teeth *See* DECIDUOUS TEETH.

milli- (m) A prefix indicating a unit is to be multiplied by 10^{-3}. For instance, one millimetre (mm) is equal to one thousandth of a metre.

mineralocorticoid Any one of a group of CORTICOSTEROID hormones, secreted by the CORTEX of the ADRENAL GLAND, that is concerned with the METABOLISM of minerals. An example is ALDOSTERONE.

mini-pill *See* PILL.

miscible (*adj.*) Describing two liquids that can mix together to form a single liquid. Most organic liquids are miscible with one another, but not with water. Water and ethanol are miscible. *See also* IMMISCIBLE.

mitochondria (*sing.* ***mitochondrion***) Rod-like (1–10 μm in length and 0.25–1 μm in width) or spherical membrane-bound bodies (ORGANELLES) within the CYTOPLASM of EUKARYOTIC cells. Mitochondria occur in varying numbers and contain the enzymes responsible for energy production, in the form of ATP, during aerobic RESPIRATION. It is thought that mitochondria are derived from free-living bacteria that once invaded larger cells to become symbiotic (*see* SYMBIOSIS).

Each mitochondrion contains its own mitochondrial DNA, and division of existing mitochondria yields new ones. They have a double outer membrane, the inner component of which is folded inwards and projects into the matrix as cristae. These are covered in particles associated with OXIDATIVE PHOSPHORYLATION. PLASTIDS of plants may be related to mitochondria.

mitosis A process in which a cell nucleus divides, usually prior to CELL DIVISION, to create two new cells each with the same number of CHROMOSOMES as the parent cell.

Mitosis can be divided into a number of phases: prophase, metaphase, anaphase and telophase. During prophase, the chromosomes become untangled and visible and a system of protein tubules called the spindle forms, which span the cell from one pole to the other, and the nuclear membrane disintegrates. During the subsequent phases, the two strands (CHROMATIDS) that constitute chromosomes are separated and move to opposite poles of the cell. Unlike during MEIOSIS, homologous (like) chromosome pairs do not associate during mitosis. A new nuclear envelope forms around each group of chromatids, they become invisible again and the spindle disintegrates. The cell itself can divide after nuclear division but does not always (so multinucleate cells exist). It is during a further phase, INTERPHASE, that the DNA content of the cell doubles and cell ORGANELLES double, so that the two daughter cells formed contain the same number of chromosomes as the parent itself.

Only a specialized group of cells in plants can undergo mitosis (*see* MERISTEM), but most animal cells can. During growth or repair of tissue, cell division is by mitosis because the new cells must be the same as the existing cells. ASEXUAL REPRODUCTION by mitosis yields organisms identical to the parent, which are suited to the same environment as their parents but are unable to adapt to new challenges.

mitral valve *See* BICUSPID VALVE.

mode The value in a set of data that occurs most often, or the range for which the frequency is greatest.

Stages of mitosis.

a) Interphase

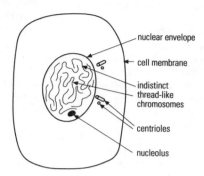

nuclear envelope

cell membrane

indistinct
thread-like
chromosomes

centrioles

nucleolus

b) Prophase

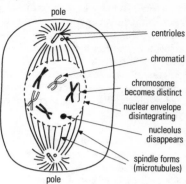

pole

centrioles

chromatid

chromosome
becomes distinct

nuclear envelope
disintegrating

nucleolus
disappears

spindle forms
(microtubules)

pole

c) Metaphase

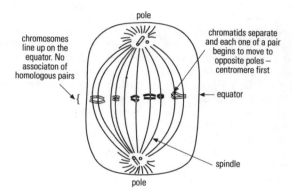

pole

chromosomes
line up on the
equator. No
associaton of
homologous pairs

chromatids separate
and each one of a pair
begins to move to
opposite poles –
centromere first

equator

spindle

pole

d) Anaphase

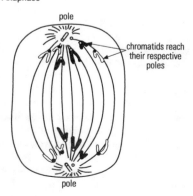

pole

chromatids reach
their respective
poles

pole

e) Telophase

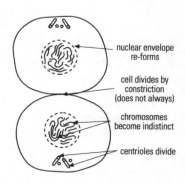

nuclear envelope
re-forms

cell divides by
constriction
(does not always)

chromosomes
become indistinct

centrioles divide

molar tooth A large, broad tooth at the back of the mouth of mammals used for crushing and grinding food. Humans have three molars on each side of the upper and lower jaws, behind the PREMOLARS, making a total of twelve. The molar teeth have several roots and a ridged surface. In contrast to the premolars, the molars are not preceded by DECIDUOUS TEETH. The third molar on each side is termed the wisdom tooth and does not appear until early adulthood.

molecular biology The study of the molecular components of biological systems. Molecular biology particularly refers to the study of NUCLEIC ACIDS and PROTEINS.

molecule The smallest part of a chemical COMPOUND that can exist without it losing its chemical identity. Molecules are made of one or more atoms held together by chemical bonds. *See also* MACROMOLECULE.

mollusc A member of the phylum MOLLUSCA.

Mollusca The second largest phylum in the animal kingdom, consisting of about 100,000 species, including snails, slugs, squids, octopuses, oysters, mussels and winkles. The members are all invertebrates whose bodies are divided into a head, muscular foot and a visceral mass. The body is soft with no internal skeleton and no limbs. In many species it is covered by a hard shell. Most molluscs are marine but some are freshwater and a few are terrestrial. The shells are variable and can be univalve, as in the snail, or bivalve, as in the mussel, as well as other forms. In squid, the shell is internal. Reproduction in molluscs is sexual, eggs are laid, and many are HERMAPHRODITES.

Molluscs are important to humans as a food source and for pearls. Snails are the intermediate hosts for a number of parasitic diseases of humans and livestock, for example schistosomiasis (*see* TREMATODA). *See also* GASTROPODA, PELECYPODA.

molybdenum (Mo) A metallic element that is a MICRONUTRIENT for plants but not vital for most animals. In plants, molybenum is required for the conversion of NITRATES to NITRITES in the formation of amino acids. It is also needed by some PROKARYOTES for NITROGEN FIXATION. Deficiency causes a reduction in crop yield.

monoclonal antibody (MAb) An ANTIBODY produced in the laboratory by fusing an antibody-producing B CELL (from mice immunized with an ANTIGEN) with MYELOMA tumour cells, which do not exhibit the usual regulation of growth and CELL DIVISION. The resulting cell is called a hybridoma, and continues to divide and produce antibodies indefinitely. This allows large quantities of antibody with a single specificity to be made.

The technique was invented in 1975 by César Milstein (1927–) at Cambridge University, UK, and has been a major breakthrough in science. Monoclonal antibodies can be linked to markers to locate specific antigen targets, for example sources of disease, or specific cell types, or linked to CYTOTOXIC drugs, which (if the antibody is directed at tumour cell antigens) could locate and hopefully destroy the tumour. One widely used technique using monoclonal antibodies is the ENZYME-LINKED IMMUNOSORBANT ASSAY (ELISA).

monocotyledon Any flowering plant that has one COTYLEDON or seed leaf in the embryo. These contrast to DICOTYLEDONS, which are flowering plants with two cotyledons. Most monocotyledons are small plants, such as orchids, lilies, grasses and cereals, but some are large, such as palms. The leaves are usually narrow with parallel veins and smooth edges, the flower parts are grouped in threes and the VASCULAR BUNDLES (XYLEM, PHLOEM) are arranged irregularly. POLLINATION is usually by wind. In most monocotyledons, the cotyledons remain below ground following GERMINATION and are called hypogeal.

monocyte A type of vertebrate LEUCOCYTE that can differentiate into a MACROPHAGE.

monoecious Plants that have separate male and female flowers on the same individual plant. This arrangement favours cross-fertilization. *Compare* DIOECIOUS, HERMAPHRODITE.

monoestrus (*adj.*) Describing non-human mammals who experience a single OESTROUS CYCLE each year. They therefore have a well-defined breeding season. *Compare* POLYOESTRUS.

monohybrid cross In genetics, a cross between two animals or plants that are genetically identical except for one GENE. The one gene could be, for example, for seed colour, with one individual HOMOZYGOUS for the DOMINANT ALLELE (where the seed is green) and one homozygous for the RECESSIVE allele (where the seed is yellow). The offspring of such a cross (F_1

GENERATION) will all have green seeds; they are monohybrids (HYBRIDS for one gene only) and are identical to one another and resemble one parent. The recessive allele for yellow seeds is hidden but will be expressed in the next, F_2, generation. In this F_2 GENERATION there will be on average three plants with green seeds and one plant with yellow seeds – a ratio of 3:1. In reality there are too many genetic differences to see this simple inheritance. *See also* DIHYBRID CROSS, MENDEL'S LAWS.

monokine A soluble substance secreted by MACROPHAGES that is involved in cell communication in the IMMUNE RESPONSE. *See* CYTOKINE.

monomer A simple chemical compound that, under suitable conditions, can join with other identical monomers to form a long chain POLYMER.

monosaccharide A single sugar with the general formula $(CH_2O)_n$ that cannot be split into smaller CARBOHYDRATE units. When n is 3, the sugar is called a triose sugar, when n is 5 it is a pentose sugar, and when n is 6 it is hexose sugar. Monosaccharides are either aldoses (aldo-sugars), which have an ALDEHYDE group (CHO), or ketoses (keto-sugars), which have a KETONE group (C=O). Both GLUCOSE and FRUCTOSE have the formula $C_6H_{12}O_6$ but glucose is an aldose and fructose is a ketose, so their properties are different. Both can easily form ring structures; glucose usually has a six-sided ring and fructose a five-sided ring, although both can form either ring structure. Most monosaccharides can form ISOMERS. Monosaccharides are sweet, soluble crystalline molecules. *See also* DISACCHARIDE, POLYSACCHARIDE.

monosodium glutamate A white crystalline salt with a taste similar to meat that is widely used as a food ADDITIVE.

monotreme The least evolved mammal, subclass PROTOTHERIA, where the young hatch from an egg outside the mother's body and are then nourished with milk from simple MAMMARY GLANDS inside an abdominal pouch on the mother. Only a few species of monotreme still exist, for example the duck-billed platypus and the spiny anteater, because they have been displaced by other, more advanced species.

morning after pill A contraceptive PILL taken up to 72 hours after unprotected intercourse that prevents implantation of a fertilized egg. It contains high levels of hormones and is not for regular use.

moss Any member of the class MUSCI.

motor nerve A NERVE that carries an impulse away from the CENTRAL NERVOUS SYSTEM to an EFFECTOR muscle or gland. It is part of the EFFECTOR SYSTEM and causes voluntary and involuntary actions.

motor system *See* EFFECTOR SYSTEM.

mould A general name for superficial growth of a fungus on foodstuffs such as fruit or bread.

mouth, *buccal cavity* The cavity forming the entrance to the DIGESTIVE SYSTEM. In mammals, it is enclosed by the jaws, cheeks and palate. Digestion of food begins in the mouth, where it is chewed (mastication) and mixed with SALIVA. The TONGUE is a muscular structure, attached to the floor of the mouth, that contains nerves and tastebuds. It aids the chewing process and pushes chewed food to the back of the mouth and into the PHARYNX. A number of reflexes exist to ensure that food goes down the OESOPHAGUS and air down the TRACHEA. *See also* EPIGLOTTIS, RESPIRATORY SYSTEM.

mouthpart An appendage around the mouth of ARTHROPODS that is adapted for feeding. Mouthparts are found in pairs, the number varying between groups, for example CRUSTACEANS have three pairs. In insects, the mouthparts can be adapted for piercing, sucking, biting or chewing. *See also* MANDIBLE.

MRI *See* MAGNETIC RESONANCE IMAGING.

mucin A sticky, jelly-like GLYCOPROTEIN that provides lubrication, intercellular bonding or binding of, for example, food (*see* SALIVA).

mucous membrane A thin layer of EPITHELIUM that lines all animal body cavities and canals that come into contact with the air. In particular, mucous membranes are found in the DIGESTIVE SYSTEM, RESPIRATORY SYSTEM and URINO-GENITAL SYSTEM. The epithelium usually contains GOBLET CELLS that secrete MUCUS.

mucus A slimy secretion, containing MUCINS, that is produced by GOBLET CELLS of MUCOUS MEMBRANES in various parts of the body. In the RESPIRATORY SYSTEM, mucus helps to trap airborne particles for expulsion. In the DIGESTIVE SYSTEM, mucus helps to lubricate the food and protect the stomach from attack by digestive enzymes.

multicellular (*adj.*) Organisms or their parts that consist of more than one cell. *Compare* ACELLULAR, UNICELLULAR.

multinucleate cell A CELL with more than one NUCLEUS. *See also* CELL DIVISION.

Musci A class of the phylum BRYOPHYTA consisting of the mosses. Mosses are small, non-flowering plants with no true roots. The sexual organs are at the tips of the leaves (*see* ANTHERIDIUM, ARCHEGONIUM). They thrive in damp conditions and show ALTERNATION OF GENERATIONS.

muscle Animal cells or fibres, derived from embryonic MESODERM, that have the ability to contract, causing movement of joints and other body movements. There are three types: voluntary, involuntary and cardiac.

Voluntary muscle (also called striated, striped, skeletal muscle) is activated by MOTOR NERVES under voluntary control and is concerned with locomotion and joint movement. It is composed of large, long fibres that consist of multinucleate cells, or syncytia (*see* SYNCYTIUM), with many nuclei held together by CONNECTIVE TISSUE and surrounded by a membrane. Within the SARCOPLASM of each fibre are longitudinal MYOFIBRILS, each with a distinctive pattern of bands caused by the distribution of the proteins ACTIN and MYOSIN. The bands form repeating units called sarcomeres.

Sliding filament theory of muscular contraction.

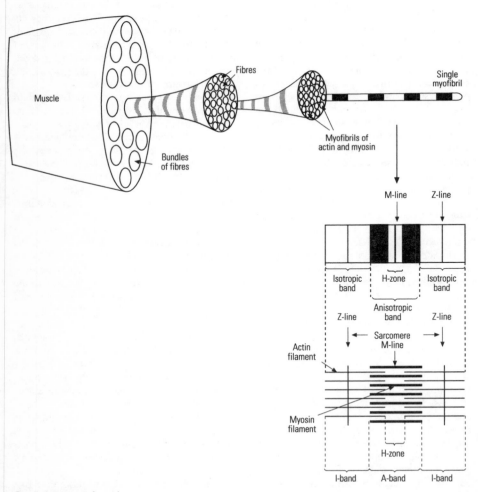

Detailed structure of muscle.

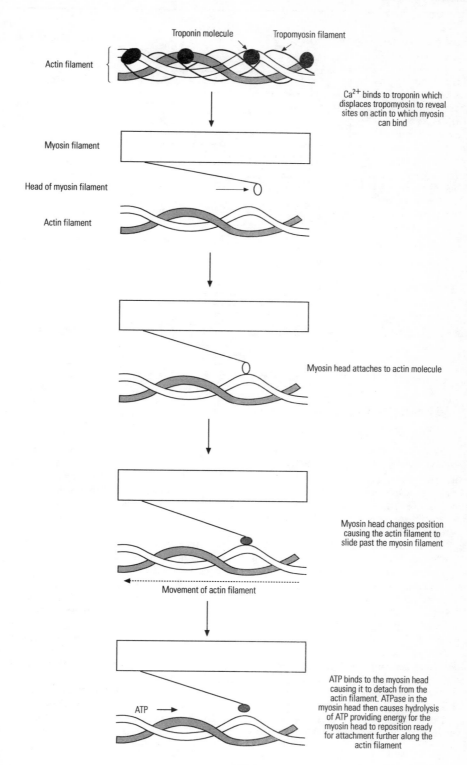

Troponin molecule   Tropomyosin filament

Actin filament

Ca^{2+} binds to troponin which displaces tropomyosin to reveal sites on actin to which myosin can bind

Myosin filament

Head of myosin filament

Actin filament

Myosin head attaches to actin molecule

Myosin head changes position causing the actin filament to slide past the myosin filament

Movement of actin filament

ATP

ATP binds to the myosin head causing it to detach from the actin filament. ATPase in the myosin head then causes hydrolysis of ATP providing energy for the myosin head to reposition ready for attachment further along the actin filament

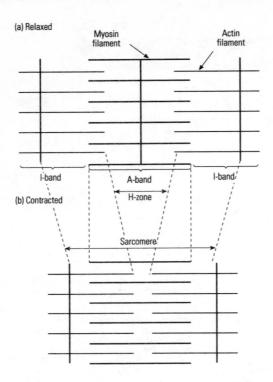

(a) Relaxed

Myosin filament

Actin filament

I-band   A-band   I-band
(b) Contracted   H-zone

Sarcomere

Changes in the myofibril banding pattern during muscular contraction.

Involuntary muscle (also called unstriated, smooth muscle) consists of spindle-shaped cells arranged in sheets or bundles bound by connective tissue and is under the control of motor nerves from the AUTONOMIC NERVOUS SYSTEM (involuntary nervous system). It is characterized by its ability to contract slowly and rhythmically over a long period of time. Therefore, involuntary muscle is important in the DIGESTIVE SYSTEM, by allowing PERISTALSIS, in the walls of blood vessels and in the tubes of the URINO-GENITAL SYSTEM.

CARDIAC MUSCLE is a specialized muscle found only in the heart, and is under involuntary control.

See also MUSCULAR CONTRACTION.

muscular contraction The response of MUSCLE cells to stimulation, causing a force in one direction. The muscle may shorten during a contraction or remain the same length.

The contraction of voluntary, or striated, muscle is understood best and involves the 'sliding filament theory', although the contraction of involuntary and CARDIAC MUSCLE is thought to be similar. Muscle fibres consist of several MYOFIBRILS, which themselves consist of thin and thick filaments of the proteins ACTIN and MYOSIN respectively. These form a distinctive banding pattern across the myofibril (visible under a microscope) that alters during a contraction, showing that the filaments slide past one another.

In order for contraction of voluntary muscle to occur, a number of conditions must be met. Firstly, the muscle needs stimulation by an impulse from a MOTOR NERVE. Secondly, the actin and myosin filaments must make contact to form a complex called ACTOMYOSIN. This complex can only be formed in the presence of calcium ions. The point at which the nerve meets the muscle is called the NEUROMUSCULAR JUNCTION, and there are many of these spread throughout a muscle to ensure rapid contraction of all the fibres simultaneously. Each

NERVE IMPULSE releases a jet of ACETYLCHOLINE, which diffuses across to the outer membrane of the muscle and depolarizes it, generating an ACTION POTENTIAL. Calcium ions are then released from the SARCOPLASMIC RETICULUM, where they are stored, where this comes into contact with infoldings of the muscle fibre's outer membrane, called transverse tubules. The calcium ions bind to protein molecules called troponin, which are attached to actin filaments. This in turn displaces tropomyosin molecules, also attached to actin, to reveal sites on the latter to which myosin can bind, forming an actomyosin complex. The myosin heads bind to the actin filament (following the HYDROLYSIS of ATP by the enzyme ATPase within the myosin heads), forming bridges, and the heads change position, so causing the actin filament to slide past the stationary myosin filament. This is the muscle contracting. ATP then binds to the myosin head, causing it to detach from the actin filament and reposition ready for reattachment further along the filament. After contraction, calcium ions are pumped back into the sarcoplasmic reticulum and tropomyosin once again hides the myosin binding sites on the actin filament. The muscle is now relaxed. The GLYCOGEN store found in muscles provides the supply of glucose needed to regenerate the ATP.

mutagen A factor, for example ultraviolet radiation, IONIZING RADIATION (such as X-rays, alpha particles, beta particles and gamma radiation) and chemicals, that increases the natural spontaneous rate of MUTATION.

mutation An alteration of a GENE in an organism caused by an alteration of the genetic material (DNA or RNA). A mutation usually occurs when DNA is replicated and mistakes are made. Most mutations are undesirable. Any beneficial mutations would be favoured by NATURAL SELECTION. There is a natural, low spontaneous rate of mutation but certain factors called MUTAGENS increase this rate.

Most mutations occur in body cells (not in the formation of GAMETES) and are therefore not passed on. The mutation may be as small as the omission, insertion or substitution of a single base (*see* NUCLEOTIDES) in the DNA, which is called a point mutation. Although small, a point mutation can have serious effects because it can create STOP CODONS (nonsense mutations) or alter the reading

frame of the DNA (frame shift mutations) so that protein synthesis is affected. An example of a disease resulting from a point mutation is SICKLE-CELL DISEASE.

Mutations can be much larger than those at the gene level, affecting CHROMOSOME structure or number. Whole sets of chromosomes can be duplicated or one chromosome can be deleted or added. An example of this in humans is in DOWN'S SYNDROME, where there are three copies of chromosome 21. Similar examples involving the SEX CHROMOSOMES are Klinefelter's syndrome, where individuals have the GENOTYPE XXY, XXXY or XXXXY and a male PHENOTYPE with some female development, and Turner's syndrome, where there is an X chromosome missing and individuals have the genotype XO and a small, sexually immature female phenotype. Chromosome structure can be mutated during MEIOSIS by a deletion of a portion, or by INVERSION, TRANSLOCATION (which is different from CROSSING-OVER because it occurs between non-homologous chromosomes) or duplication of a portion of chromosome.

A mutation does not always affect the organism because it may not be translated into protein, or if it is it may be within a non-functional part of the protein. These are a neutral mutations, and can eventually be important as they build up with time. In the laboratory, mutagenesis is a process that can be used to modify existing gene products.

Predictions can be made about the likelihood of offspring having a particular condition, taking account of the family genetic history. Some defects, particularly chromosomal, can be detected in early pregnancy by examining foetal cells taken from AMNIOTIC FLUID.

See also ONCOGENE, POLYMORPHISM.

mutualism A relationship between two different species in which neither partner suffers, or both benefit. Mutualism is another term for true SYMBIOSIS, and does not include the variations of symbiosis such as COMMENSALISM and PARASITISM.

mycelium The main body of a fungus that consists of a mass of thread-like HYPHAE.

mycorrhiza Structures formed by the association of a fungus with a root of a higher plant, such as a pine, oak, beech or birch tree.

myelin sheath An insulating layer around some nerve AXONS that speeds up the passage of NERVE IMPULSES. The myelin sheath is made up of fats and proteins contained within many layers of membrane laid down by specialized SCHWANN CELLS. The myelin sheath is discontinuous and is absent at intervals, called nodes of Ranvier, along the axons. A nerve impulse travels faster along these axons because the myelin sheath acts as an electrical insulator and the impulse therefore jumps from node to node.

myeloma A malignant TUMOUR of BONE MARROW. *See also* MONOCLONAL ANTIBODY.

myofibril A structural component of striated (voluntary) MUSCLE fibres. There are several myofibrils to each fibre and they have a distinctive banding pattern due to the distribution of the proteins ACTIN and MYOSIN. Actin is made of thin filaments, which form a light (isotropic) I-band across the muscle fibril. Myosin is made of thick filaments, and where these overlap with actin filaments a darker (anisotropic) A-band is seen. These bands alternate across the myofibril, causing its striated appearance. Within each light band is a central Z-line, and within each dark band is a lighter region called the H-zone, which may have a central dark M-line. During MUSCULAR CONTRACTION, the actin filaments slide over the myosin filaments as they are pulled together. This has the effect of shortening the I-band and H-zone while the A-band remains unchanged.

myogenic (*adj.*) Originating in or forming MUSCLE tissue. The contractions of the CARDIAC MUSCLE are said to be myogenic since they are stimulated within the heart itself.

myoglobin A GLOBULAR PROTEIN found in vertebrate muscle that is closely related to HAEMOGLOBIN and binds oxygen. Myoglobin has a single HAEM group (human haemoglobin has four) and a greater affinity for oxygen than haemoglobin. Myoglobin stores oxygen until it is needed in situations of extreme exertion, when the blood oxygen supply from haemoglobin is not sufficient to keep up with demands of muscle cells. Myoglobin is a red colour and is responsible for the coloration of meat.

During really vigorous exercise, if sufficient oxygen cannot be provided, muscle cells switch from AEROBIC RESPIRATION to ANAEROBIC RESPIRATION, which requires no oxygen but causes a build up of LACTIC ACID in the muscle, which is felt as cramp. This physiological state is known as OXYGEN DEBT. When exercising is stopped, the lactic acid is broken down and the oxygen debt is paid off.

See also RESPIRATORY PIGMENT.

myopia *See* SHORT-SIGHTEDNESS.

myosin A protein found in most EUKARYOTIC cells. Two classes of myosin exist; both consisting of a head and tail region. Myosin I is involved in cell locomotion, and myosin II involved in MUSCULAR CONTRACTION. Myosin II forms the thick filaments of muscle myofibrils, along with the thin filaments of ACTIN. The head regions of myosin contain actin-binding sites and it is the interaction of myosin with actin to form ACTOMYOSIN that is important in muscular contraction.

N

NAD (nicotinamide adenine dinucleotide) A COENZYME derived from the vitamin nicotinic acid (*see* VITAMIN B), which is an electron carrier in the ELECTRON TRANSPORT SYSTEM and in the KREBS CYCLE in RESPIRATION. When reduced NAD receives a hydrogen atom it becomes NADH, which carries the electrons. In its phosphorylated form NAD is NADP.

NADP (nicotinamide adenine dinucleotide phosphate) The phosphorylated form of NAD (*see* PHOSPHORYLATION). NADP is important as an electron carrier in PHOTOSYNTHESIS. When reduced, NADP receives a hydrogen atom to become NADPH. NADP is not as abundant in animal cells as NAD.

nano- (n) A prefix indicating that a unit is to be multiplied by 10^{-9}. For instance, one nanometre (nm) is one billionth of a metre.

nasopharynx The upper part of the PHARYNX.

nastic movement A plant movement in response to an external stimulus that is unrelated to the direction of the stimulus (unlike TROPISM). The movement can be due to growth or changes in TURGOR pressure. Examples of stimuli are light (PHOTONASTY), temperature (THERMONASTY) and chemicals (CHEMONASTY). *See also* THIGMONASTY.

natural killer cell (NK cell) In higher vertebrates, a type of WHITE BLOOD CELL that can recognize alterations to the surface of virally infected and cancerous cells and then kill them. They are distinct from T CELLS and B CELLS and act without the need for the foreign ANTIGENS of the MAJOR HISTOCOMPATIBILITY COMPLEX. *Compare* CYTOTOXIC T CELL.

natural selection The process by which individuals within a POPULATION, with characteristics that favour survival in their environment, reproduce more efficiently and so are selected (non-randomly) in preference to individuals without those characteristics. The frequency of a favourable characteristic therefore increases in a population, and the frequency of an undesirable one decreases. The process of natural selection was recognized in 1858 by Charles Darwin (1809–1882) and Alfred Wallace (1823–1913).

It is now known that natural selection is a slow process relying upon random gene MUTATIONS (some of which are favourable and passed on to future generations) and on the genetic RECOMBINATION resulting from SEXUAL REPRODUCTION. It takes many generations for a particular trait, for example beak shape, limb shape and fur coat, to become an adaptation, and at any one time there will be a range of individuals with respect to any one character. The effect of the environment on this selective gene transmission is called selection pressure. Sometimes more than one distinct form of a characteristic can coexist within a population (*see* POLYMORPHISM).

The process of natural selection can cause problems for humans, for example the development of ANTIBIOTIC RESISTANCE by certain bacteria. Over many generations, through the inheritance of many favoured VARIATIONS, natural selection will eventually lead to the formation of new species (although other factors affect EVOLUTION).

See also ADAPTIVE RADIATION, ARTIFICIAL SELECTION, GENETIC DRIFT.

nectar A sugary liquid secreted by some plants that attracts insects, birds or other animals to the flower for POLLINATION. Nectar consists of sugars, amino acids and other nutrients. Bees use nectar to make honey.

nectary A specialized GLAND, near the base of some flowers, that produces NECTAR.

negative feedback Where the end-product of a pathway inhibits the ENZYME at the start of the pathway, as occurs in many metabolic reactions. *See also* HOMEOSTASIS.

nekton The swimming animals of the PELAGIC zone of a mass of water, for example whales and fish. *See also* PLANKTON.

Nematoda A phylum consisting of worms that have an unsegmented, cylindrical body with two openings to the gut (mouth and anus) and no CILIA or FLAGELLA. There are 10,000 known species, which are thought to represent only about 2 per cent of the phylum, living in a variety of habitats.

The triploblastic (three-layered) body is pointed at both ends and has a tough outer CUTICLE of protein. Most nematodes are free-living but some are PARASITES. Nematodes move by a series of longitudinal contractions down their body. Food is pumped into the intestine by the PHARYNX and is varied, including algae, animals and organic debris. Reproduction is sexual; male and female individuals are separate and fertilization is internal. Parasitic nematodes may have a complex life cycle with several hosts. Roundworms, threadworms and eel worms cause diseases of humans, such as elephantiasis, and attack plant roots, such as the potato.

Compare FLATWORM.

nematode Any member of the phylum NEMA-TODA.

neo-Darwinism The current theory of EVOLU-TION. It is a combination of Darwin's original theory of NATURAL SELECTION and Mendel's theories on genetics (*see* MENDEL'S LAWS), together with modern knowledge of GENES and CHROMOSOMES and incorporates the concept of genetic variation upon which natural selection can work.

neoplasm A TUMOUR.

nephron The functional unit of the KIDNEY. A nephron is a filtering unit forming URINE. There are over a million nephrons in the human kidney.

A nephron is made up of a tight knot of blood capillaries called the GLOMERULUS, which is surrounded by a cup-shaped structure called the BOWMAN'S CAPSULE. This capsule forms part of a long tubule that has several clearly defined regions. The tubule extends from the Bowman's capsule along the proximal convoluted tubule, to a long narrow collecting tubule called the LOOP OF HENLE. The distal convoluted tubule links the loop at the other side to a collecting duct. Blood containing waste materials passes over the Bowman's capsule and useful minerals and water are reabsorbed back into the blood. Over 80 per cent of reabsorption occurs within the proximal convoluted tubule. The cells here are well suited to reabsorption because they have MICROVILLI, to increase the surface area, and contain many MITOCHONDRIA, to provide the

ATP needed for ACTIVE TRANSPORT. Food substances, water and sodium are reabsorbed by the proximal convoluted tubule. Further reabsorption and concentration of the filtrate occurs in the loop of Henle. The remaining fluid passes through the distal convoluted tubule to the collecting duct and then to the URETER as urine.

The distal convoluted tubule controls the pH of the blood and urine by adjusting excretion and retention of hydrogen (H^+) and hydrogen carbonate (HCO^{3-}) ions. ANTI-DIURETIC HORMONE controls the permeability of the walls of the distal convoluted tubule and the collecting duct, and hence the degree to which water is drawn out from them by OSMO-SIS, which occurs due to the high salt concentration in the inner (MEDULLA) region of the kidney.

nerve A bundle of NEURONES and GLIAL CELLS together with their associated CONNECTIVE TISSUE and blood vessels surrounded by a connective tissue sheath. Nerves can contain either sensory or motor neurones or a mixture of both (mixed nerves). Each neurone conducts independently of its neighbour. Nerves are of varying lengths but can be nearly as long as the whole animal.

nerve cell *See* NEURONE.

nerve fibre The AXON of a NEURONE and its MYELIN SHEATH if present.

nerve impulse A wave of chemical and electrical changes, affecting the membrane of a NEU-RONE, which passes along NERVE FIBRES to relay information rapidly between different parts of the body.

All living cells have a POTENTIAL DIFFERENCE across their membranes (the membrane potential) that is caused by the distribution of four ions, sodium (Na^+), potassium (K^+), chloride (Cl^-) and organic ANIONS (COO^-). In a normal state, the membrane of a neurone is negatively charged inside with respect to the outside (resting potential) and the membrane is said to be polarized (*see* POLARIZE). Following appropriate stimulation above a threshold value, sodium and potassium ions move across the membrane by DIFFUSION or CATION PUMPS, and the membrane becomes positively charged. This is called an ACTION POTENTIAL and the membrane is said to be depolarized. This is an all or none response. Once an action potential has been received by

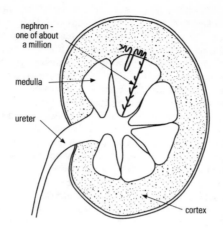

Above: Mammalian kidney showing the position of a nephron.
Below: Regions of the nephron.

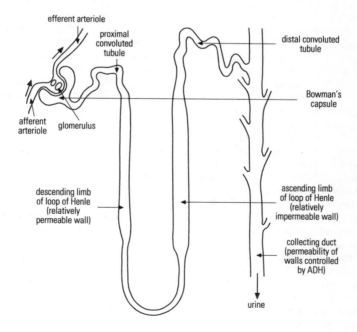

the nerve cell body, it travels quickly along the AXON to a SYNAPSE, which is linked either to other neurones or EFFECTOR cells (e.g. muscles). Here a NEUROTRANSMITTER is released that diffuses across the synapse and stimulates either an impulse through another nerve cell or the action of an effector cell. Very quickly, the membrane returns to its resting potential and is thus repolarized.

The speed of transmission of an impulse can vary from 1 to 100 cms⁻¹. There is a refractory period after an action potential, which separates a second impulse from the first.

nerve net A network of NERVE cells that forms a primitive type of NERVOUS SYSTEM in some simple invertebrates

nervous system All of the cells forming nervous tissue, the NEURONES, NERVES, RECEPTORS and GLIAL CELLS, that detect and relay information about an animal's internal and external environment and co-ordinate a response. All animals have a nervous system derived from embryonic ECTODERM, although the system can vary from being very simple, as in the 'nerve net' of jellyfish and sponges, to very complex, as in humans.

The nervous system of humans can be divided into a number of parts. (i) The SENSORY SYSTEM is formed from receptors collecting information from the internal and external environment and the sensory neurones carrying this information to the CENTRAL NERVOUS SYSTEM, where it is processed. (ii) The central nervous system (CNS) receives information from sensory neurones, interprets this information and sends messages to motor neurones to stimulate the appropriate action. In vertebrates, the CNS consists of a BRAIN and SPINAL CORD enclosed and protected by the spinal column surrounded by three membranous coverings called the MENINGES. (iii) The EFFECTOR SYSTEM transmits the information received from the CNS to effectors, which act upon it.

See also AUTONOMIC NERVOUS SYSTEM, REFLEX.

neural network An artificial network of processors that tries to mimic the transmission of impulses in the human brain. A neural network may be an electronic or optical construction or a computer simulation.

neurilemma The outer covering of a NERVE FIBRE.

neuroendocrine cell *See* NEUROSECRETORY CELL.

neurogenic (*adj.*) Originating in or stimulated by the NERVOUS SYSTEM or NERVE IMPULSES.

neuromuscular junction The point at which a MOTOR NERVE meets a MUSCLE and a SYNAPSE forms. The muscle membrane under the nerve is called the endplate. A NERVE IMPULSE releases a jet of ACETYLCHOLINE, which depolarizes the endplate and generates an ACTION POTENTIAL that travels along the muscle, causing it to contract (*see* MUSCULAR CONTRACTION).

neurone, *nerve cell* A major cell type of the NERVOUS SYSTEM specialized to transmit information rapidly between different parts of the body, in the form of NERVE IMPULSES.

The cell body of a neurone consists of a nucleus and cytoplasm and one or more short projections called DENDRITES (branching from a DENDRON), which conduct impulses towards the neurone. A longer extension (usually only one) called the AXON conducts impulses away from the cell body of the neurone to the SYNAPSE, which links with another nerve cell or an EFFECTOR cell such as muscle. Sensory neurones carry impulses towards the CENTRAL NERVOUS SYSTEM; relay neurones are found within the central nervous system and connect with motor neurones, which carry impulses away from the central nervous system to an EFFECTOR cell (muscles or glands). Neurones are bundled together by CONNECTIVE TISSUE which, with associated blood vessels, forms NERVES.

neurosecretory cell, *neuroendocrine cell* Any cell of the NERVOUS SYSTEM that can both conduct NERVE IMPULSES and secrete HORMONES. Neurosecretory cells are found, for example, in the HYPOTHALAMUS. *See also* ENDOCRINE SYSTEM.

neurotransmitter A chemical of low RELATIVE MOLECULAR MASS that is released at a SYNAPSE and transmits impulses between NEURONES or between neurones and EFFECTORS. Neurotransmitters are stored in vesicles located at the end of a nerve AXON and are released upon arrival of a NERVE IMPULSE. The neurotransmitter then diffuses across the synapse to RECEPTORS on the postsynaptic cell (*see* SYNAPSE). Neurotransmitters can be excitatory, generating an ACTION POTENTIAL, or they can be inhibitory. Neurotransmitters are inactivated, often by enzymes, to ensure that impulses do not merge at the synapse.

About 50 neurotransmitters are known, including ACETYLCHOLINE, NORADRENALINE,

ADRENALINE, ENDORPHINS and ENCEPHALINS. A number of drugs exist that can mimic neurotransmitters, for example amphetamines mimic the action of noradrenaline; nicotine mimics natural neurotransmitters. Other drugs affect the release of neurotransmitters, for example caffeine increases release, whereas beta-blockers inhibit release.

neutral mutation *See* MUTATION.

neutrophil A type of GRANULOCYTE (blood cell) with cytoplasmic granules that do not take up acid or basic dyes. The majority of granulocytes are of this type.

newton The SI UNIT of force. One newton is defined as the force that will make a mass of one kilogram accelerate at one metre per second per second.

niacin *See* VITAMIN B.

nicotinamide adenine dinucleotide *See* NAD.

nicotinamide adenine dinucleotide phosphate *See* NADP.

nicotinic acid *See* VITAMIN B.

nitrate Any salt containing the nitrate ion, NO_3^-. Most nitrates are soluble and many are important fertilizers since the nitrate ion is an important source of fixed nitrogen (*see* NITROGEN FIXATION), which is needed by plans to make proteins and NUCLEIC ACIDS. Nitrates are readily leached from the soil by rain, causing high levels in rivers and lakes. This leads to EUTROPHICATION. *See also* NITROGEN CYCLE.

nitrification The process occurring in the soil by which ammonia (from urea, urine and the breakdown of protein by AMMONIFICATION) is oxidized by bacteria to form NITRATES. The free-living bacterium *Nitrosomonas* oxidizes ammonium ions (NH_4^+) to NITRITES (NO_2^-), which are toxic but quickly oxidized to nitrates (NO_3^-) by *Nitrobacter*. These processes release energy, which the bacteria use for their own respiratory processes.

Nitrification is reduced if the temperature or pH of the soil is low. Because nitrates are soluble, they can easily leach out of the soil, causing nitrogen deficiency. Artificial fertilizers are therefore often added to prevent growth limitation.

See also NITROGEN CYCLE, DENITRIFICATION.

nitrite Any salt containing the nitrite ion, NO_2^-. Nitrites are easily oxidized (*see* OXIDATION) to NITRATES. Nitrites are used as preservatives and as colouring agents in cured meats, such as bacon and sausages.

nitrogen (N) A colourless, odourless, tasteless gas. Nitrogen makes up 78 per cent of the atmosphere, but is chemically fairly unreactive.

Nitrogen is an essential element for life, being present in all proteins and nucleic acids. Some bacteria are able to 'fix' nitrogen from the air and incorporate it into the growth of certain plants (*see* NITROGEN CYCLE, NITROGEN FIXATION). NITRATES are often used as a fertilizer since this is a form of nitrogen that can be readily used by many organisms. Nitrogen deficiency in plants causes chlorosis (yellowing of the leaves) and stunted growth.

Nitrogen is obtained commercially from the fractional distillation of liquefied air and is used to manufacture AMMONIA. Liquid nitrogen is commonly used to cool objects to low temperatures, for example to store cells or tissue for subsequent culture, or eggs fertilized by *IN VITRO* FERTILIZATION for further use.

nitrogen cycle *(See also diagram on following page)* The circulation of nitrogen, mostly by living organisms, through the ECOSYSTEM. Nitrogen is an essential mineral for all organisms because it is used to make proteins and other organic compounds. Although the atmosphere is 78 per cent nitrogen, this cannot be readily used by most organisms (*see* NITROGEN FIXATION). Plants obtain nitrogen from NITRATES in the soil, by absorption through the roots, and convert them to proteins. The proteins are passed to HERBIVORES and CARNIVORES in the FOOD CHAIN, and nitrogen is eventually returned to the soil as excrement or when organisms die.

There are several important groups of bacteria involved in the processes of the nitrogen cycle. The free-living chemosynthetic (*see* CHEMOSYNTHESIS) bacteria *Nitrosomonas* and *Nitrobacter* are important in NITRIFICATION, which is the process by which ammonia is oxidized to form nitrates. Bacteria that can utilize atmospheric nitrogen by nitrogen fixation to make nitrogenous compounds are called nitrogen-fixing bacteria.

Anaerobic bacteria, such as *Pseudomonas denitrificans,* and *Thiobacillus denitrificans* are also important because they convert nitrates in the soil back to atmospheric nitrogen in the process of DENITRIFICATION. DECOMPOSERS (organisms capable of feeding on excrement and other dead organisms) are crucial in the nitrogen cycle for breaking down proteins,

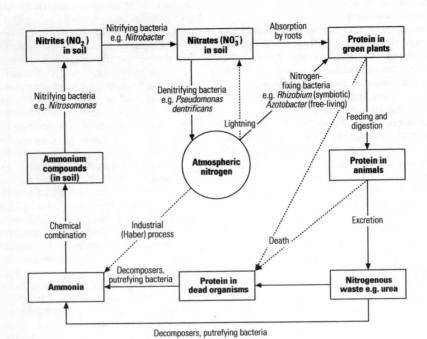

A summary of the nitrogen cycle.

amino acids and other nitrogenous compounds to form ammonia, ammonium ions (by AMMONIFICATION) and amines (by putrefaction), which are used by the nitrifying bacteria.

nitrogen fixation The process by which atmospheric nitrogen is converted to nitrogenous compounds by the action of nitrogen-fixing micro-organisms. These organisms can be free-living bacteria, such as *Azotobacter* and *Clostridium*, or CYANOBACTERIA, for example *Nostoc*. They convert atmospheric nitrogen to ammonia, which they use to make amino acids.

Nitrogen fixation is also carried out by symbiotic bacteria such as *Rhizobium* (*see* SYMBIOSIS) that live in specialized ROOT NODULES on the roots of leguminous plants, such as beans and peas, and a few non-leguminous plants. LEGUMES are therefore important crops for improving soil fertility.

See also NITROGEN CYCLE.

NK cell *See* NATURAL KILLER CELL.

NMR An abbreviation for NUCLEAR MAGNETIC RESONANCE.

node A swelling or lump. In zoology, an example is a LYMPH NODE; in botany, an example is a region on a plant stem from which leaves develop.

node of Ranvier One of several areas along a nerve AXON where the insulating MYELIN SHEATH is absent. The NERVE IMPULSE jumps from node to node along these axons and therefore travels faster than along a non-myelinated axon.

non-disjunction The failure of one or more pairs of CHROMOSOMES to separate at MEIOSIS. *See also* DOWN'S SYNDROME.

non-reducing sugar A sugar that cannot act as a REDUCING AGENT in solution, as indicated by a negative BENEDICT'S TEST or FEHLING'S TEST. *See also* CARBOHYDRATE.

non-renewable (*adj.*) A term used to describe FOSSIL FUELS and other energy sources that are being consumed at a rate that far exceeds the production of new reserves. *See also* RENEWABLE RESOURCE.

nonsense codon *See* STOP CODON.

nonsense mutation *See* MUTATION.

noradrenaline, *norepinephrine* A HORMONE and NEUROTRANSMITTER that is, like ADRENALINE, derived from the amino acid tyrosine. It is secreted by the MEDULLA of the ADRENAL GLAND and by modified NEURONES of the SYMPATHETIC NERVOUS SYSTEM. Noradrenaline maintains arousal in the brain, for example in response to external stress, dreaming and emotion.

norepinephrine *See* NORADRENALINE.

normal distribution, *Gaussian curve* In statistics, a bell-shaped curve obtained when the frequency distribution for a characteristic that shows continuous variation is plotted on a graph. An example is the height of individuals in a population; most individuals are of intermediate height, with a few at each extreme. *See also* AVERAGE.

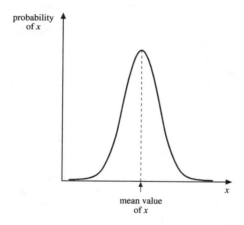

Normal distribution.

Northern blotting A technique similar to SOUTHERN BLOTTING except that it is used to detect RNA fragments. A DNA probe is used to hybridize with complementary RNA fragments on the filter.

nose The SENSE ORGAN for smell and an opening of the respiratory tract. There are numerous olfactory RECEPTORS detecting smell that are found in the roof of the nasal cavity within the MUCOUS MEMBRANE lining the whole cavity. This membrane moistens and warms the air entering the nose and traps dirt. Small hairs inside the nostrils also prevent the entry of foreign objects. A septum of cartilage divides the external part of the nose.

notochord A stiff, but flexible, rod of tissue that exists between the DIGESTIVE SYSTEM and CENTRAL NERVOUS SYSTEM of all CHORDATES at some stage in their life. In most vertebrates, the notochord occurs in the embryo and is replaced in adults by the VERTEBRAL COLUMN.

nucellus A mass of cells within the OVULE of a SEED PLANT. The nucellus is completely surrounded by the INTEGUMENT except for a small hole (MICROPYLE) at the tip. The cells of the nucellus differentiate and divide to form the EMBRYO SAC and the egg cell (the female GAMETE). The nucellus also provides nutrition for the developing ovule.

nuclear magnetic resonance (NMR) A technique widely used in organic chemistry to identify organic molecules and to determine their structures. The technique can also be used to examine living organs without destroying them, which has revolutionized the diagnosis of disease.

Some of the nuclei in a molecule possess the property of spin, which can be in one of two orientations. If a magnetic field is applied to the molecule these differences in spin cause a splitting of the nuclear energy levels. The molecule is then subjected to an additional weak, oscillating magnetic field. At a precise frequency the nuclear magnets resonate and it is this that is recorded and amplified.

The resonance frequencies of a particular element depend on its environment. Thus using NMR it is possible, for example, to detect three different types of hydrogen atoms in ethanol – those in CH_3, CH_2 and OH groups in ethanol.

See also MAGNETIC RESONANCE IMAGING.

nuclease A general term for an ENZYME that degrades NUCLEIC ACID. *See* RESTRICTION ENDONUCLEASE.

nucleic acid The complex organic acid present in the cells of all organisms that is responsible for their genetic make-up. The two types of nucleic acid are DNA and RNA, and each is made of long chains of NUCLEOTIDES.

nucleoid The region of a PROKARYOTIC cell that contains the genetic material in the form of CHROMOSOMES or PLASMIDS. In contrast to a NUCLEUS, a nucleoid is not bounded by a membrane.

nucleolus (*pl.* *nucleoli*) A spherical body within the nucleus of a non-dividing EUKARYOTIC cell.

They contain RNA and protein and are concerned with the synthesis of RIBOSOMES. There may be one or more nucleoli present. Nucleoli stain with basic dyes and their size reflects their level of activity.

nucleoside The organic base and PENTOSE sugar part of a NUCLEOTIDE; that is, a nucleotide without the PHOSPHATE group. In RNA, the sugar is RIBOSE and the common nucleosides are adenosine, guanosine, cytidine and uridine. In DNA, the sugar is DEOXYRIBOSE and the nucleosides found are the same as RNA except that uridine is replaced by thymidine.

nucleotide The constituent unit of the nucleic acids DNA and RNA, consists of an organic base, a PENTOSE sugar (RIBOSE $C_5H_{10}O_5$ or DEOXYRIBOSE $C_5H_{10}O_4$) and a PHOSPHATE group. There are five organic bases: adenine, guanine, cytosine, thymine and uracil. Adenine and guanine are PURINES, and have double rings (one with six sides, one with five). Cytosine, thymine and uracil are PYRIMIDINES, and have two single, six-sided rings. The organic bases are abbreviated to A, G, C, T and U, and the order in which they are placed in the nucleic acid strand contains the GENETIC CODE. The three components of a nucleotide join together. Links then form similarly between the sugar and phosphate groups of two or more nucleotides, to form dinucleotides or polynucleotides. Although the main role of nucleotides is in the formation of nucleic acids, they are also found in other molecules, such as AMP, ADP, ATP, NAD, NADP and FAD.

nucleus (*pl. nuclei*) A central dense body surrounded by a nuclear membrane or envelope, found in almost all EUKARYOTIC cells (human red blood cells have no nucleus). The nucleus contains the genetic material of the cell in the form of CHROMOSOMES within a liquid nuclear sap. The nucleus controls all the activities of the cell, including cell division by MITOSIS or MEIOSIS. The nuclear membrane is a double membrane containing many pores, which allow large molecules such as RNA to pass between the nucleus and the CYTOPLASM. One or more nucleoli (*see* NUCLEOLUS) are present, within the nucleus of a non-dividing cell concerned with the synthesis of RIBOSOMES.

nut A dry, INDEHISCENT (does not split open) fruit with a single seed surrounded by a hard woody wall (PERICARP). Examples are the hazelnut and chestnut. A nut is formed from more than one CARPEL but only one seed develops, and the others abort (*compare* ACHENE). The term nut is used to describe not only true nuts, such as hazelnuts and acorns, but also hard-shelled fruits, for example almonds and walnuts, which are really DRUPES, and seeds, for example Brazil nuts and peanuts.

Nutrasweet *See* ASPARTAME.

nutrient A chemical substance required by plants and animals for their normal growth and development. In animals, nutrients are largely obtained through the diet and include CARBOHYDRATES, PROTEINS, LIPIDS, VITAMINS and certain minerals. Plants obtain their nutrients through the soil and air. *See also* MACRONUTRIENT, MICRONUTRIENT.

nutrition The processes by which living things take in food and use it. There are two types of nutrition: AUTOTROPHIC NUTRITION and heterotrophic nutrition (*see* HETEROTROPH).

O

obesity *See* DIET.

objective *See* OBJECT LENS.

object lens, *objective* In a microscope, the lens that collects light from the object being viewed.

obligate parasite A PARASITE that cannot live independently of its host.

oesophagus The tube carrying food from the MOUTH to the STOMACH. It is a muscular tube, 23 cm long in humans, that begins in the lower part of the PHARYNX. The oesophagus contains glands in its lining that secrete MUCUS to lubricate the food. *See also* DIGESTIVE SYSTEM.

oestrogen A STEROID HORMONE produced by the GRAAFIAN FOLLICLE in the OVARY of mammals. Oestrogen actually refers to a group of hormones, including synthetic ones; the main synthetic oestrogen used by humans is oestradiol. Oestrogens cause the development of secondary sexual characteristics, such as breasts and fat deposition, help prepare the UTERUS for pregnancy and maintain the pregnancy if it follows. Oestrogens stimulate production of LUTEINIZING HORMONE and repair the uterine lining following MENSTRUATION if pregnancy does not follow.

Synthetic oestrogens are a major component of the contraceptive PILL, and are thought to be responsible for the side-effects. It has been suggested recently that environmental levels of oestrogens, for example in plastics, are too high and could be responsible for feminization of males and a possible reduction in male fertility.

oestrous (*adj.*) Describing OESTRUS or the cycle during which it occurs. *See* OESTROUS CYCLE.

oestrous cycle The hormonal cycle occurring in many non-human mammals that is equivalent to the MENSTRUAL CYCLE of humans. There are four phases. During pro-oestrus (follicular phase), GRAAFIAN FOLLICLES develop in the OVARY and secrete OESTROGENS. The next phase is oestrus, during which OVULATION occurs and the female's sexual desire is heightened so mating is most likely to occur (this is termed 'on heat'). In metoestrus (luteal phase), the CORPUS LUTEUM develops, and in dioestrus, PROGESTERONE is secreted by the corpus luteum to prepare the UTERUS for IMPLANTATION. There may be long periods between consecutive oestrous phases with several each year (polyoestrus) or a single oestrous period each year (monoestrus). The oestrous cycle often occurs at a time that will favour the survival of the offspring.

oestrus (*n.*) The phase in the OESTROUS CYCLE during which OVULATION occurs.

oil One of many types of naturally occurring HYDROCARBONS, which can be solid (FATS, WAXES) or liquid. Oils are flammable and usually insoluble in water. Mineral oils are those obtained from refining petroleum and are used as fuels and lubricants. Essential oils are those obtained from plants which possess pleasant odours used in perfumes and flavourings. Fixed oils are LIPIDS found in animals and plants, such as fish and nuts, and are used, for example, in foods, soaps and paints.

oleic acid An unsaturated FATTY ACID with one double COVALENT BOND. Oleic acid is a major constituent of animal and plant fats. It is found, for example, in lard, groundnut oil, soya-bean oil and butterfat.

olfaction The sense of smell or the action of smelling.

olfactory (*adj.*) Relating to smell.

olfactory receptor A RECEPTOR cell found in olfactory organs, such as the nose, that is associated with the detection of smell. *See* SENSE ORGAN.

Oligochaeta A class of the phylum ANNELIDA that includes the earthworm (*Lumbricus*). Earthworms move by contraction and relaxation of muscles with the aid of chaetae (bristles), by which they anchor themselves. They have no distinct head and are HERMAPHRODITE. The earthworm is of particular economic importance to humans because it contributes to soil formation and improvement, by improving aeration and drainage, mixing vegetation and soil, and neutralizing acid soil with their gut secretions. *Compare* POLYCHAETA and HIRUDINEA.

oligodendrocyte A type of GLIAL CELL within the CENTRAL NERVOUS SYSTEM that deposits the MYELIN SHEATH.

ommatidium (*pl.* **ommatidia**) One of many units of the compound eye of an insect. Each ommatidium consists of a CORNEA, a LENS, a CONE and a group of RECEPTOR cells with light-sensitive pigments linked to a NERVE FIBRE.

omnivore An animal, for example humans, apes and ants, that eats both plants and animal meat in its diet. Omnivores have gut bacteria that help digestion of a range of substances.

oncogene A tumour-inducing GENE identified in DNA that arises from MUTATIONS in normal genes. Cellular oncogenes are sometimes called proto-oncogenes. Oncogenes usually originate from a host cell and are carried by a virus, often a RETROVIRUS. They may undergo mutation and cause cancer in host cells when they become infected by the virus. Most viruses that are known to be capable of transforming a normal cell to a tumour cell are found to have oncogenes that are inserted into the host cell DNA and induce abnormal growth and division (in several different ways). Viral oncogenes are generally associated with cancer, but chemicals and radiation can activate cellular oncogenes to cause cancer as well.

ontogeny The whole course of an organism's development, from fertilized egg to maturity. *Compare* PHYLOGENY.

oocyte In animals, an immature female GAMETE that gives rise to an OVUM. *See* OVARY.

oogenesis In animals, the process of ova production. *See* OVARY.

oogonium In animals, an immature female GAMETE that gives rise to OOCYTES. *See* OVARY.

Oomycete A member of the phylum OOMYCOTA.

Oomycota A phylum of the kingdom PROTOCTISTA. consisting of water moulds and related organisms. Oomycetes have HYPHAE, but are distinct from fungi, and they FLAGELLA. Oomycetes can be SAPROTROPHS or PARASITES and they can reproduce asexually (by ZOOSPORES) or sexually. The best example is *Phytophthora infestans,* the cause of potato blight and the Irish famine of 1845.

operculum 1. In botany, a covering over the SPORANGIUM of some fungi and mosses that opens to allow the release of mature SPORES.

2. In zoology, a bony flap covering the GILL slits of bony fish, which protects the gills and aids their ventilation.

operon A group of adjacent GENES on a CHROMOSOME that act together to produce a POLYPEPTIDE or ENZYME. Operons were discovered in bacteria by the French biochemists Jacob (1920–) and Monod (1910–76) in 1961.

An operon is made up of structural genes (CISTRONS) that are responsible for the production of the polypeptide, and operator genes that regulate the switching on and off of the structural genes. Another gene, further away from the group, called the regulator gene, codes for a protein called the repressor, which can bind to the operator gene and prevent it switching on the structural genes. Operons are found mostly in PROKARYOTES and are less common in EUKARYOTES, where regulatory mechanisms are more complex.

See also GENE EXPRESSION.

opiate A term referring to drugs derived from OPIUM. The term opiate is also used to refer to natural pain-relieving chemicals produced by the body, such as ENDORPHINS and ENCEPHALINS.

opium A drug that is obtained from unripe seeds of the opium poppy *Papaver somniferum.* It contains a number of ALKALOIDS, including the pain-killing substances morphine and codeine, and also a highly poisonous substance called thebaine. Morphine is a strong but addictive pain-killing drug and heroin is an even stronger synthetic derivative of morphine.

opsonin A protein, often an ANTIBODY, that coats the surface of foreign substances, making them more vulnerable to ingestion by MACROPHAGES.

optical fibre A very fine, pure glass fibre through which light can be transmitted with very little escaping through the sidewalls. The light is reflected to carry information or an image from one end to the other. Optical fibres are used in ENDOSCOPES to examine inaccessible parts of the body.

optic nerve A large nerve between the eye and the brain. It carries information from the sensory cells in the RETINA to the visual centres in the brain. The optic nerve develops as part of the brain wall.

order One of the subdivisions of CLASS in the CLASSIFICATION of organisms. The names of orders for birds and fish end in '*-formes*', for mammals, amphibians and reptiles '*-a*' and for fungi and plants '*-ales*'. Orders consist of groups of FAMILIES.

organ A structural and functional unit of an animal or plant. An organ consists of more than one type of TISSUE and is co-ordinated to perform usually one main function. Organs often work together in organ systems, such as the DIGESTIVE SYSTEM, which includes organs such as the STOMACH, LIVER and PANCREAS. Some organs are also GLANDS and can therefore belong to more than one organ system, for example the pancreas is also part of the ENDOCRINE SYSTEM. Examples of organs in plants include roots, stems and leaves.

organelle A discrete structure found in living cells that is specialized for a particular function. Organelles include the NUCLEUS, which contains the DNA, and cytoplasmic organelles such as MITOCHONDRIA, RIBOSOMES, CHLOROPLASTS, LYSOSOMES, ENDOPLASMIC RETICULUM, and GOLGI APPARATUS. *See also* PLASTID.

organic A term that was used in the late eighteenth century to refer to compounds obtained from living material in contrast to those derived from minerals (which were termed inorganic). Now 'organic' refers to compounds containing both carbon and hydrogen. Most organic compounds also contain other elements, such as oxygen, nitrogen, sulphur and phosphorus.

Organic compounds are more numerous than inorganic compounds and form the basis of life. Although many organic compounds are made only by living organisms, such as PROTEINS and CARBOHYDRATES, it is now possible to manufacture many synthetically. Many organic compounds are derived from petroleum, the remains of microscopic marine organisms.

Examples of organic compounds include ALCOHOLS, ALDEHYDES, AMINO ACIDS, CARBOHYDRATES, ESTERS, KETONES and PROTEINS.

organism A living individual plant or animal. *See also* MICRO-ORGANISM.

organ of Corti *See* COCHLEA.

organogenesis *See* EMBRYONIC DEVELOPMENT.

origin of life The beginning of life on Earth about 4,000 million years ago. There are a number of theories regarding how life began. The spontaneous generation theory (abiogenesis) suggests that life arose from non-living matter on a number of separate occasions. This does not seem to occur now, but could have played a role in the origin of life. The

creation theory suggests that God created the Earth and all its life forms. The cosmozoan theory suggests that life began elsewhere in the universe and later arrived on earth. The biochemical EVOLUTION theory suggests that atoms combined into simple molecules, which subsequently combined into complex ones and then into cells.

The biochemical EVOLUTION theory is the one most widely accepted by scientists today. It is thought that simple molecules, such as amino acids and MONOSACCHARIDES, were formed by the combination of gases present in the Earth's early atmosphere (carbon dioxide, methane, hydrogen, ammonia and water). The energy for this came from ultraviolet radiation from the sun and electrical energy from lightning storms. Such molecules probably floated on the surface of the oceans as a 'primeval soup', and when concentrated enough formed POLYMERS, such as starch, and eventually developed into cells.

ornithine ($H_2N(CH_2)_3CH(NH_2)COOH$) An AMINO ACID that is an intermediate in the UREA CYCLE and in the synthesis of ARGININE. It is not a constituent of proteins.

ornithine cycle *See* UREA CYCLE.

ornithology The study of birds.

ornithophily POLLINATION of flowers by birds. Flowers adapted for ornithophily, for example tropical plants. They are large and brightly coloured (red or orange) with large amounts of NECTAR. They are often unscented, because birds do not respond well to smell.

osmoregulation The maintenance of a constant level of water and salt concentration in living organisms. Osmoregulation is necessary for the normal functioning of vital body processes.

In mammals, the KIDNEY controls the balance of water and salts by adjusting water reabsorption before URINE production. In animals living in dry climates where water has to be conserved, the kidney is specially adapted by having a longer LOOP OF HENLE along which water and salts are reabsorbed. This reabsorption is controlled by hormones, mostly of the ADRENAL GLAND. In birds, the kidney is again the main organ of excretion, but as they need to conserve water they excrete URIC ACID as their main nitrogenous waste, which uses less water than urine to eliminate.

In freshwater species, osmoregulation must counteract the tendency for water to enter by OSMOSIS, which would make them gain excess fluid. This is achieved by having many large glomeruli (*see* GLOMERULUS) in their kidneys, so they produce large amounts of dilute urine. Salts are selectively reabsorbed. In marine (salt water) species, the tendency is to lose too much water by osmosis. This is overcome by having kidneys with few glomeruli and short tubules, which allows greater excretion of salts but prevents excess loss of water. These species only produce small amounts of urine.

In protozoans (*see* PROTOZOA) osmoregulation is by a CONTRACTILE VACUOLE.

osmosis The movement of a liquid solvent, usually water, from a less concentrated SOLUTION to a more concentrated one through a SEMIPERMEABLE MEMBRANE (permeable in both directions to water but varying in permeability to the SOLUTE) until the two concentrations are equal, or isotonic.

Osmosis is a passive process requiring no energy. If external pressure is applied to the more concentrated solution, osmosis is prevented. This provides a measure of the OSMOTIC PRESSURE of the more concentrated solution, which is measured in pascals (Pa). The osmotic pressure is greater the more concentrated the solution. In animal cells, the more dilute solution is called hypotonic, and the more concentrated solution is called hypertonic. These terms are no longer used with reference to plant cells. The passage of water by osmosis will occur across a semipermeable membrane from any solution of weaker osmotic pressure to one of higher osmotic pressure, regardless of whether the dissolved substance on both sides of the membrane is the same or not.

Osmosis can also be explained in terms of WATER POTENTIAL, which is now recommended for the use of plant studies. In this case, water moves from an area of high (less negative) water potential to an area of low (more negative) water potential.

Osmosis is vital in controlling the distribution of water in living organisms, for example in the transport of water from the roots up to the stems of plants, and in maintaining a constant water balance (OSMOREGULATION) to prevent the concentration of salts becoming too high or too low and affecting vital functions. Fish have a protective mechanism to counteract osmosis. Without such a mechanism, salt water fish would lose fluid by osmosis and freshwater fish would gain excess fluid.

See also REVERSE OSMOSIS.

osmotic potential *See* OSMOTIC PRESSURE.

osmotic pressure, *osmotic potential* The pressure difference that can occur across a SEMIPERMEABLE MEMBRANE as a result of OSMOSIS. It is defined as the pressure that needs to be applied across a semipermeable membrane to prevent osmosis. This term is no longer used with regard to solutions in plant cells. *See also* WATER POTENTIAL.

ossification The process by which BONE develops from CARTILAGE. Specialized cells called osteoblasts secrete EXTRACELLULAR MATRIX onto the surface of existing cartilage, and calcium phosphate crystals deposit within the matrix to form bone. Osteoblasts that become included within the bone structure during its development become osteocytes (bone cells) and cease to divide and form bone matrix.

Osteichthyes A class of VERTEBRATES consisting of the bony fish (which have a bony skeleton). Examples include the salmon, herring and stickleback. Osteichthyes forms the largest class in the phylum CHORDATA. *Compare* CHONDRICHTHYES.

osteoarthritis A degenerative joint disease resulting from destruction of the joint CARTILAGE. It is associated with much pain and reduced mobility of the affected joints. Treatment (but not cure) is with anti-inflammatory drugs such as CORTICOSTEROIDS. *See also* RHEUMATOID ARTHRITIS.

osteoblast A specialized cell responsible for the formation of BONE by the process of OSSIFICATION.

osteoclast A multinucleate cell in BONE that allows remodelling of bone shape during growth by breaking down the calcified matrix under the regulation of PARATHORMONE.

osteocyte A bone cell. *See* OSSIFICATION.

osteoporosis A human condition in which bones become brittle and weak and are easily broken. Osteoporosis occurs in elderly people, particularly in women past the MENOPAUSE, and is possibly linked to a drop in OESTROGEN levels.

otolith In the MACULA of the inner EAR, a deposit of calcium carbonate embedded in a jelly-like substance. There are many such deposits within the gelatinous mass and they move in response to gravity, thereby pulling on the gelatinous mass. This movement displaces sensory hairs that are also embedded in the gelatinous layer, and a message is then transmitted to the brain so that it is able to respond to vertical and lateral movements of the head.

outbreeding The mating between unrelated or distantly related individuals of a species. Outbreeding populations show greater genetic variability than INBREEDING populations, which is preferable since harmful GENES are likely to be RECESSIVE and masked by dominant ALLELES. Outbreeding populations have a greater potential for adaptation to environmental changes than do inbreeding populations.

oval window A membrane-covered opening between the middle EAR and the inner ear, which detects vibrations of the EARDRUM. *See also* EAR OSSICLES, COCHLEA.

ovary 1. In animals, the female GONAD that produces the OVUM. In humans, there are two ovaries, about 25 by 35 mm, in the lower abdomen, close to the FALLOPIAN TUBES. Each month, a mature ovum is released from the ovary, a process called OVULATION, and is either fertilized (*see* FERTILIZATION) or lost as part of the menstrual flow (*see* MENSTRUAL CYCLE).

The outer layer of the ovary contains germinal EPITHELIAL cells that begin to divide to form ova while in the FOETUS, before birth. The process of ova production is called oogenesis, and begins with oogonia dividing by MITOSIS to form primary oocytes, which then divide by MEIOSIS. Oogenesis stops in the newborn at the primary oocyte stage. At this stage many oocytes are organized in structures called primary follicles. The newborn female contains all the genetic information it needs to provide for its offspring.

At puberty, a hormone produced by the PITUITARY GLAND, called FOLLICLE-STIMULATING HORMONE, restarts oocyte development within the follicle. At any one time there will be oocytes and follicles at different stages of maturity in the ovary. A GRAAFIAN FOLLICLE is the largest, mature stage and contains the secondary oocyte (produced by meiotic division of a primary oocyte), which is released as the mature ovum. After the ovum has been released, the Graafian follicle remains as a CORPUS LUTEUM that secretes the steroid hormone PROGESTERONE. The ovaries also produce the steroid hormone OESTROGEN.

2. For plant ovaries, *see* CARPEL.

overfishing Fishing at a rate that exceeds the sustainable yield and causes the depletion of fishing stock. Overfishing is a result of modern fishing methods, which use huge factory ships and specialized equipment to locate shoals of fish. In the North Sea, cod and haddock have suffered through overfishing and herring are almost extinct. Overfishing is also a serious problem in the developing world because stocks used by local people have been depleted.

There are now restrictions on fishing, although these are a cause of much controversy between countries. The mesh size of fishing nets is regulated to allow smaller, immature fish to fall through so that they can reach sexual maturity and breed to replenish the stocks.

oviduct *See* FALLOPIAN TUBE.

ovipary Reproduction by which the female lays eggs that develop outside her body. Ovipary is the most common form of reproduction. *See also* OVOVIVIPARY, VIVIPARY.

ovovivipary Reproduction by which fertilized eggs develop within the female but gain no nutrition from her. Ovovivipary occurs in fish, reptiles and many insects. *See also* OVIPARY, VIVIPARY.

ovulation In female animals, the release of a mature OVUM (egg cell) from the OVARY. Ovulation is under hormonal control as part of the MENSTRUAL CYCLE, and occurs monthly in women from puberty until the MENOPAUSE.

ovule In SEED PLANTS, the structure containing the female GAMETE that develops into the SEED after fertilization (*see* DOUBLE FERTILIZATION). The ovule consists of an EMBRYO SAC containing several nuclei, one of which is the egg nucleus or female gamete, surrounded by nutritive tissue called the NUCELLUS. There are one or two protective layers called the INTEGUMENTS that develop into the TESTA after fertilization. In flowering plants (ANGIOSPERMS) the ovule is within an ovary (*see* CARPEL) fixed to the wall by a stalk, but in conifers (CONIFEROPHYTA) it is not enclosed by an ovary and is found on an ovule-bearing scale within a CONE.

ovum (*pl.* **ova**) The female GAMETE before FERTILIZATION. In higher organisms, such as humans, it is called an egg and is produced in the OVARY. It is larger than the male SPERM and non-motile. It consists of a nucleus, a large CYTOPLASM containing yolk grains as a food source, and a thick outer membrane surrounded by cumulus cells, which aid movement of the ovum toward the UTERUS and provide nutrients. Once fertilized, the ovum is called a ZYGOTE. In some species, such as birds, the egg is covered by a shell.

In plants, the ovum is called an egg cell (see OVULE).

oxaloacetic acid A colourless, crystalline acid that combines with ACETYL COENZYME A at the beginning and end of the KREBS CYCLE. It is also produced in C_4 PLANTS in the light-independent stage of PHOTOSYNTHESIS. See also CALVIN CYCLE.

oxidase See OXIDOREDUCTASE.

oxidation In simplest terms, the addition of oxygen to a substance. The term also refers to the removal of hydrogen from a substance or the removal of ELECTRONS to form a positive ION. It is the opposite of REDUCTION.

oxidative phosphorylation The process by which ATP is formed through AEROBIC RESPIRATION, by the transfer of electrons in the ELECTRON TRANSPORT SYSTEM, which results in the oxidation of hydrogen atoms. Oxidative phosphorylation occurs in the MITOCHONDRIA, within particles found on the folded inner layer of the double membrane.

oxidizing agent A material that readily brings about an OXIDATION, itself being reduced (see REDUCTION).

oxidoreductase One of a group of ENZYMES that transfer oxygen and hydrogen atoms between substances, such as dehydrogenases, oxidases.

oxygen (O) A colourless, odourless gas, vital for life. Oxygen makes up 20 per cent of the atmosphere, mostly as the molecule O_2 though some occurs as OZONE, O_3. It is the most common element in the Earth's atmosphere and is very reactive, combining with most elements.

Oxygen is vital to organisms that carry out AEROBIC RESPIRATION. These organisms absorb oxygen directly from the atmosphere or make use of dissolved oxygen in water. Oxygen also supports combustion – many materials will burn very rapidly in pure oxygen.

oxygen debt The oxygen needed to break down LACTIC ACID produced by a fatigued muscle during anaerobic RESPIRATION. During vigorous exercise, if the lungs cannot supply enough oxygen to a muscle, the cells will switch from aerobic to anaerobic respiration

in order to produce energy. This results in a build up of lactic acid, which causes cramp. After resting, the body will use extra oxygen (obtained by the automatic panting response) to break down the lactic acid, so paying off the oxygen debt. See also FERMENTATION.

oxygen dissociation curve An S-shaped curve obtained when the percentage saturation of a RESPIRATORY PIGMENT such as HAEMOGLOBIN is plotted against the partial pressure of oxygen (or oxygen tension), which is a measure of oxygen concentration in the surroundings. The haemoglobin curve rises sharply, indicating a high affinity for oxygen. A small increase in oxygen tension (such as in the lungs) causes haemoglobin to be rapidly saturated with oxygen. Conversely, a small drop in oxygen tension (such as in body tissues using oxygen at a high rate) results in a rapid dissociation of oxygen from haemoglobin.

Different pigments have different dissociation curves. For example, the curve for foetal haemoglobin is displaced to the left of that for adult haemoglobin, indicating a greater affinity of the former for oxygen. This enables the foetus to obtain oxygen from the mother's haemoglobin. MYOGLOBIN is also displaced to the left of adult haemoglobin, which means it has a greater affinity for oxygen and is thus able to store oxygen during periods of strenuous exertion. In the presence of a high concentration of carbon dioxide, the dissociation curve for haemoglobin is displaced to the right, indicating a reduced affinity for oxygen. Thus oxygen is released in the presence of high concentrations of carbon dioxide.

See also BOHR EFFECT.

oxyhaemoglobin HAEMOGLOBIN combined with oxygen. It occurs in regions of high oxygen concentration, such as the lungs or GILLS. Oxygen is transported in the blood in this form to the body tissues, and is easily released where oxygen is at low concentration. This dissociation of oxyhaemoglobin is encouraged by the presence of carbon dioxide. See also BOHR EFFECT.

oxytocin A PEPTIDE hormone, produced by the PITUITARY GLAND of birds and mammals, that is involved in the contraction of the smooth muscle of the UTERUS during birth and, in mammals, the muscular contractions causing expression of milk from the MAMMARY GLANDS during LACTATION.

ozone (O_3) A colourless gas with a distinctive odour. Ozone can be made by the action of an electric discharge on oxygen:

$$3O_2 \rightarrow 2O_3$$

Ozone is produced in the upper atmosphere and plays an important part in protecting the Earth's surface from ultraviolet radiation. It is a powerful OXIDIZING AGENT. *See also* OZONE LAYER.

ozone hole An area of lower than usual ozone concentration in the OZONE LAYER above the Earth's poles.

ozone layer A protective layer consisting of the gas ozone, O_3, 15–40 km above the Earth's surface. It is is formed by the effect of ultraviolet (UV) radiation on oxygen molecules. UV light splits oxygen (O_2) molecules into two atoms, one of which then combines with oxygen to create ozone.

The ozone layer prevents harmful UV radiation reaching the Earth's surface, but in recent years it has become clear that the layer is being damaged by human activities. *See also* CHLOROFLUOROCARBON, GREENHOUSE EFFECT.

PQ

pacemaker, *sinoatrial node, SA node* In vertebrates, a group of muscle cells in the wall of the HEART that controls the basic rate of contractions of the heart. The pacemaker is found in the wall of the right ATRIUM. The cells of the pacemaker spontaneously and rhythmically contract, and this rate can be adjusted according to demand by the AUTONOMIC NERVOUS SYSTEM.

An artificial pacemaker can be implanted in a person whose heart beats irregularly. This operates by stimulating the heart muscles with minute electric shocks at regular intervals to restore the normal heartbeat.

See also PURKINJE FIBRES.

PAGE See POLYACRYLAMIDE GEL ELECTROPHORESIS.

palaeontology The study of FOSSILS. Palaeontology looks at the structure and evolution of extinct organisms and their environment, as can be determined from their fossil remains. It also has applications in geology, as fossils help date the rock strata in which they are found and help identify potential sources of FOSSIL FUELS.

palisade mesophyll *See* MESOPHYLL.

palynology *See* POLLEN ANALYSIS.

pancreas A GLAND in all vertebrates (except some fish) that is part of the DIGESTIVE SYSTEM and located near the DUODENUM. In humans the pancreas is behind and below the stomach and is about 18 cm long. The pancreas has two major roles, as both an EXOCRINE GLAND and an ENDOCRINE GLAND.

As an endocrine gland, specialized groups of cells within the pancreas called the ISLETS OF LANGERHANS secrete the hormones INSULIN and GLUCAGON directly into the blood, which act antagonistically to regulate blood sugar levels.

As an exocrine gland, the pancreas produces an alkaline PANCREATIC JUICE, when stimulated by the hormones SECRETIN and CHOLECYSTOKININ, which is secreted into the duodenum where it aids digestion.

pancreatic juice A neutral secretion produced by the PANCREAS, when stimulated by the hormones SECRETIN and CHOLECYSTOKININ made by the SMALL INTESTINE. Pancreatic juice is secreted into the DUODENUM, where it aids digestion. The juice contains AMYLASES, PROTEASES, LIPASES and NUCLEASES. In order for these enzymes to function correctly, mineral ions (such as sodium hydrogencarbonate) present in the pancreatic juice neutralize the acid mix coming from the stomach. The enzymes are made by acinar cells in the tubules of the pancreas.

pandemic (*adj.*) Describing a disease that occurs over a wide geographical area.

Paneth cell *See* CRYPT OF LIEBERKÜHN.

pantothenic acid *See* VITAMIN B.

paper chromatography A CHROMATOGRAPHY technique in which the stationary phase is absorbent paper. A spot of the mixture to be separated is placed near the base of a paper strip, the bottom edge of which is placed in a solvent. The solvent moves up the paper by the CAPILLARY EFFECT, and the different components in the mixture are carried along at different rates. Colourless components may be made visible by 'developing' the CHROMATOGRAM – spraying it with some material that reacts to make the components visible. Colourless components that fluoresce may be viewed in ultraviolet light. Paper chromatography is used for analysing mixtures rather than for purifying particular components.

paracetamol, *4-acetamidophenol* ($C_8H_9NO_2$) A popular pain-killing drug that also reduces fever. It is less damaging to the stomach than ASPIRIN, but liver damage can be caused by excess doses.

parapodium In POLYCHAETA, any of the paired fleshy appendages with numerous chaetae (bristles), used for locomotion.

parasite An organism living on another organism (the host) and dependent on it for nourishment at the expense of the host. A parasitic organism must spend some time on the host (so a fly bite is not a parasitic relationship). Parasites that live inside the host, for example the liver tapeworm, are endoparasites, and parasites that live outside the host, for example fleas and lice, are ectoparasites.

Obligate parasites cannot survive without their host, while facultative parasites can, and partial parasites, for example mistletoe, can photosynthesize and also be parasitic. It is crucial for the parasite to be able to reproduce in a host. The host a parasite reproduces in (or lives in when it becomes sexually mature) is a primary (definitive) host. A secondary (intermediate) host is one in which the parasite does not reproduce but it used for some stages of its development. Parasites often have complex life cycles with one or two hosts interspersed with periods of free-living. The host can mount an IMMUNE RESPONSE to a parasite, but parasites are very resistant to immune attacks.

Examples of parasites include *Phytophthora infestans*, which causes potato blight (causing serious crop loss), the malarial parasite *Plasmodium* (*see* MALARIA) and parasitic flatworms (tapeworms and flukes). Liverfluke causes epidemics of 'liver rot' in sheep and cattle. The pork tapeworm *Taenia solium* uses the pig as its intermediate host and then attaches to the intestinal mucous membrane of humans as its primary host. Its eggs are passed out in the host's faeces. Parasites cause a great deal of damage, as diseases in humans and domestic animals and as a cause of crop damage.

parasitism An association between two different species in which one partner (the PARASITE) gains considerably at the expense of the other (the host). Parasitism can be considered as a variation of SYMBIOSIS.

parasympathetic nervous system Part of the AUTONOMIC NERVOUS SYSTEM that is important when the body is resting. It slows down the heart rate, decreases blood pressure and stimulates the digestive system. The parasympathetic nervous system usually opposes the SYMPATHETIC NERVOUS SYSTEM.

parathormone, *parathyroid hormone* A hormone produced by the PARATHYROID GLANDS. It regulates the level of calcium in the blood to ensure that it is high enough to permit normal muscle and nerve activity. It does this in three ways: by reducing calcium excretion by the kidneys; by releasing more calcium from bone (by increasing breakdown of the bone matrix by OSTEOCLASTS); and by converting VITAMIN D to its active form, which increases calcium absorption in the intestine. Another hormone called CALCITONIN works antagonistically to parathormone; that is, it reduces levels of calcium ions in the blood and together they maintain the correct balance.

parathyroid gland Either one of a pair of ENDOCRINE GLANDS found near to or embedded within the THYROID GLAND. Parathyroid glands produce PARATHORMONE, which regulates the level of calcium in the blood to ensure that it is high enough to permit normal muscle and nerve activity. Another hormone called CALCITONIN works antagonistically with parathormone – calcitonin reduces levels of calcium ions in the blood, and together they maintain the correct balance.

parathyroid hormone *See* PARATHORMONE.

parenchyma A simple (one cell-type only) plant tissue, consisting of loosely packed, almost spherical, cells with thin CELLULOSE walls that provide the main packing tissue within a plant. Parenchyma cells may also store food, and intercellular spaces allow gas exchange to occur. Photosynthetic parenchyma is called CHLORENCHYMA and is found in some stems and the MESOPHYLL of leaves. EPIDERMIS is a specialized parenchyma. *See also* GUARD CELL, STOMA.

parotid gland A large SALIVARY GLAND.

parthenogenesis The process by which a female egg (OVUM) develops without being fertilized by a male GAMETE. It is a modified form of SEXUAL REPRODUCTION, although because the offspring are from a single parent it is ASEXUAL REPRODUCTION. In some cases, fertilization with male sperm is the stimulus for parthenogenesis, even though the male chromosomes are not taken into the nucleus. Parthenogenesis can be induced artificially, for example in rabbits, by stimulating the egg. Some plants, for example dandelions, and certain fish reproduce by parthenogenesis, and others use it at some stage during their life cycle, for example aphids use it in the summer to build up large population numbers. In the honey bee, unfertilized eggs develop into male bees and fertilized eggs develop into female bees.

partial parasite An organism that is capable of photosynthesis but is also parasitic. *See* PARASITE.

parturition In mammals, the birth of a FOETUS at the end of a full-term PREGNANCY. The precise control of the onset of labour and the subsequent birth is not fully understood, but the foetus seems to play a role by stimulating release of PROSTAGLANDIN from the UTERUS, which causes it to contract. Synthetic prostaglandins can be used to induce labour. The PEPTIDE hormone OXYTOCIN also plays a part in parturition.

passive immunity *See* IMMUNITY.

pasteurization A process used to kill some disease-causing bacteria present in food without spoiling the food. The process was discovered by Louis Pasteur (1822–1895) in the 1850s. He first used it on wine and beer by heating it to slow down the multiplication of bacteria and hence the souring process. The most well-known pasteurized food product is pasteurized milk, which is heated to 72°C for 15 seconds and rapidly cooled to 10°C. This kills the organisms causing tuberculosis, typhoid, diphtheria and dysentery. Some beneficial bacteria are also killed, which reduces the nutritional value of milk.

pathogen A disease-causing agent, usually a MICRO-ORGANISM.

peat A brownish compact mass of partly decayed vegetation, saturated with water, that has accumulated in wetland areas. Peat is used as a fertilizer and, when dried, as fuel.

pectic substance One of a group of POLYSACCHARIDES that are important in plant CELL WALLS and the MIDDLE LAMELLA between adjacent cell walls. In the latter, pectic substances are present in the form of calcium pectate. Another example is pectin, which is used in jam making since it forms a gel with sucrose.

pectoral fin Either one of a pair of FINS that are attached to the shoulder girdle of a FISH. They are for steering and breaking.

pectoral girdle, *shoulder girdle* In vertebrates, the skeletal support, within the THORAX, for the trunk that consists of the bones and muscles necessary for attachment of the upper limbs or fins. It is attached to the sternum (breast bone) at the front of the chest. The scapula (shoulder blade) and clavicle (collar bone) form part of the pectoral girdle.

pedicel In botany, a small stalk of an individual flower of an INFLORESCENCE.

peduncle In botany, the main stalk of an INFLORESCENCE, or a stalk bearing a single flower. It contains XYLEM and PHLOEM to provide nourishment for the flower.

pelagic (*adj.*) Describing animals and plants that live in the mass of water of the sea or a lake, rather than the bottom of the water (benthos). Pelagic organisms can be divided into PLANKTON and NEKTON.

Pelecypoda A class of the phylum MOLLUSCA (the bivalves), including mussels. Bivalves have shells that are in two halves and they have an indistinct head with no tentacles. They are filter feeders that trap organic matter from the water around them in CILIA located in GILLS.

pellicle The outer layer of protein that protects and maintains the shape of certain unicellular organisms, such as *Euglena*, of the phylum EUGLENOPHYTA.

pelvic fin Either one of a pair of FINS that are attached to the pelvic (hip) girdle of a FISH. They are for steering and breaking.

pelvic girdle, *hip girdle* In vertebrates, a set of bones in the PELVIS providing support and a site for muscle attachment to enable movement of the legs, hindlimbs or back fins.

pelvis In vertebrates, the lower region of the ABDOMEN that is bounded by a set of bones called the PELVIC GIRDLE to which the lower limbs and the muscles needed for their movement are attached.

penicillin The first ANTIBIOTIC, discovered in 1929 by Alexander Fleming (1881–1955), that now forms a family of antibiotics obtained from moulds of the genus *Penicillium*. It can also be produced synthetically.

penis The male sex organ, used to insert sperm into the female VAGINA during sexual intercourse (*see* SEXUAL REPRODUCTION). The penis contains erectile tissue and a good blood supply that allows it to remain stiff when filled with blood (upon sexual stimulation). In most mammals (but not humans), the penis is stiffened by a bone. The penis itself consists of spongy tissue covered by an elastic skin and the end is expanded to form the sensitive region (the glans penis), which is protected by a retractable skin called the foreskin.

pentadactyl limb The basic limb found in mammals, birds, reptiles and amphibians, consisting of an upper part with one bone, a lower part with two bones and a hand or foot with five digits (fingers or toes). This basic pattern has been modified by loss or fusion of bones to suit the lifestyle of different vertebrate species; sometimes it has been greatly altered, for example in birds and bats for flying and in whales for swimming.

pentose A MONOSACCHARIDE containing five carbon atoms in the molecule. Examples are RIBOSE and DEOXYRIBOSE, the sugar components in NUCLEIC ACIDS.

pepo A type of berry with a hard exterior, for example cucumber fruit.

pepsin A PROTEASE enzyme found in the stomach that breaks down proteins during digestion. Pepsin is secreted as its precursor, pepsinogen, which is activated by the stomach acidity to form pepsin.

pepsinogen The inactive precursor of the PROTEASE enzyme PEPSIN. Pepsinogen is secreted by the lining of the stomach and is converted to pepsin by the action of hydrochloric acid in the stomach.

peptide Two or more AMINO ACIDS joined together by a peptide bond (–CO-NH–) that forms between the amino group (NH$_2$) of one amino acid and the carboxyl group (COOH) of another amino acid with the loss of a water molecule. A long chain peptide (more than three amino acids) is called a polypeptide, which can fold or twist to form a PROTEIN. Some peptides are HORMONES, for example OXYTOCIN and ANTI-DIURETIC HORMONE.

peptidoglycan A MACROMOLECULE that gives rigidity to the CELL WALL of BACTERIA. It is not found in EUKARYOTES. GRAM'S STAIN, which is used to differentiate bacteria, relies upon differences in the peptidoglycan layer. Gram-positive bacteria have a thick layer of peptidoglycan in their cell walls, whereas Gram-negative bacteria have a thinner layer of peptidoglycan surrounded by another outer layer.

perennation The ability of plants to survive for more than one year by means of underground storage organs such as a CORM, RHIZOME, BULB, ROOT or TUBER.

perennial plant A plant that continues to grow year after year. In woody perennials, such as trees, the stems remain above ground during winter and new growth starts from these in the spring. In herbaceous perennials, the above-ground growth dies in the winter and new growth begins from an underground storage organ such as a RHIZOME or CORM. *Compare* ANNUAL PLANT, BIENNIAL PLANT.

perianth A collective term for the outer layers of a flower, consisting of the PETALS and SEPALS. Petals form the inner layer of the perianth. Collectively, the petals are called the corolla and their function is to attract pollinators such as insects or birds. The sepals form the outer layer of the perianth. The sepals are collectively known as the calyx. Sepals are usually green and are a site of PHOTOSYNTHESIS. The main function of sepals is to protect the other floral parts when the flower is a bud.

In most DICOTYLEDONS, the perianth is composed of two whorls, the calyx and corolla, but in MONOCOTYLEDONS these two are indistinguishable and are called tepals. Dicotyledons usually have sepals and petals in groups of five, and monocotyledons usually have them in groups of three.

pericarp The wall of a fruit. In fleshy fruits such as berries, the pericarp is composed of three layers, a tough outer exocarp, a middle mesocarp that is often fleshy, and an inner endocarp that surrounds the seeds. In fruits such as the nut, the pericarp is dry and hard.

pericycle A layer of PARENCHYMA cells in the ROOTS of plants, that lies between the vascular tissue (XYLEM and PHLOEM) and the ENDODERMIS. Lateral roots originate from the pericycle.

periderm *See* CORK, CAMBIUM.

perilymph Fluid found in the inner EAR. *See* COCHLEA.

periosteum A tough fibrous membrane surrounding bones, to which muscles and tendons attach.

peripheral nervous system (PNS) The combined SENSORY SYSTEM and EFFECTOR SYSTEM.

peristalsis Involuntary waves of movement, caused by contraction of smooth muscle, that pass along tubular organs, for example in the intestines and earthworms. Peristalsis helps to move material along the tube.

peritoneal cavity The body cavity enclosing the DIGESTIVE SYSTEM, surrounded by an EPITHELIAL membrane called the peritoneum and containing peritoneal fluid.

peritoneal fluid The fluid contained in the PERITONEAL CAVITY.

peritoneum The SEROUS MEMBRANE lining the PERITONEAL CAVITY.

pernicious anaemia A deficiency of VITAMIN B$_{12}$ where RED BLOOD CELLS are malformed, resulting in bruising and slow recovery from minor injuries.

peroxisome, *microbody* A small, spherical, membrane-bound body within a EUKARYOTIC cell that contains the enzyme CATALASE. Peroxisomes have very little internal structure other than being granular. Catalase breaks down hydrogen peroxide, which is a toxic by-product of several biochemical reactions. Peroxisomes are found particularly in cells that are metabolically active, such as in the liver.

pest Any living organism that has a detrimental effect on human welfare, profit or convenience. Pest does not include organisms that are the direct cause of disease. Most pests damage crops, livestock or buildings and may also spread disease. *See also* PESTICIDE.

pesticide A poisonous chemical, used to kill PESTS that may be hazardous to health or simply unwanted by humans. Pesticides are named after the organisms they kill: insecticides kill insects, fungicides kill fungi, herbicides kill plants considered to be weeds, and rodenticides kill rodents. Pesticides are used in farming and gardening but because of their toxic nature they can cause a pollution problem. A good pesticide is one that is specific to the organism to which it is directed, is easily broken down to harmless substances so it does not persist in the environment, and does not accumulate within an organism to be passed along the FOOD CHAIN. *See also* BIOMAGNIFICATION.

petal The part of a flower forming the inner layer of the PERIANTH. The collective term for petals is the corolla. The main function of petals is to attract pollinators, such as insects and birds, and they are therefore often brightly coloured, large and scented and may produce a sweet NECTAR at their base. Petals are derived from modified leaves and may be absent or small in wind-pollinated plants.

petiole The stalk of a LEAF.

Peyer's patches Small areas of secondary LYMPHOID TISSUE in the gut of mammals, birds and reptiles.

pH A measure of the acidity or alkalinity of a solution. pH 7 indicates a neutral solution, a value smaller than this indicate acidic solutions, whilst larger values indicate alkaline solutions.

Phaeophyta A phylum from the kingdom PROTOCTISTA that consists of brown ALGAE. Brown algae are diverse, and most are marine. They include all the larger seaweeds and some smaller ones. No brown alga is unicellular. Brown algae contain CHLOROPHYLL a and c and other pigments, which give them their colour. Both ASEXUAL REPRODUCTION and SEXUAL REPRODUCTION can occur.

phage *See* BACTERIOPHAGE.

phagocyte A type of WHITE BLOOD CELL that can ingest particles (foreign micro-organisms or food) from its surroundings to form internal VACUOLES consisting of the particle surrounded by the CELL MEMBRANE (a process known as PHAGOCYTOSIS). The vacuoles may fuse with LYSOSOMES and the particle is then destroyed or broken down. MACROPHAGES are phagocytic but are larger than other phagocytes and have a longer life span.

phagocytosis The process by which cells ingest particles too large to enter by DIFFUSION or ACTIVE TRANSPORT. The cell invaginates to form a cup-shaped depression that contains the particle. This is then pinched off to form a VACUOLE (or PHAGOSOME), and a LYSOSOME fuses with it to become a heterophagosome (a type of secondary lysosome; *see* PHAGOSOME). Enzymes are then released to break down the particle and useful elements are absorbed by the cell. This process occurs in *Amoeba* as a means of feeding and in a few specialized cells (PHAGOCYTES) such as WHITE BLOOD CELLS and MACROPHAGES, which ingest harmful bacteria or other invading bodies. *See also* PINOCYTOSIS, POTOCYTOSIS.

phagosome A type of secondary LYSOSOME. Heterophagosomes result from the fusion of primary lysosomes with substances entering the cell for digestion, for example bacteria. Such substances are then digested by the enzymes contained within the lysosome. Some of the products pass out through the lysosomal membrane and others remain as residual bodies.

Autophagosomes result from the fusion of a primary lysosome with cell ORGANELLES wrapped in a membrane (probably originating from smooth ENDOPLASMIC RETICULUM). This may occur during cell reorganization or to remove worn-out organelles.

pharynx The cavity in the throat at the back of the mouth. The upper part of the pharynx (nasopharynx) is an airway but the rest is for the passage of food. Air enters from the nostrils and into the pharynx for passage, then to the TRACHEA (windpipe). Food is pushed to the pharynx by the tongue for passage into the OESOPHAGUS, leading on to the stomach. To ensure that food goes down the oesophagus and air down the trachea, a number of reflexes exist. The EPIGLOTTIS closes the opening to the trachea and a soft palate closes the opening to the nasal cavity. In addition to the trachea and oesophagus meeting at the pharynx, the EUSTACHIAN TUBES from each middle ear also enter the pharynx. *See also* RESPIRATORY SYSTEM.

phase contrast microscopy A modification of a light MICROSCOPE. It utilizes the different refractive index of features within an object so that a transparent object can be seen in detail without the need for a coloured stain.

phenotype The actual visible traits of an organism, for example eye or hair colour, that is determined by the GENOTYPE and the effects of dominance (*see* ALLELE) and environmental factors. *See also* VARIATION.

phenylalanine An ESSENTIAL AMINO ACID found in many proteins. It is normally converted to TYROSINE in the body and failure to do this results in a condition called PHENYLKETONURIA. Phenylalanine is obtained in the diet. It is a main component in ASPARTAME (Nutrasweet).

phenylketonuria A genetic disorder in humans in which there is a defect in protein metabolism. The amino acid PHENYLALANINE, which is taken in with the diet, is normally converted in the body to TYROSINE. In phenylketonuria, the enzyme causing this conversion is defective and there is a build up of phenylalanine, which is eventually excreted in the urine. This build up can cause mental retardation, epilepsy and lighter hair and skin colour. Symptoms can be controlled by reducing the dietary intake of phenylalanine. Many foods display phenylalanine as a content for this reason.

pheromone A chemical signal (such as an odour), considered to be a 'social hormone', that operates in a hormone-like manner between individuals of a species rather than within an individual. Pheromones are often used to attract mates, for example they allow female silk moths to attract male mates many kilometres away. Ants produce pheromones that warn other ants of danger, and bees produce several pheromones each with its own effects.

phloem A compound plant tissue (made of a number of cell types) that transports sugars and other food materials from the leaves to other parts of the plant. Phloem consists of PARENCHYMA cells, SCLEREIDS and fibres (*see* SCLERENCHYMA), and specialized SIEVE ELEMENTS with associated COMPANION CELLS. The sieve elements are long, thin-walled cells joined end to end, whose cross-walls break down to form SIEVE TUBES with SIEVE PLATES. This results in the formation of large pores in the cross-wall, which allow the continuous passage of nutrients. The sieve tube elements are living but have no nucleus and very little CYTOPLASM, so depend on adjacent companion cells to sustain them. Phloem is usually found associated with XYLEM in structures (or veins) called VASCULAR BUNDLES. *See also* MASS FLOW, TRANSLOCATION.

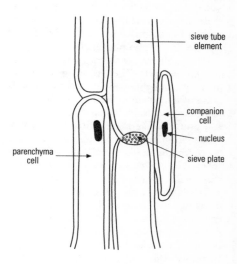

Longitudinal section of the phloem.

phosphate Any salt containing the phosphate ion, PO_4^{3-}. Phosphates are components of NUCLEOTIDES, ATP and some proteins. They are found in some cell membranes as PHOSPHOLIPIDS. Phosphates are involved in many biological processes. In plants, deficiency of phosphates leads to stunted growth, especially of the root system and dull, dark green leaves. In animals, deficiency of phosphates causes RICKETS. Phosphates, particularly ammonium phosphate, are widely used as fertilizers.

phosphoenolpyruvate (PEP) A METABOLITE that plays an important role in the light-independent stage of the PHOTOSYNTHESIS of C_4 PLANTS. *See* CALVIN CYCLE.

phospholipid A LIPID based on GLYCEROL and comprising a PHOSPHATE group, with one or more FATTY ACIDS. Phospholipids are found throughout living systems because they are found in CELL MEMBRANES. The phosphate end is hydrophilic (water-loving) and the other end is hydrophobic (water-repelling). *See also* FLUID MOSAIC MODEL.

phosphorus (P) A non-metallic element occurring in several forms. The most common are white phosphorus, which is toxic and highly reactive (it ignites spontaneously on exposure to air), and red phosphorus, which is non-toxic and less reactive. Phosphorus is essential for life. It is required for the formation of NUCLEIC ACIDS and certain energy-carrying molecules (for example ATP). Phosphorus is also an important constituent of bones and teeth. Salts of phosphorus, PHOSPHATES, are used as fertilizers. *See also* PHOSPHORUS CYCLE.

phosphorus cycle The recycling of phosphorous throughout the ECOSYSTEM. A wide range of important biological chemicals contain phosphorus, including NUCLEOTIDES, ATP and proteins. The main source of phosphorous is from rocks, which release it into the ecosystem through erosion. Plants absorb dissolved PHOSPHATES, which are passed to other animals in the FOOD CHAIN. Phosphates are recycled through plant and animal waste and decaying organisms, including their bones and shells, which can themselves be eroded to provide more dissolved phosphates or be deposited in rocks.

phosphorylation The transfer of a PHOSPHATE group by the enzyme phosphorylase, often from inorganic phosphate ions, to an organic compound. The conversion of AMP and ADP to ATP occurs by phosphorylation reactions in two main metabolic pathways: PHOTO-PHOSPHORYLATION (using light energy) and OXIDATIVE PHOSPHORYLATION (occurring during cellular RESPIRATION in aerobic cells). Many enzymes are also activated or deactivated by phosphorylation.

photoautotroph Any organism that uses light energy from the sun to synthesize organic compounds from inorganic molecules. This is in contrast to CHEMOAUTOTROPHS, which use chemical energy. All green plants and some bacteria are photoautotrophs obtaining their energy by the process of PHOTOSYNTHESIS, which uses light to convert carbon dioxide and water into sugars. *See also* AUTOTROPH.

photoheterotroph Any organism that uses light energy to synthesize organic compounds (similar to a PHOTOAUTOTROPH), but which relies on some organic compound as a nutrient material (and is therefore a HETEROTROPH). A few bacteria are photoheterotrophs, such as the purple non-sulphur bacteria, which rely on organic acids as their energy source instead of hydrogen sulphide. Most other heterotrophs, all animals and fungi, do not use light as an energy source.

photonasty The NASTIC MOVEMENT of plants in response to light. For example, many flowers open in the day and close at night.

photoperiodism The mechanism in plants by which they can respond to changes in daylength and therefore regulate some of their activities. Photoperiodism is regulated by the light-sensitive pigment PHYTOCHROME.

The ratio of light and dark controls the flowering of many plants, which fall into three groups. (i) In long-day plants (LDP), flowering is triggered by a dark period of less than a critical length. In other words, flowering occurs mostly in the summer. Examples include the radish, clover and petunia. (ii) In short-day plants (SDP), flowering is triggered when the dark period exceeds the critical length; that is, mostly in the winter, for example the chrysanthemum. (iii) In day-neutral plants, flowering is unaffected by the length of day, for example the cucumber and begonia. The categories are not absolute and some plants need a combination of long and short days before flowering. The stimulus is detected by the leaves and thought to be transmitted by a growth substance (as yet unidentified) called 'florigen'.

Another factor affecting flowering is vernalization, when exposure of the plant to a period of low temperatures stimulates flowering. Vernalization is thought to be affected by a growth substance, vernalin, that is a GIBBERELLIN. This, together with photoperiodism, ensures that flowering occurs at specific times of the year. Both can be controlled artificially, with commercial implications.

Animals are also affected by photoperiodism (*see* BIORHYTHM).

photophosphorylation PHOSPHORYLATION that uses light energy. Photophosphorylation occurs during PHOTOSYNTHESIS in which ATP is formed from ADP and inorganic phosphate. This takes place on the THYLAKOID membranes within the CHLOROPLAST. There are two pathways; cyclic and non-cyclic (*see* PHOTOSYNTHESIS).

photoreceptor A receptor cell that detects light and some forms of electromagnetic radiation. *See* CONE, EYE, ROD, SENSE ORGAN.

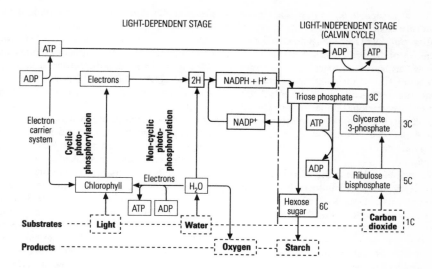

Summary of photosynthesis.

photorespiration A very active type of RESPIRA-
TION that occurs outside the MITOCHONDRIA in
plants, when light intensity is high, carbon
dioxide levels low and oxygen levels high.
Photorespiration involves the reoxidation of
some of the carbohydrates formed during
photosynthesis to carbon dioxide. No energy is
generated, so the process seems wasteful and
reduces the efficiency of photosynthesis. It
only occurs in C_3 PLANTS (temperate plants;
about 85 per cent of plants) and not in C_4
PLANTS (tropical plants).

photosynthesis The use of light energy by green
plants to convert carbon dioxide and water to
carbohydrates and oxygen. Photosynthesis is a
form of AUTOTROPHIC NUTRITION or self-feeding
and all animals depend on it because it is the
means by which basic food (sugar) is created
and oxygen is released for use by aerobic organisms. Photosynthesis occurs in the CHLORO-
PLASTS of higher plants and algae by means of
various light-trapping pigments, the most common of which is CHLOROPHYLL. There are two
stages to photosynthesis: the light-dependent
stage (photolysis) and the light-independent
stage (formerly called the dark reaction).

The light-dependent stage occurs in the
grana of the chloroplast, and involves the
splitting of water by light to yield oxygen,
hydrogen ions (protons) and electrons with
the simultaneous conversion of ADP to ATP by
PHOTOPHOSPHORYLATION, so converting light
energy to chemical energy. This can be non-
cyclic photophosphorylation, in which electrons are lost from the chlorophyll to be
passed into the light-independent stage and
are replaced by electrons from a water molecule (with the production of ATP). Thus the
same electrons are not recycled through the
chlorophyll. Cyclic photophosphorylation also
occurs, in which electrons from the chlorophyll are returned to it via an electron carrier
system. NADPH is also formed during the
light-dependent stage by the reduction of NADP
to NADPH and H^+.

Electrons and protons from the light-
dependent stage (NADPH and H^+) are used in
the light-independent stage (along with the
ATP generated) to convert carbon dioxide
from the air (entering by DIFFUSION through
the STOMATA and eventually into the chloroplast STROMA) into carbohydrates. The light-
independent stage occurs in the stroma of
chloroplasts and is also called the CALVIN CYCLE.
This latter stage differs between C_3 PLANTS and
C_4 PLANTS. Most of the products of photosynthesis are derived from intermediates of the
Calvin cycle.

A plant is capable of synthesizing all the organic materials necessary for its survival, including carbohydrates, LIPIDS and PROTEINS. A number of factors affect photosynthesis: light intensity, carbon dioxide concentration, temperature (the light-independent stage is temperature-dependent), and chlorophyll and oxygen concentrations. Under certain conditions, PHOTORESPIRATION occurs in some plants at the expense of photosynthesis. As light intensity increases, so does the rate of photosynthesis.

The point at which as much carbon dioxide is being used in photosynthesis as is evolved from RESPIRATION is called the compensation point. Beyond this the rate of photosynthesis increases until light saturation is reached – that is, where an increase in light intensity has no effect on photosynthesis.

Compare CHEMOSYNTHESIS.

phototaxis The directional movement of an organism in response to light. For example the green alga CHLAMYDOMONAS swims towards light to increase the rate of PHOTOSYNTHESIS. *See* TAXIS.

phototropism The directional growth of a plant in response to an external light stimulus. Shoots are positively phototropic and bend towards a light source; some roots are negatively phototropic and bend away from a light source; leaves position themselves at right angles to light and are called diaphototropic. Phototropism can be explained in terms of AUXIN, a PLANT GROWTH SUBSTANCE. Auxin moves away from the light source, causing cell expansion and therefore bending of a shoot towards the light. As higher concentration of auxins can inhibit growth, and roots are more sensitive to auxin than shoots, the same concentration that causes positive phototropism in shoots will cause inhibition of growth in roots (on the side away from the light) therefore causing the root to bend away from the light.

See also TROPISM.

phylogeny The evolutionary development of an organism or species. *Compare* ONTOGENY.

phylum (*pl. **phyla***) The major division within a KINGDOM. An example is the phylum CHORDATA, which includes mammals, birds, reptiles, amphibians, fish and tunicates. In plant classification the term 'division' is sometimes used instead of phylum. The subdivisions of phyla are CLASSES. *See* CLASSIFICATION.

physiology The study of the functional processes and activities that occur within living organisms. In higher organisms, physiology includes the study of interactions between cells, tissues and organs. Although usually used in relation to animals, physiology can also refer to the processes occurring within plants.

phytochrome A light-sensitive protein pigment that, by absorbing light of different wavelengths, enables a plant to detect changes in daylength, so controlling PHOTOPERIODISM. There are two forms of phytochrome: phytochrome 660 (P_{660}), which absorbs red light with a wavelength of 660 nm; and phytochrome 730 (P_{730}), which absorbs light at the far red end of the visible spectrum, at wavelength 730 nm. These two forms are interconvertible and even short exposure to light at the appropriate wavelength causes a switch. Daylight and darkness also have an effect; during daylight P_{660} is converted to P_{730} and in the dark P_{730} is converted to P_{660} (although this is slower).

Phytochrome is distributed in small amounts all over the plants but mostly at the growing tips. The two forms have different (usually ANTAGONISTIC) effects on plants. P_{730} stimulates GERMINATION of some seeds, causes an increase in leaf area, stimulates flowering in long-day plants (*see* PHOTOPERIODISM) and induces formation of other plant pigments. P_{660} inhibits these effects but stimulates flowering in short-day plants. The ratio of the two forms can be important in breaking the DORMANCY of some seeds.

phytomenadione *See* VITAMIN K.

phytoplankton The plant constituent of PLANKTON.

pia mater The innermost of the three membranes that cover the BRAIN and SPINAL CORD. *See* MENINGES.

pico- (p) A prefix placed in front of a unit to denote that the size of that unit is to be multiplied by 10^{-12}.

pigment Any substance, for example HAEMOGLOBIN and CHLOROPHYLL, that gives a characteristic colour to animal or plant tissue.

piliferous layer The layer of cells on the surface of a plant ROOT that produces the ROOT HAIRS.

pill A general term for a variety of contraceptive tablets (*see* CONTRACEPTION) taken orally that use the female hormones to interfere with the

MENSTRUAL CYCLE. The combined pill contains synthetic hormones that mimic OESTROGEN and PROGESTERONE produced in pregnancy. It prevents OVULATION and alters the CERVIX, making it hostile to SPERM. It is the most effective form of birth control (99 per cent) but has some side-effects, for example thrombosis due mainly to the oestrogens, headaches and high blood pressure. The mini-pill works in a similar way but contains progesterone only and prevents ovulation and the IMPLANTATION of a fertilized egg; it has fewer side-effects but a slightly lower success rate.

The morning-after pill can be taken up to 72 hours after unprotected intercourse and prevents implantation of a fertilized egg, but contains high levels of hormones and is not for regular use.

pilus (*pl. pili*) A small hair-like outgrowth on the surface of many bacteria that assist them in attaching to certain surfaces. DNA is exchanged or donated between bacteria through a pilus. *See also* CONJUGATION.

pineal body *See* PINEAL GLAND.

pineal gland, *pineal body, epiphysis* A small mass of nervous tissue within the vertebrate MIDBRAIN. The nervous connection to the brain is lost but a nerve supply comes from the SYMPATHETIC NERVOUS SYSTEM. The pineal gland secretes two products that are linked to body rhythms, a hormone-like substance called melatonin (which is inhibited by light) and a NEUROTRANSMITTER called SEROTONIN. Serotonin causes general inhibition of activity (opposes NORADRENALINE).

In lower vertebrates, such as amphibia and some snakes, the pineal gland develops a basic lens and retina to form an eye that is situated on top of the head. In fish, the pineal gland detects the surrounding light and controls colour change involved in camouflaging. In birds, it detects changes in length of daylight and stimulates breeding behaviour as spring approaches.

See also SENSE ORGAN.

pinna 1. In mammals, part of the outer EAR.

2. In botany, the primary division of a pinnate LEAF.

pinnate leaf A type of compound LEAF.

pinocytosis A process similar to PHAGOCYTOSIS that is used by living cells for the uptake of liquids not solids.

pit A region in a plant CELL WALL that is not secondarily thickened and therefore allows the movement of substances between adjacent cells. Pits usually develop in areas in the primary cell wall where PLASMODESMATA are numerous. Pits in adjacent cells coincide to form a pit pair. A pit consists of a pit membrane (the primary cell wall and the MIDDLE LAMELLA between cells) and a pit cavity (a depression in the secondary thickening). *See also* CELL WALL.

pith The soft, spongy tissue in the centre of stems or roots of VASCULAR PLANTS. Pith is usually PARENCHYMA tissue. Pith also refers to the white, fibrous tissue between the rind and pulp of fruits such as oranges.

pituitary gland, *hypophysis* A small ENDOCRINE GLAND found in the centre of the vertebrate brain that secretes a number of hormones that have control over the activities of the other endocrine glands. The pituitary gland is itself largely under the control of the HYPOTHALAMUS, to which the posterior pituitary is attached.

The posterior pituitary is of nervous origin and communicates with the hypothalamus via nerves. It stores and secretes (under hypothalamic control) ANTI-DIURETIC HORMONE (ADH) and OXYTOCIN (made by the hypothalamus), concerned, respectively, with the salt/water balance and the functioning of the MAMMARY GLANDS and UTERUS.

The other region of the pituitary is the ANTERIOR pituitary, which produces five hormones called trophic hormones (which affect other endocrine glands) and a non-trophic hormone, GROWTH HORMONE, which controls body METABOLISM and growth generally. The trophic hormones are ADRENOCORTICOTROPHIC HORMONE, FOLLICLE-STIMULATING HORMONE, LUTEINIZING HORMONE, PROLACTIN and THYROID-STIMULATING HORMONE.

See also THYROID.

placenta 1. The structure in most mammals that attaches the developing FOETUS to the UTERUS and which links the blood supply of the mother and baby so that exchange of oxygen, nutrients and waste products can sustain the foetus throughout PREGNANCY. The placenta is connected to the foetus by an UMBILICAL CORD.

The blood supply of the mother and baby are separate (and therefore AGGLUTINATION, or clumping, is prevented) and exchange of nutrients, for example glucose, amino acids, vitamins, fats and oxygen, and waste products, for example UREA and carbon dioxide, occurs between the capillaries of mother and baby by DIFFUSION. Some maternal antibodies can cross the placenta, and provide protection against some diseases during the early months of life. However, some PATHOGENS can also cross, for example rubella (the German measles virus), causing physical and mental damage to the baby, or HIV causing AIDS. Many organisms do not cross the placenta, however, and it allows considerable protection against disease. The placenta also prevents many chemicals or high levels of maternal hormones entering the foetal blood but some, for example alcohol, nicotine, caffeine and heroine can cause damage to the foetus. The placenta also has an essential immunological role in the protection of the foetus.

Once fully developed, at 3 months, the placenta takes over the role of hormone production necessary to maintain the pregnancy from the earlier TROPHOBLAST. At this time it also begins to secrete OESTROGEN, PROGESTERONE, HUMAN CHORIONIC GONADOTROPHIN and HUMAN PLACENTAL LACTOGEN. Many other proteins are secreted by the placenta that are of considerable interest but the function of many is as yet unknown. The placenta is shed at birth as the afterbirth.

See also ENDOMETRIUM.

2. The part of the ovary of a flowering plant to which the OVULES are attached. See CARPEL.

plankton Small organisms living on the surface of fresh or salt water. Plankton is an important food source for larger animals such as fish and whales. The plants (algae) of plankton are called phytoplankton, and they are nonmotile, simple organisms of varying sizes that move with the water currents. The animals of the plankton are called zooplankton, and they feed on phytoplankton and are themselves food for larger fishes. Zooplankton are mostly able to move by FLAGELLAE. See also NEKTON, PELAGIC.

plant An immobile multicellular, EUKARYOTIC organism capable of PHOTOSYNTHESIS. Plants are the primary producers in all FOOD CHAINS and therefore all animal life is dependent on them. They contain CELLULOSE in their cell walls. Plants make up the kingdom Plantae.

One common method of plant classification is into the phyla: BRYOPHYTA (mosses and liverworts); LYCOPODOPHYTA (club mosses);

SPHENOPHYTA (horsetails), FILICINOPHYTA (FERNS); CONIFEROPHYTA (conifers); and ANGIOSPERMOPHYTA (angiosperms; flowering plants). The study of plants is called botany.

See also SEED PLANT, VASCULAR PLANT.

Plantae The kingdom consisting of PLANTS.

plant growth substance, *plant hormone* A chemical substance (natural or artificial) that modifies plant growth. Plant growth substances are not produced by a particular area of the plant (in contrast to animal HORMONES) and may affect different areas of growth or a related process (for example, leaf fall, flowering or fruit ripening). There are five groups of growth substances: AUXINS, GIBBERELLINS, CYTOKININS, ABSCISSIC ACID (inhibitor) and ETHENE.

plant hormone See PLANT GROWTH SUBSTANCE.

plasma The liquid part of BLOOD (without the cells) that is a watery yellowish fluid containing many dissolved sugars, fats and proteins, including antibodies, hormones and salts, as well as blood-clotting substances (*see* BLOOD CLOTTING CASCADE). Plasma is obtained by centrifugation of blood to sediment the cells.

plasma cell A mature B CELL that produces large amounts of specific ANTIBODY following exposure of B cells to the specific ANTIGEN that causes them to divide. Plasma cells die after a few days of antibody production. See also MEMORY CELL.

plasmagel See ECTOPLASM.

plasma membrane See CELL MEMBRANE.

plasma protein One of many PROTEINS found in blood PLASMA. Examples include ALBUMIN, GLOBULIN and blood-clotting factors (PROTHROMBIN, FIBRINOGEN).

plasmasol See ENDOPLASM.

plasmid A small circular loop of DNA found in bacterial cells that is separate from the bacterial chromosome but replicates with it. Usually only one copy is present but several can be found. Plasmids can confer useful properties onto the cell, such as ANTIBIOTIC RESISTANCE. If a plasmid is used as a carrier for foreign DNA, it is called a vector (vectors can also be BACTERIOPHAGES). By growing cultures of the bacteria containing the plasmid with its foreign DNA fragment, many copies of the foreign DNA can be obtained. Plasmids are an invaluable tool in GENETIC ENGINEERING.

plasmodesma (*pl. plasmodesmata*) A fine cytoplasmic tube (30–60 nm in diameter) that is present in the CELL WALL of a plant, at intervals or in regions. There is no LIGNIN at these sites and so 'pits' are formed, through which water and dissolved minerals can pass from one cell to another.

plasmolysis The shrinkage of a cell CYTOPLASM away from the cell wall as a result of water loss by OSMOSIS. Water leaves the cell VACUOLE, causing the cytoplasm to shrink without affecting the overall cell shape or size. The cell is then called 'flaccid'. Plasmolysis can be induced in the laboratory by immersing plant cells in a saline or sugar solution, so water leaves the cells by osmosis. It is unlikely to occur naturally except in extreme conditions. *See also* TURGOR.

plastid The general name for a cell ORGANELLE of plants and algae, containing a series of internal membranes bounded by a double membrane. Plastids contain a small amount of DNA, and some RIBOSOMES of the PROKARYOTIC type may be present. They develop by division of existing plastids and there can be one or more per cell of various shapes. Plastids mature into CHLOROPLASTS or CHROMOPLASTS, which contain coloured pigments, or LEUCOPLASTS, which are colourless.

platelet, *thrombocyte* In vertebrates, a small cell-like fragment that buds from larger cells in the BONE MARROW and is involved in blood clotting following injury. Platelets adhere to injured ENDOTHELIAL cells (lining blood vessels), and aggregate, releasing a number of factors needed for prevention of bleeding and repair of the damaged tissue. In the BLOOD-CLOTTING CASCADE, platelets release an enzyme called thrombokinase, which converts the inactive enzyme prothrombin into the active thrombin. This in turn converts the soluble plasma protein FIBRINOGEN into the insoluble FIBRIN, which forms a clot over the wound. *See also* BLOOD.

platelet-derived growth factor (PDGF) A GROWTH FACTOR that plays a role in wound healing.

Platyhelminthes A phylum consisting of the flatworms, which are free-living or parasitic invertebrates of great economic importance. Flatworms have a head region and a flattened shape showing bilateral symmetry. Their body walls are triploblastic (three layers), and there is one opening to the gut. Many are HERMAPHRO-DITES that can self-fertilize, but usually cross-fertilization occurs. The life cycle of parasitic flatworms can be complex and often involves several intermediate hosts (*see* PARASITE).

Examples are the liverfluke (*Fascida hepatica*) of sheep and cattle (the intermediate host is a snail) that causes liver rot; the blood fluke (*Schistosoma*) of humans (the intermediate host is a snail) that causes schistosomiasis (bilharzia), damaging the lungs and liver; the pork tape worm (*Taenia solium*) that infects humans (the intermediate host is the pig), causing anaemia, diarrhoea, weight loss, intestinal pain and blockage.

Some flatworms move by secreting MUCUS and gliding over this by means of CILIA, others produce muscular contractions down the length of their body, while parasitic flatworms are usually attached by hooks or suckers to their host. Flatworms are carnivores that either trap prey in the mucus they produce, or suck blood and cells into their bodies. Tapeworms absorb pre-digested food from their host by diffusion through their body surface (*see* CESTODA, TREMATODA, TURBELLARIA).

pleura, *pleural membrane* A double membrane covering the exterior surface of the LUNGS in mammals. It is a SEROUS MEMBRANE and contains a small space between the two layers called the PLEURAL CAVITY.

pleural cavity The space between the double pleural membrane (*see* PLEURA) surrounding each LUNG of mammals. A fluid (pleural fluid) enters the pleural cavity to lubricate the lungs to allow for their expansion.

pleural membrane *See* PLEURA.

pleurisy An infection of the PLEURAL CAVITY.

plumule The part of the embryo of a SEED PLANT that develops into the shoot and bears the first true leaves (*see* GERMINATION).

PMN *See* POLYMORPHONUCLEAR LEUCOCYTE.

pneumatophore An erect plant root that rises above the soil surface, for example in swamps, to aid gas exchange.

PNS *See* PERIPHERAL NERVOUS SYSTEM.

poikilotherm, *ectotherm* An animal that exhibits POIKILOTHERMY. A poikilotherm is also known as a cold-blooded animal.

poikilothermy, *ectothermy* (*poikilo* = various, *thermo* = heat) A condition where the body temperature of an animal fluctuates depending on the external temperature. Poikilothermy is characteristic of all animals except birds and mammals, which maintain a constant body temperature (HOMEOTHERMY). Invertebrates,

fish, amphibians and reptiles are all poikilotherms. Despite being called cold-blooded,
poikilotherms can have body temperatures as
high as homeotherms, or higher. Poikilotherms
have behavioural means of temperature control, such as cooling in water, shivering, lying in
the sun and sheltering.

point mutation The omission, insertion or substitution of a single base (*see* NUCLEOTIDES) in
DNA. *See* MUTATION.

polarity 1. The property of cells, tissues or
organisms of being functionally or morphologically different at opposite ends of their
longitudinal axis. For example, shoots/roots in
plants and head/tail of animals.

2. A property of molecules in which there
is an uneven distribution of electrons, so that
one region is more positively charged and
another more negatively charged. These types
of molecules are termed polar molecules.
Water is an example.

polarization The difference in charge across a
nerve or muscle membrane. This is due to the
distribution of various ions across the membrane. *See* NERVE IMPULSE, DEPOLARIZATION.

polar nucleus Either one of the two nuclei
within the OVULE of a flowering plant that
migrate from each pole of the EMBRYO SAC
towards the centre. Polar nuclei fuse with one
of the male nuclei carried to the ovule by the
POLLEN TUBE, and form the PRIMARY ENDOSPERM
NUCLEUS. This is part of DOUBLE FERTILIZATION.

pollen The grain of SEED PLANTS carrying the
male GAMETE. Pollen is produced in POLLEN
SACS of the ANTHERS of flowering plants and
within male cones in CONIFEROPHYTA. Pollen
grains are usually yellow with a hard outer wall
and can be light, for wind dispersal, or heavier,
larger, sticky and spiny for insect dispersal.
Pollen grains are extremely resistant to decay,
which makes them a useful tool for gathering
information on the abundance of past species.
This is termed POLLEN ANALYSIS or palynology.
See also DOUBLE FERTILIZATION, POLLEN TUBE.

pollen analysis, *palynology* The study of fossil
POLLEN and SPORES in order to gain an understanding of past floras and climates. Pollen
and spores are very resistant to decay and may
therefore be preserved in peat or sedimentary
rock. Details of their shape, internal and external structure can provide information on the
dominant flora and the climate and conditions
when they were produced.

pollen sac A structure within the ANTHER of
flowering plants containing POLLEN. There are
usually four pollen sacs (two in each lobe of
the anther).

pollen tube An outgrowth from a POLLEN grain
when it lands on a STIGMA of a flowering plant
and begins to germinate. The pollen tube
transports two male GAMETES to the female
gamete (egg nucleus) to allow DOUBLE FERTIL-
IZATION. The tube passes through the STYLE and
towards the OVULE, where it enters through the
MICROPYLE. It penetrates the EMBRYO SAC and
releases the male gametes, one of which then
fuses with the egg nucleus and the other with
two polar nuclei (*see* POLAR NUCLEUS). The
pollen tube then disintegrates.

pollination The transfer of POLLEN from the
male to female parts of plants. In flowering
plants (ANGIOSPERMS) this transfer is from
ANTHERS to STIGMAS, and in CONIFERS it is from
male to female cones. In cross-pollination the
transfer is between two plants, and in self-
pollination it is within one flower or another
flower on the same plant. Although self-
pollination does occur, it is less favoured than
cross-pollination because it reduces the variability of the population. To reduce self-pollination, the STAMEN and stigma may mature at
different times. If the stamen develops first, the
plant is termed protandrous; in protogynous
plants the stigma and OVULES develop first.

Transfer of pollen can be by wind, water,
insects, birds, bats or other small animals, and
flowers are usually adapted for one form of
pollination. Pollination is not the same as fertilization, which occurs after pollination (*see*
DOUBLE FERTILIZATION).

See also ANEMOPHILY, ENTOMOPHILY, ORNI-
THOPHILY.

pollution Contamination of the environment
by the by-products of human activity, mostly
from industrial and agricultural processes. Air
pollution is largely due to the burning of FOS-
SIL FUELS in the home, industry or the combustion engine of vehicles. Such pollutants
include smoke (tiny particles of carbon), sulphur dioxide, carbon dioxide, carbon monoxide, nitrogen oxides and lead, particularly from
car exhaust emissions. These contribute to the
GREENHOUSE EFFECT and ACID RAIN. Other
important air pollutants are CHLOROFLUORO-
CARBONS, which contribute to the thinning of
the OZONE LAYER.

Water pollution can be the result of rain polluted by contaminants in the air, but is also due to the release of toxic chemicals, such as copper, zinc, lead, mercury and cyanide, into rivers and seas, killing fish and plant species. Oil pollution of water (usually accidental) is localized, but devastating to seabirds, shellfish and seaweed. Oil spillages can be degraded by some micro-organisms. Sewage is treated (*see* SEWAGE DISPOSAL) but still contains large amounts of PHOSPHATES from washing powders and detergents that remain a problem as a source of water pollution.

Land (terrestrial) pollution comes from the dumping of solid wastes, for example slag heaps from ore digging, metal refining and coal mining, which are unsightly and often cannot sustain any vegetation. Domestic rubbish is usually disposed of by burning; some plastics are now being used that can be degraded by micro-organisms to avoid the dangerous gases given off by burning. Other land pollutants are pesticides and noise.

Radioactive pollution can be from medical waste, televisions, watches, waste from the nuclear power industry, or from testing nuclear weapons. Disposal of RADIOACTIVE WASTE has been at sea or by burial on land, but decay can take thousands of years and presents problems of safety, pollution and security.

See also EUTROPHICATION, GLOBAL WARMING.

polyacrylamide gel electrophoresis (PAGE) A type of ELECTROPHORESIS, widely used to determine the size and composition of protein mixtures. In this technique, proteins are placed on a gel matrix of polyacrylamide and an electric field is applied. The proteins migrate towards the positive pole of the electric field at a rate determined by their size as they pass through the pores of the gel. Various stains can be used to visualize the proteins. It is possible to purify some proteins by this technique, although it is more frequently used to analyse mixtures.

Polychaeta A class of the phylum ANNELIDA, including the lugworm (*Arenicola*) and ragworm (*Nereis*). Polychaetes are marine, free-swimming and can move slowly on structures called 'parapodia', with which they can burrow. They are clearly segmented, with a distinct head, and they possess many chaetae (bristles). Polychaetes have separate sexes and FERTILIZATION is external. *Compare* HIRUDINEA, OLIGOCHAETA.

Polychaete A member of the phylum POLYCHAETA.

polygene Any one of a group of GENES that interact together to have an effect on a phenotypic characteristic (*see* PHENOTYPE). Individually, the genes have little effect but combined their effects are marked. Polygenes affect characteristics that show continuous variation, such as the height and weight of an organism. An individual organism receives a range of genes from any polygenic complex because of the random assortment of genes during MEIOSIS. This usually ensures that the characteristic is intermediate, for example neither very tall nor very small. A characteristic that is determined by polygenes is called a polygenic character.

polymer A large molecule made up of two or more similar or identical repeated units joined together to form a chain or branching matrix. There are many naturally occurring polymers, including PROTEINS, POLYSACCHARIDES and NUCLEIC ACIDS and also many synthetic polymers including polythene and nylon.

polymerase An enzyme, for example DNA polymerase and RNA polymerase, that joins MONOMERS together to form POLYMERS.

polymorphism In genetics, the existence of more than one distinct form of a particular characteristic within a species population that cannot be explained by MUTATION. Examples include the BLOOD GROUP SYSTEM in humans and different colour forms of the peppered moth (which depends on their environment).

polymorphonuclear leucocyte (PMN) A type of WHITE BLOOD CELL. *See* GRANULOCYTE.

polynucleotide A long-chain molecule of NUCLEOTIDES.

polyoestrus (*adj.*) Describing non-human mammals who have many OESTROUS CYCLES each year. *Compare* MONOESTRUS.

polyp The sedentary structural form in the life cycle of the CNIDARIA.

polypeptide A PEPTIDE consisting of three or more AMINO ACIDS. Polypeptide chains can fold or twist to form a PROTEIN.

polyploid A nucleus, cell or organism that has three or more sets of CHROMOSOMES. Polyploidy is rare in animals but common in plants. Polyploidy can arise spontaneously, if diploid GAMETES (with two sets of chromosomes each) self-fertilize or if chromosome numbers double after fertilization. Polyploidy can also be induced by treatment with a chemical called colchicine. This prevents spindle formation during MEIOSIS and therefore chromosomes cannot separate.

Some polyploids are sterile, depending on whether HOMOLOGOUS chromosomes can pair during meiosis. Sterile polyploids can still undergo ASEXUAL REPRODUCTION.

An example of polyploidy in plants is wheat, where varieties with four or six sets of chromosomes exist. Many flowering plants are polyploid, including bananas, potatoes, apples and sugar cane. Polyploid varieties usually have an advantage, such as larger fruit.

polysaccharide A CARBOHYDRATE consisting of a variable number of MONOSACCHARIDES joined together in chains that can be branched or not and can fold for easy storage. Polysaccharides can be broken down into their constituent DISACCHARIDES or monosaccharides for use by an organism. In chains, they are insoluble. Examples include CELLULOSE, CHITIN, STARCH and GLYCOGEN.

polyunsaturate A FATTY ACID with two or more double COVALENT BONDS.

pome A type of PSEUDOCARP, for example apples and pears.

pons A broad band of nerve fibres that links the MEDULLA OBLONGATA to the MIDBRAIN.

population A group of animals of one species living together in a particular area. The actual size of a population is determined by the balance between the birth rate and the death rate, although other factors such as immigration and emigration also influence population size. The human population is increasing exponentially and it was suggested as far back as 1798 by Thomas Malthus (1766–1834) that future food supplies would not support our population size; this problem is still true today. *Compare* COMMUNITY. *See also* DENSITY-DEPENDENCE.

population growth The increase in a population that occurs when the birth rate exceeds the death rate, or when immigration exceeds emigration. For most populations, if the growth is plotted against time on a graph, an S-shaped (sigmoid) growth curve is obtained. This represents initial slow growth followed by a period of fast growth or EXPONENTIAL GROWTH. The curve then levels off when the CARRYING CAPACITY of the environment has been reached. This is when food, space or

other factors are sufficient only to support a given number of individuals with no further increase. When such factors become limiting the population size will decrease as the death rate exceeds the birth rate. Sometimes a J-shaped growth curve is obtained, which represents initial exponential growth that ceases abruptly followed by a decrease in populations numbers. There could be a number of reasons for such a decrease, for example some environmental resistance. *See also* GROWTH, BACTERIAL GROWTH CURVE.

Porifera A phylum consisting of the sponges. Sponges are simple invertebrate animals that are usually marine and possess no organs or tissues. The body is hollow but is lined by outer contractile cells and inner choanocytes (collar cells) that bear FLAGELLA. The movement of the flagella circulates water through the sponge, thereby providing a constant supply of food particles and oxygen. Between the outer and inner cell types are amoebocytes, which store food and give rise to reproductive cells. ASEXUAL REPRODUCTION or SEXUAL REPRODUCTION can occur in sponges; they are HERMAPHRODITE but cross-fertilization does occur. There are many openings all over the body wall that can be strengthened with protein, silica or calcium carbonate.

positive feedback Where the end-product of a pathway stimulates the ENZYME at the start of the pathway. *Compare* NEGATIVE FEEDBACK.

posterior (*adj.*) The rear (hind) end of the body of an organism. In humans it is the back side. Posterior also refers to the back of a structure such as an organ or gland. In plants, it refers to buds or flowers that are nearest to the main stem. *Compare* ANTERIOR.

post-synaptic neurone The NEURONE at a SYNAPSE that receives the NERVE IMPULSE across the synaptic cleft from the transmitting neurone (the PRE-SYNAPTIC NEURONE).

potassium (K) A soft, light, silver-white metallic element that is highly reactive. It oxidizes rapidly on exposure to air and reacts violently with water. It is widespread in nature and essential to all forms of life. In animals it helps to maintain the electrical, osmotic and cation/anion balance across cell membranes, along with sodium. Potassium therefore has many important roles in animals, including in the transmission of NERVE IMPULSES. In plants it is a constituent of the sap vacuole and therefore helps to maintain turgidity; deficiency leads to premature death.

potential difference (pd) The potential difference between two points is a measure of the amount of energy converted from electrical energy to other forms when one unit of charge flows between the two points. The unit of potential difference is the volt.

potocytosis A cellular process in which small molecules and ions are concentrated before entering the CYTOSOL at special CELL MEMBRANE sites called caveolae. It is similar to PHAGOCY-TOSIS.

precipitate A solid or solid-phase separated from a solution, which may subsequently settle. In chemistry, a precipitate is a finely divided powder suspended in a liquid, which is formed when a reaction between two soluble salts in solution produce an insoluble salt.

precipitation The formation of a PRECIPITATE.

precipitin An ANTIBODY that combines with a soluble ANTIGEN to form an antigen–antibody complex, which then precipitates and can be removed (or separated if produced experimentally).

precursor Any compound produced in some intermediate step in a series of chemical reactions, and which is then involved in further reactions leading to the desired final product.

predation A relationship in an ecological COMMUNITY in which one animal species (the predator) captures, kills and feeds on another (the prey). The level of predation influences the size of POPULATIONS. *See also* FOOD CHAIN.

pregnancy The duration of EMBRYO development within the UTERUS of mammals, beginning with conception and ending at birth. The duration varies between species, for example it lasts 40 weeks in humans, 18–22 months in elephants, and 60 days in cats.

Some of the early signs of pregnancy in humans are cessation of menstruation, tenderness or enlargement of breasts, and the detection of the hormone HUMAN CHORIONIC GONADOTROPHIN in the urine, which forms the basis of pregnancy testing. Levels of the steroid hormones OESTROGEN and PROGESTERONE prepare for and maintain the pregnancy, and there are changes during pregnancy in the circulating levels of many other constituents of blood. In humans and other primates, the pregnancy length seems to be determined by the FOETUS, although the precise control of the

onset of labour and birth is not fully understood (*see* PARTURITION).

One in five pregnancies fail, usually very early and often unnoticed (except by a late period) but sometimes later as a miscarriage. These failures are often due to an abnormality in the foetus. Foetal death after 24 weeks is called a stillbirth, when the baby is born dead. Factors such as drugs, alcohol and certain PATHOGENS can affect foetal development (*see* PLACENTA).

The term GESTATION is more commonly used for pregnancy in animals other than humans.

See also BLASTOCYST, ECTOPIC PREGNANCY, EMBRYONIC DEVELOPMENT, IMPLANTATION.

premolar In mammals, a large TOOTH behind the INCISOR teeth that has two cusps (ridges) and a single or double root, and is used for grinding food. Humans have eight premolars and they are present in the milk teeth (*see* TOOTH).

pressure flow *See* MASS FLOW.

pressure potential (φ_p) A measure of the HYDROSTATIC PRESSURE pushing outwards on a cell wall. Pressure potential is a component of WATER POTENTIAL. In turgid plant cells, in which water has entered, the pressure potential has a positive value, since the water pushes outwards against the cell wall. In XYLEM cells the pressure potential has a negative value due to TRANSPIRATION, which pulls water up the plant. *See also* OSMOSIS.

pre-synaptic neurone The NEURONE at a SYNAPSE that transmits the NERVE IMPULSE across the synaptic cleft to the receiving neurone (the POST-SYNAPTIC NEURONE) or to a muscle. The pre-synaptic neurone does this by releasing NEUROTRANSMITTER into the synaptic cleft.

primary endosperm nucleus A nucleus within the EMBRYO SAC of a flowering plant that results from the fusion of one of the male nuclei carried by the POLLEN TUBE with the two polar nuclei (*see* POLAR NUCLEUS) present in the embryo sac. The primary endosperm nucleus is TRIPLOID and it divides further to form the ENDOSPERM. *See also* DOUBLE FERTILIZATION.

primary growth In plants, the increase in the size (length) of shoots and roots derived from apical MERISTEMS. This is in contrast to SECONDARY GROWTH, which is derived from lateral meristems and causes an increase in girth.

primate A member of the order Primates in the

class MAMMALIA, including apes, monkeys and humans (called anthropoids) and lemurs, bushbabies, lorises and tarsiers (called prosimians). Features of primates include a large brain relative to the body, forward-directed eyes with good colour vision, five digits on limbs, nails instead of claws, opposable thumbs and big toes, gripping hands and feet, good mobility of limbs and shoulder joints, only two MAMMARY GLANDS, young that are usually born singularly and are nourished by a PLACENTA. Many of these characteristics are adaptations to a climbing way of life in trees.

primordium In botany, a cell or group of cells that is immature but will develop into a specific structure, for example a leaf primordium.

prion A protein fragment thought to be responsible for causing BOVINE SPONGIFORM ENCEPHALOPATHY (BSE) in cattle, SCRAPIE in sheep and CREUTZFELDT–JACOB DISEASE (CJD) in humans. Prions are produced normally in brain cells by a gene known as the PrP gene, which is located on chromosome 20 in humans. It may be that prions are able to direct their own replication without the need for nucleic acids.

There are two normal forms of prions, spiral molecules and pleated sheets. A theory to explain why prions should cause the damage to the brain seen in victims of BSE, scrapie and CJD is that an abnormal form of the prion exists, which fails to fold properly and therefore grows at angles that cause damage to the brain cells. Some cases of CJD have been shown to be associated with mutations of the PrP gene. Furthermore, many mutations are already known to exist in the area on chromosome 20 where the PrP gene is located. The fact that these diseases are infectious means that the abnormal prion protein may be transmitted by injection or ingestion of tissue containing the prion. It is thought to be only transmissible between closely related species. It is possible that another agent switches on, or interferes with, the PrP gene.

Unlike other infectious diseases, no evidence of an IMMUNE RESPONSE being mounted to the prion can be found (i.e. no antibodies have been detected) in the victims. This makes early detection very difficult. Also, prions are resistant to PROTEASE enzymes, which normally destroy proteins and are therefore difficult to destroy.

procambium Tissue derived from the MERISTEM in VASCULAR PLANTS that gives rise to the VASCULAR BUNDLES. The procambium consists of elongated cells that are grouped into strands just behind the growing points of stems and roots. As these divide, cells towards the inside of a shoot form PROTOXYLEM and those to the outside form PROTOPHLOEM. Later, the procambium gives rise to METAXYLEM and METAPHLOEM. In roots, there is only a single strand of procambium, which differentiates to form protophloem, protoxylem and later metaphloem and metaxylem.

productivity The quantity of carbon compounds formed by the primary producers (plants) in a FOOD CHAIN during photosynthesis, that can be used by the consumers (animals). The gross productivity is the total quantity of carbon compounds produced. The net productivity is the gross productivity minus the quantity of carbon compounds used by the plants themselves in their respiration. Thus the net productivity represents the amount of food available to consumers in an ECOSYSTEM.

progesterone In mammals, a STEROID HORMONE that regulates the MENSTRUAL CYCLE and prepares the UTERUS for PREGNANCY. Progesterone is secreted by the CORPUS LUTEUM in the OVARY and if FERTILIZATION of an OVUM occurs, this provides the progesterone needed for the maintenance of early pregnancy until the PLACENTA takes over at 3 months. If fertilization does not occur the corpus luteum regresses and dies, and levels of progesterone fall dramatically, causing menstruation. Progesterone also inhibits production of FOLLICLE-STIMULATING HORMONE.

Prokaryotae The kingdom consisting of PROKARYOTES.

prokaryote (*Pro* = before, *karyote* = nucleus) A simple organism in which the genetic material DNA is not contained within a nucleus and no membrane-bound ORGANELLES exist. Prokaryotes are thought to be the first forms of life on Earth. BACTERIA and CYANOBACTERIA are prokaryotes; all other organisms are EUKARYOTES.

Prokaryotes are all unicellular but can be found in filaments and clusters. The DNA of prokaryotes forms a coiled structure called a nucleoid; often more than one nucleoid exists within a single cell because nucleoids replicate faster than the cell divides. There are no

MITOCHONDRIA or CHLOROPLASTS in prokaryotic cells, but there are structures thought to function similarly – MESOSOMES and CHROMATOPHORES. RIBOSOMES in prokaryotes are smaller than in eukaryotes and there is no MITOSIS or MEIOSIS.

prokaryotic (*adj.*) Describing an organism that is a PROKARYOTE.

prolactin, *luteotrophic hormone, luteotrophin* In vertebrates, a protein hormone secreted by the PITUITARY GLAND that stimulates LACTATION (milk production) and promotes secretion of PROGESTERONE by the CORPUS LUTEUM.

propagate (*vb.*) To breed (of animals) or multiply (of plants, by cuttings, grafts, etc.).

prophase The first stage of MITOSIS and MEIOSIS.

prop root A form of ADVENTITIOUS ROOT that grows from the lower part of a plant stem or trunk to the ground to provide extra support, for example in maize and some woody plants.

prorennin The inactive precursor of RENNIN.

prostaglandin In mammals, any one of a group of complex FATTY ACIDS synthesized continuously by most nucleated cells. Prostaglandins act in a similar way to HORMONES as chemical messengers between cells; they can be released directly into the blood but usually only act locally. A rich source of prostaglandins is SEMEN, where they were first discovered, but they are made all over the body and may act as an intermediary between a hormone binding to its RECEPTOR on the target cell and the activation of the second messenger. Their effects include stimulating the contraction of smooth muscle (for example the UTERUS during PARTURITION), regulation of the production of stomach acid, modifying other hormonal activity, assisting in blood clotting by causing PLATELET aggregation, and being responsible for INFLAMMATION following injury or infection.

In excess, prostaglandins may be involved in causing inflammatory disorders such as arthritis. Pain-relieving drugs, such as aspirin, act by inhibiting prostaglandins. Prostaglandins are of great potential importance in the alteration of blood pressure and broncodilation and constriction.

prostate gland In male mammals, a GLAND at the base of the BLADDER that opens into the URETHRA. The prostate gland secretes an alkaline fluid that forms up to a third of the volume of SEMEN. In humans it is a rich source of PROSTAGLANDINS. In later life in humans, the prostate gland often enlarges to block the urethra and has to be removed. *See also* SEXUAL REPRODUCTION.

prosthetic group A non-protein group that firmly attaches to a PROTEIN or ENZYME to create a functional complex, in contrast to a COENZYME. Examples include HAEMOGLOBIN, which contains iron as its prosthetic group, and GLYCOPROTEINS, which contain CARBOHYDRATES as the prosthetic groups. Metal ions such as Zn^{2+}, K^+ and Na^+ are often prosthetic groups for enzymes, providing a charge needed in an active site. *See also* COFACTOR.

protandrous (*adj.*) Describing a plant whose STAMENS develop before the CARPELS, ensuring that self-pollination does not occur. *See* POLLINATION.

protease A general term for an ENZYME, for example TRYPSIN and PEPSIN, that digests PROTEINS to PEPTIDES.

protein Any one of a group of complex organic compounds, essential to all living organisms, that have a large RELATIVE MOLECULAR MASS and consist of AMINO ACIDS linked together. Proteins always contain carbon, hydrogen, oxygen and nitrogen, usually sulphur and sometimes phosphorus.

The amino acids link together to form PEPTIDES or POLYPEPTIDES, and the sequence of amino acids in a polypeptide chain is its primary structure. There is an amino end (NH_2) and carboxyl end (COOH) to any protein molecule. The ultimate shape of the protein molecule depends on the types of bonds that exist within it, and the shape of the polypeptide chain is called the secondary structure. HYDROGEN BONDS can form that often result in a polypeptide chain coiling into an ALPHA HELIX or BETA-PLEATED SHEETS. Other types of bonding can occur, including IONIC BONDS and HYDROPHOBIC interactions between different groups on the same molecule, which cause folding of the protein to shield these groups from water. Together these bonds cause folding and twisting of the constituent polypeptide chains of a protein into a three-dimensional structure called the tertiary structure. A large, complex protein molecule has many polypeptide chains combined and incorporates non-protein groups (*see* PROSTHETIC GROUP, COFACTOR) vital to its function into its structure, which is then referred to as the quaternary structure. The shape of a protein is crucial to its functioning, for example in providing ENZYME-binding sites.

There are a limitless number of proteins and, unlike CARBOHYDRATES, they vary from species to species. Their functions are numerous. FIBROUS PROTEINS, for example COLLAGEN, provide a structural role, while GLOBULAR PROTEINS, for example enzymes, provide a metabolic role. CONJUGATED PROTEINS incorporate non-protein groups into their structure that play a vital role in their functioning, for example the haem group in HAEMOGLOBIN.

Proteins can be denatured, for example by heat, which breaks up the three-dimensional structure and prevents it from functioning. Protein is an essential requirement of the human diet to provide energy; it is not usually stored in the body, so needs to be included regularly in the diet (60 g per day).

See also PROTEIN SYNTHESIS.

protein synthesis The manufacture of all the PROTEINS needed by an organism. Protein synthesis is ultimately controlled by the DNA contained within the cell nucleus. A copy of the genetic code contained within the DNA is carried, in the form of RNA, to the cell cytoplasm, where it determines which AMINO ACIDS are to be linked in which order. This in turn determines the type of protein made.

Animals obtain the ESSENTIAL AMINO ACIDS from their diet and synthesize the others. Plants synthesize their own amino acids from NITRATES in the soil and CARBOHYDRATE products from, for example, the KREBS CYCLE. In the cell cytoplasm, the amino acids combine with a specific TRANSFER RNA (tRNA) by which they are carried to the RIBOSOMES. In the ribosomes, the amino acids meet MESSENGER RNA (mRNA), which is synthesized in the cell nucleus by a process called TRANSCRIPTION. Ribosomes then attach to one end of the mRNA (where there is a START CODON) and attract a specific amino acid carried by the tRNA in a process called TRANSLATION. In this way amino acids are linked in a specific order to form POLYPEPTIDES. Protein synthesis ends when a STOP CODON occurs on the mRNA. The polypeptides are then assembled into proteins.

prothrombin The precursor of the enzyme THROMBIN. See also BLOOD CLOTTING CASCADE.

protoctist A member of the kingdom PROTOCTISTA.

Protoctista A CLASSIFICATION kingdom made of single-celled EUKARYOTIC organisms with varied characteristics, including PROTOZOA, slime mould and ALGAE. Some, but not all, protoctists feed by photosynthesis, most have sexual reproduction and some have FLAGELLA. Important protoctists include *Plasmodium* (the cause of MALARIA), *Trypano-soma* (the cause of sleeping sickness), AMOEBA, *Phytophthora infestans* (the cause of potato blight and the Irish famine of 1845) and *Euglena*. See also OOMYCOTA, EUGLENOPHYTA.

protogynous (*adj.*) Describing a plant whose CARPELS develop before the STAMENS, ensuring that self-pollination does not occur. See POLLINATION.

proton A positively charged elementary particle found in the nucleus of an atom. The number of protons in a nucleus is called the atomic number and determines the number of ELECTRONS needed to produce a neutral atom, which in turn determines the chemical properties of the element. A proton has a charge equal but opposite to that on an electron.

protophloem In plants, primary PHLOEM tissue formed from the PROCAMBIUM at the expanding region of a shoot or root, whilst it is still elongating. Protophloem precedes METAPHLOEM.

protoplasm The CYTOPLASM and NUCLEUS of a CELL, including all the structures in the cytoplasm but excluding the CELL WALL, if present.

Prototheria A subclass of MAMMALIA comprising the MONOTREMES, the least advanced of the mammals. *Compare* EUTHERIA, METATHERIA.

protoxylem In plants, primary XYLEM tissue formed from the PROCAMBIUM at the expanding region of a shoot or root, whilst it is still elongating. Protoxylem vessel walls are thickened with rings or spirals of LIGNIN, which allow them to continue expanding. Protoxylem is succeeded by METAXYLEM.

protozoa (*sing. protozoan*) A collective term for a group of four phyla of the kingdom PROTOCTISTA, consisting of single-celled organisms without rigid cell walls. Protozoa are all aquatic, EUKARYOTIC organisms that lack chlorophyll. Most can be free-living or parasitic but some, for example *Plasmodium* (the cause of MALARIA), are solely parasitic.

The protozoa are divided into phyla mostly by their mode of locomotion. Some protozoa move by CILIA, for example *Paramecium*, some by FLAGELLA, for example

Trypanosoma (the cause of sleeping sickness), and some by pseudopodia (*see* PSEUDO-PODIUM), for example AMOEBA. Reproduction is commonly by BINARY FISSION. Protozoa form a large proportion of PLANKTON and are therefore an important food for fish and other aquatic animals. *See also* APICOMPLEXA, CILIO-PHORA, RHIZOPODA, ZOOMASTIGINA.

provirus A viral GENOME that is integrated into the CHROMOSOME of a host cell and remains latent there for long periods of time before being expressed. *See also* RETROVIRUS, VIRUS.

proximal convoluted tubule In the KIDNEY, the part of the NEPHRON between the BOWMAN'S CAPSULE and the LOOP OF HENLE along which useful minerals and water are reabsorbed back into the blood.

pseudocarp A soft fruit that results from parts other than the OVARY being incorporated into the fruit structure. An example is the strawberry, which develops from the RECEPTACLE (flower stalk) and the true fruits are the pips on the outer surface. Other examples are apples and pears, in which the outer skin and fleshy tissue develop from the receptacle after FERTILIZATION and the CARPELS (true fruits) form the core that surrounds the seeds. This grouped pseudocarp is called pome. The pineapple is a multiple pseudocarp that is composed of fleshy tissue from the receptacles of many flowers.

pseudopodium (*pl. pseudopodia*) An extension of PROTOPLASM that forms in some cells, for example AMOEBA and MACROPHAGES, and is used as a method of locomotion (amoeboid movement) or for engulfing food or foreign bodies.

Pteridophyta A division of the PLANT kingdom that is no longer used. It included the ferns, club mosses and horsetails, which are now classified into phyla of their own: FILICINO-PHYTA, LYCOPODOPHYTA and SPHENOPHYTA, respectively.

ptyalin An AMYLASE found in the SALIVA of some mammals, including humans.

pulmonary (*adj.*) Relating to the lungs.

pulse In vertebrates, a rhythmic throbbing that can be felt where ARTERIES are close to the surface, for example in the wrist. The pulse is caused by the contractions of the heart muscle forcing blood into the AORTA, which results in a sudden increase in pressure in the arteries that, because of their elastic walls, causes a swelling

or throbbing to pass through them. An average human pulse rate is about 70 beats per minute.

pupa, chrysalis The stage between LARVA and adult in INSECTS undergoing METAMORPHOSIS. The protective covering made by the larva to surround it during its pupal stage is called a cocoon.

pupil An aperture at the centre of the IRIS of the EYE. The size of the pupil can be adjusted by the muscles of the iris to control the amount of light entering the eye.

purine A type of organic base occurring in NUCLEIC ACIDS and NUCLEOTIDES that consists of a double ring, one with six sides and one with five. Adenine and guanine are the most common.

Purkinje fibres Specialized fibres of CARDIAC MUSCLE that conduct waves of excitation from the PACEMAKER (in the wall of the right ATRIUM) to the apex of the VENTRICLES. The Purkinje fibres form a network throughout the wall of the ventricles and ensure simultaneous contractions of both ventricles.

pus The yellowy fluid consisting of dead cells and bacteria that accumulates at the site of an INFLAMMATION.

putrefaction The largely anaerobic (*see* ANAER-OBE) decomposition of organic matter by micro-organisms. Foul-smelling AMINES are often produced.

pyloric sphincter A ring of muscles at the exit of the STOMACH that relax and contract to allow food to leave the stomach.

pyramid of biomass A diagrammatic representation of the dry mass of the organisms at each level of a FOOD CHAIN. Pyramids of biomass overcome some of the problems of a PYRAMID OF NUMBERS, but they are difficult to achieve. They are largely dependent on the time span used, for example the biomass of crops in summer can be very different to the biomass in winter. *See also* PYRAMID OF ENERGY.

pyramid of energy A diagrammatic representation of the energy used at each level in a FOOD CHAIN. A pyramid of energy is determined over a given time period, which provides an accurate picture of the food chain but is difficult to obtain. *See also* PYRAMID OF BIOMASS, PYRAMID OF NUMBERS.

pyramid of numbers A diagrammatic representation of the numbers of individuals at each level in a FOOD CHAIN. Each level is shown as a horizontal bar. Usually the numbers of primary consumers at the bottom of the pyramid exceeds the number of secondary consumers at the top of the pyramid, and so the overall shape is of a classic pyramid. However, the pyramid of numbers does not take into account the size of an organism, so a tree counts as one organism, as does an insect, which sometimes results in an inverted pyramid or bulging in the middle. *See also* PYRAMID OF BIOMASS, PYRAMID OF ENERGY.

pyrenoid A region of small protein bodies embedded in the CHLOROPLASTS of many algae that is concerned with converting the early products of PHOTOSYNTHESIS into compounds for storage.

pyridoxine *See* VITAMIN B.

pyrimidine A type of organic base occurring in NUCLEIC ACIDS and NUCLEOTIDES that consists of two single, six-sided rings. The most common are cytosine, thymine and uracil.

pyruvate Any salt or ester of PYRUVIC ACID.

pyruvic acid, *2-oxopropanoic acid* ($CH_3COCOOH$) A colourless, pleasant-smelling organic acid. It is an important product in the metabolism of carbohydrates and proteins. *See* KREBS CYCLE.

quadrat A folding square frame (usually 1 m^2) placed on the ground and used to study numbers and types of plants and animals within it, as a representation of an area as a whole. Sampling can be random or systematic (placed at regular intervals). Usually a number of quadrats are studied in order to reach a valid conclusion about the distribution, type and number of species within an area.

A point quadrat or point frame consists of two vertical poles with a horizontal bar across the top of them. A long pin is placed through holes at regular intervals along the horizontal bar and every species touched by the pin is recorded. The point quadrat is useful where there is dense vegetation because it samples at different levels.

Quadrats are not reliable for more mobile animals that move away when disturbed. To gain information on population sizes of such animals, MARK, RELEASE, RECAPTURE methods are used. *See also* TRANSECT.

R

radial symmetry A type of structure of an organ or organism in which a cut through the centre in any direction produces two halves that are mirror images of each other. It is a characteristic of the bodies of COELENTERATES and ECHINODERMS and also of many flowers. *Compare* BILATERAL SYMMETRY.

radiation In general, the emission of rays, waves or particles from a source. In particular, the term radiation is used for IONIZING RADIATION and ELECTROMAGNETIC RADIATION. *See also* BACKGROUND RADIATION, IRRADIATION, RADIOACTIVITY.

radiation sickness Sickness resulting from exposure to RADIATION. Symptoms include nausea, vomiting, diarrhoea, genetic damage to cells causing cancers and damage to germ cells resulting in subsequent birth defects. *See* IONIZING RADIATION, ELECTROMAGNETIC RADIATION.

radicle The part of the embryo in a SEED PLANT that becomes the primary ROOT. Once GERMINATION begins the radicle appears first and its apical MERISTEM pushes through the soil and may develop into the entire root system. *See also* ADVENTITIOUS ROOT.

radioactive isotope *See* RADIOISOTOPE.

radioactive tracer *See* RADIOLABELLING.

radioactive waste Any waste material that emits RADIATION in excess of the BACKGROUND RADIATION levels. Such waste comes from the nuclear power and nuclear weapons industries, mining of radioactive ores, hospitals and research laboratories.

Nuclear waste is of three types: high level waste from spent fuel, which has to be stored carefully until its RADIOACTIVITY has fallen to a safe level (this may take several thousand years); intermediate waste (reactor components and processing plant sludge), which may be long or short-lived; and low-level waste (lightly contaminated substances), which is short-lived but bulky. There is no safe method of disposing of radioactive waste but much is treated or stored until safer to dispose of and then buried in steel drums in deep chambers lined with concrete. No radioactive waste has been buried at sea since 1983.

radioactivity The spontaneous decay of unstable atomic nuclei with the emission of IONIZING RADIATION, either alpha, beta or gamma radiation. Radioactivity occurs spontaneously in many naturally-occurring radioisotopes without any external influence, and may also be induced in certain unstable nuclei by bombarding with subatomic particles.

Radioactivity can be harmful to living tissue because of the damage done to living cells by the ionizing radiation. In particular, the formation of free radicals in the vicinity of the DNA in a cell can lead to MUTATIONS that may cause cancer. However, radioactivity can also be used to kill cancerous cells: a cell is most vulnerable to genetic damage when it is dividing, and since cancer cells divide more rapidly than healthy cells, they are more easily killed.

See also RADIOTHERAPY.

radiocarbon dating A method of estimating the age of archeological specimens with a biological origin. All living organisms take in carbon from their surroundings. Some of this carbon is the radioactive isotope carbon–14, which is produced in the atmosphere by the interaction of cosmic rays with nitrogen nuclei. When an organism dies, it stops taking in carbon from its surroundings, and since carbon–14 has a half-life of about 5,700 years, measurement of the proportion of carbon–14 compared to stable carbon–12 enables the age to be determined. Radiocarbon dating assumes that the proportion of carbon–14 in the atmosphere has remained constant. This is not quite true and so there are inaccuracies in this method of dating, although consistent results are obtained for specimens up to 40,000 years old.

radiograph An X-RAY shadow picture often used in medical diagnosis, taken by placing the patient between a source of X-rays (from an X-ray tube) and a suitable detector such as a fluorescent screen or photographic film. *See* RADIOGRAPHY.

radiography The use of radiation, especially X-RAYS, to produce images on photographic film or FLUORESCENT screens. The image produced (radiograph) is the shadow cast by X-rays as they penetrate matter. Radiography is widely used in medicine to examine bone structures, since these absorb more radiation than soft tissues and so show up white on the radiograph. Another application is the use of an X-ray absorbing material, or CONTRAST ENHANCING MEDIUM, such as barium sulphate. This is introduced into the patient as a barium meal, to examine the stomach for example, or injected into an artery to produce an arteriogram (an image of the arteries). See also AUTORADIOGRAPHY.

radioimmunoassay (RIA) An IMMUNOASSAY in which a RADIOISOTOPE is used to label either the ANTIGEN or ANTIBODY involved.

radioisotope, *radioactive isotope* Any radioactive ISOTOPE. Radioisotopes are a valuable tool in science and medicine, since they can be incorporated into a compound and used as a tracer to facilitate detection of the compound (by the radiation it emits) as it passes through a system or experiment. A radioisotope replaces its stable counterpart in the compound, for example tritium atoms are often used to replace hydrogen atoms in a compound, which is then said to be tritiated. The radiolabelled compound will behave in the same way chemically and physically as the unlabelled compound.

radiolabelling The tagging of a compound with a RADIOISOTOPE to enable its path through a biological or chemical system to be monitored by its emission of radiation. Radiolabelled compounds are also known as tracers.

Radiolabelling is widely used in medicine, for example to see where new drugs go to in the body and in diagnosis of cancer, foetal abnormalities, heart disease and to monitor thyroid activity by administering radioactive iodine. Radiolabelling is also a valuable tool in the laboratory for following the course of a compound during an experiment.

radiology The study and use of IONIZING RADIATION such as X-RAYS and radioactive materials in medical diagnosis and in the treatment of diseases, including RADIOTHERAPY for cancer. See also RADIOGRAPHY.

radiotherapy The treatment of disease by RADIATION from an X-ray machine or a radioactive source. Radiation reduces the activity of dividing cells and is used particularly in the treatment of cancer. To ensure that healthy tissue does not receive a dose which may create MUTATIONS leading to further cancers, the radiation source is either implanted in the patient or is in the form of a beam aimed at the patient from several directions, overlapping to form a large dose at the location of the tumour.

The high-energy X-ray machines often used in radiotherapy are more powerful than the low-voltage machines used for taking X-rays. There can be unpleasant side-effects from radiotherapy, such as hair loss and nausea. Radioactive substances can be administered to a patient to have a more local effect, for example the use of radioactive iodine in the treatment of thyroid disease.

radula The tongue of MOLLUSCS, consisting of a horny strip bearing rows of teeth and used for rasping food.

rainforest A dense forest found in wet climates. Over half the tropical rainforests occur in Central and South America, and the rest in Africa and Southeast Asia. Temperate rainforests also exist, in New Zealand for example. Rainforests are of global importance because they contain a great diversity of species, including many useful plants (used, for example, as medicines, oils, resins and beverages), and provide much of the oxygen needed for plant and animal respiration. They also help regulate global weather patterns.

More than half of the world's rainforests have been destroyed (and not replaced) for timber or cleared to use the land for agricultural purposes. This is called DEFORESTATION, which can lead to DESERTIFICATION. Despite the destruction, rainforests still house most of our growing wood and many of the Earth's species of plants and animals.

receptacle The end of a flower stalk (PEDUNCLE) to which the flower parts are attached. The receptacle is often rounded but can be flattened or cup-shaped.

receptor 1. A cell that is specialized to receive a particular stimulus and convert this energy into a NERVE IMPULSE. The stimulus may be chemical, electrical, mechanical (movement, pressure) or temperature or light. See also SENSE ORGAN.

2. A PROTEIN in a MEMBRANE forming a RECEPTOR SITE.

3. Some HORMONES operate by intracellular receptors called nuclear receptors. These are proteins that bind to a hormone and are then able to bind a specific region of DNA to inhibit or stimulate TRANSCRIPTION of particular genes. STEROID HORMONES operate in this way.

receptor site A PROTEIN in a cell membrane with a specific function. This can be a channel-linked RECEPTOR associated with the passage of an ION, for example sodium, across the membrane. Other membrane receptor sites are associated with ENDOCYTOSIS, for example Fc receptors on MACROPHAGES. These bind to an ANTIBODY that has already bound a PATHOGEN, thus allowing endocytosis of the pathogen. In many cases, binding of a substance to a membrane receptor site triggers a cascade of secondary changes within the cell. For example, GROWTH FACTOR receptors (such as the INSULIN receptor) are linked to molecules such as ENZYMES, which are triggered upon receptor binding and control many aspects of the cell cycle, differentiation and GENE EXPRESSION.

recessive In genetics, an ALLELE that is masked by a DOMINANT allele and therefore is not expressed in the HETEROZYGOUS form. For example, the allele for blue eyes is recessive and the allele for brown eyes is dominant. A heterozygous individual with one blue and one brown allele will have brown eyes. A recessive allele will only be expressed in the HOMOZYGOUS form.

recombinant DNA DNA in which the NUCLEOTIDE sequence has been altered by its combination with another fragment of DNA. The nucleotide sequence can be altered by incorporation of, or exchange with, the new fragment of DNA. This can be a natural event occurring during MEIOSIS as a result of RECOMBINATION, or it can be a result of deliberate manipulation of genes. Recombinant DNA technology refers to deliberate manipulation. *See also* GENETIC ENGINEERING.

recombination A process in genetics in which genetic material is rearranged, resulting in genetic VARIATION of the offspring. This is evolutionarily advantageous. One of the recombination processes is CROSSING-OVER of some CHROMOSOME pairs during MEIOSIS, which results in the exchange of segments of chromosomes. The new combinations are called recombinants. Another recombination process

is random reassortment of the chromosomes when pairs are split to go into GAMETES (the gametes receive only one of each pair of chromosomes).

recombination frequency *See* CROSS-OVER VALUE.

rectum The final part of the DIGESTIVE SYSTEM in animals in which FAECES are stored before their elimination via the ANUS to the outside.

rectus muscle A muscle that is of equal width or depth throughout its length. *See also* EYE.

red blood cell, *erythrocyte* Any one of the cells carried in blood PLASMA that form almost half of the blood by volume and the majority by number. Human red blood cells are biconcave, 7 μm in diameter, have no nucleus, are made by red BONE MARROW and stored by the SPLEEN. They carry oxygen around the body complexed to HAEMOGLOBIN, which gives the cells their red colour.

Red blood cells have a life of about 4 months and are therefore constantly being destroyed in the liver and spleen and being replaced. There are about six million red blood cells per millilitre of adult human blood.

See also WHITE BLOOD CELL.

reducing agent A material that brings about a REDUCTION, being itself oxidized (*see* OXIDATION). Hydrogen is an important reducing agent.

reducing sugar A SUGAR that can act as a REDUCING AGENT in solution, as indicated by a positive BENEDICT'S TEST or FEHLING'S TEST. This depends on the presence of a free ALDEHYDE or KETONE group. Most MONOSACCHARIDES are reducing sugars, as are most DISACCHARIDES except SUCROSE. *See also* CARBOHYDRATE.

reduction In simple terms, reduction is the removal of oxygen from a substance, the addition of hydrogen or the addition of ELECTRONS to form a negative ion. It is the opposite of OXIDATION.

reductive division *See* MEIOSIS.

reflex An automatic, involuntary response in animals to a particular stimulus that is under the control of the NERVOUS SYSTEM but only involves a few NEURONES. In humans, examples of reflexes include the knee-jerk reflex, withdrawal of a hand or foot from a painful stimulus, breathing, dilation or constriction of the eye iris in response to light, sneezing, coughing and regulation of the heart rate.

A reflex arc is the pathway of neurones involved in a reflex action. It may consist simply of a stimulus sending a NERVE IMPULSE along a sensory neurone, to a single SYNAPSE (monosynaptic) in the spinal cord with an effector neurone causing a response. In vertebrates, however, there is often an intermediate (relay) neurone, involved in transmitting impulses. Reflex arcs with two or more synapses are termed polysynaptic. If a reflex arc is restricted to the spinal cord and does not involve the brain, it is a spinal reflex.

Involuntary reflexes are important, especially in primitive animals, in the avoidance of danger. Involuntary responses can be modified by experience (conditioned reflexes), where the first stimulus causing a simple reflex action becomes associated with a second stimulant that involves transmission of nerve impulses to the brain. An example of this was demonstrated by Ivan Pavlov (1849–1936), who conditioned dogs to salivate in response to the sound of a bell that indicated that food was to be delivered.

reflex arc The pathway of NEURONES involved in a REFLEX action.

relative molecular mass (rmm), *molecular weight* The mass of a molecule measured in atomic mass units. The relative molecular mass of a molecule is equal to the sum of the relative atomic masses of the atoms from which the molecule is composed.

relay neurone A NEURONE that is found within the CENTRAL NERVOUS SYSTEM that carries impulses away from here to an EFFECTOR cell (muscle or glands).

renal (*adj.*) Relating to the KIDNEY.

renal dialysis A term used to refer to the artificial removal of toxic substances from the bloodstream of patients whose kidneys have failed. Several methods of renal dialysis are used, all based on the principle of DIALYSIS. *See also* HAEMODIALYSIS.

renewable resource Natural resources that can be replaced naturally in a reasonable amount of time, for example wood, soil, water and fish. Although they are renewable, the continued supply of such resources relies on their proper use and conservation by humans. Renewable energy is power obtained from a renewable source, for example solar energy, wave power, hydroelectric power, geothermal energy and wind. NON-RENEWABLE resources cannot be replaced, for example coal, oil and metal ores. Some resources, for example used in motor

cars and tin cans, could be recycled but it is often uneconomical to do so.

renin An enzyme produced by the kidney in response to a drop in blood pressure. Renin stimulates ANGIOTENSIN.

rennin An enzyme that coagulates milk in the STOMACH and is therefore important to young animals in their digestion of milk. Rennin is secreted as the inactive prorennin, which is activated in the stomach to form rennin.

reproductive system The system of organs and tubes involved in SEXUAL REPRODUCTION.

reptile A member of the vertebrate class REPTILIA.

Reptilia A class of VERTEBRATES consisting of the reptiles. Reptiles lay hard-shelled eggs on land. The eggs are filled with yolk and fully formed young hatch from them. In some reptiles, for example lizards and snakes, the eggs remain inside the female and they give birth to live young. The skin of reptiles is covered with horny scales and they are POIKILOTHERMS, unable to maintain their body temperature. Reptiles usually have four legs and some live in water (but breathe air) and some live on land; they possess lungs. Their METABOLISM is slow and their teeth are all the same.

Many species of reptiles are now extinct, for example the dinosaurs, but surviving species include crocodiles, alligators, tortoises, lizards and snakes.

RES *See* RETICULOENDOTHELIAL SYSTEM.

resolution, *resolving power* The ability of an optical system, such as a camera, microscope or telescope, to show fine detail in an image. It is usually measured in terms of the closest separation of two objects that can just be shown to be separate by the system under consideration. The resolution is inversely proportional to the wavelength of light being used. Thus the resolution of a light microscope is limited to two points 0.2 µm apart, compared to typically 1 nm for the ELECTRON MICROSCOPE.

resolving power *See* RESOLUTION.

respiration The biochemical processes occurring within cells to break down food molecules to release energy. There are three stages to respiration: GLYCOLYSIS, the KREBS CYCLE and the ELECTRON TRANSPORT SYSTEM.

In glycolysis, glucose is broken down to pyruvate with the release of energy in the form of ATP. In the Krebs cycle, the pyruvate is broken down by a series of reactions to carbon dioxide and water, with the release of more energy. Unlike glycolysis, the Krebs cycle requires oxygen. In the electron transport system, hydrogen atoms from the Krebs cycle are carried along a chain of electron carriers with the generation of ATP, so providing energy.

Respiration is more correctly called 'internal cellular respiration' and contrasts with 'external respiration', which is the exchange of oxygen and carbon dioxide during BREATHING. Most food is converted into the sugar glucose, which is then converted into carbon dioxide and water with the release of energy. Fats and proteins can also be used as respiratory substrates (without being first converted to carbohydrates) but only during starvation (or dieting).

Respiration is usually aerobic (requiring oxygen) and occurs in MITOCHONDRIA of EUKARYOTES and MESOSOMES of PROKARYOTES. Some eukaryotic cells can function for short periods without oxygen, but most die. Some organisms, such as certain bacteria, yeast and parasites, are ANAEROBES and can use glucose to make energy without the use of oxygen. This is called anaerobic respiration and is less efficient than aerobic respiration, but because it produces alcohol and carbon dioxide it is of great use in the baking and brewing industry (see FERMENTATION).

In plants, there is a variable balance between respiration and PHOTOSYNTHESIS. The oxygen produced as a waste product of photosynthesis goes directly from the CHLOROPLAST, where it is made, to the mitochondria, where it can be used for respiration. This occurs during the night in most plants (C_3 PLANTS). During daylight, some plants engage in the apparently wasteful process of PHOTORESPIRATION. Carbon dioxide released by respiration can be used for photosynthesis.

respiratory chain See ELECTRON TRANSPORT SYSTEM.

respiratory pigment In most vertebrates and some invertebrates, one of a group of coloured proteins that can bind weakly to oxygen in the blood or other tissues to increase the uptake, transport and unloading of oxygen. HAEMOGLOBIN is the main example of a respiratory pigment in blood, which contains iron and colours the blood red.

Other examples in blood are haemocyanin, which contains copper and is blue, and

is found in the MOLLUSCS and CRUSTACEANS; haemoerythrin, which contains iron and is red/brown, and is found in ANNELIDS; chlorocruorin, which contains iron and is green, and is found in POLYCHAETES. MYOGLOBIN is found in muscles of all vertebrates but does not colour them.

Respiratory pigments are efficient because they have a high affinity for oxygen when there is a high concentration of oxygen (and so pick oxygen up easily), but a low affinity for oxygen when it is in low concentrations (so release oxygen where it is needed). Respiratory pigments can also bind to carbon dioxide or to carbon monoxide (see HAEMOGLOBIN).

The term respiratory pigment can also be used to refer to substances, such as CYTOCHROMES, involved in the ELECTRON TRANSPORT SYSTEM.

respiratory quotient (RQ) The ratio of the volume of carbon dioxide produced by an organism to the volume of oxygen consumed during aerobic RESPIRATION. The theoretical RQ value for carbohydrates as respiratory substrates is 1, for fats 0.7 and for proteins 0.8, although a mixture of these is normally utilized.

respiratory system The series of components involved in the exchange of gases (usually oxygen and carbon dioxide) by an organism. In humans, the respiratory system consists of two lungs each containing millions of air sacs called alveoli (see ALVEOLUS), where exchange of oxygen and carbon dioxide between air and the bloodstream takes place. The opening of the respiratory tract is considered to be the nose, through which air enters and passes to the mouth cavity, entering at the PHARYNX. The TRACHEA is the main airway running into the lungs, which branches into two tubes called bronchi (see BRONCHUS). The bronchi divide further into small tubes called bronchioles, and then into the air sacs or alveoli. Dust and other unwanted particles are cleared from the respiratory tract by fine hairs in the nose and by a sticky liquid called MUCUS secreted by the trachea and bronchus.

Birds, like mammals, have lungs in which gaseous exchange takes place. In fish, the respiratory system consists of a series of GILLS, and in insects gases enter and leave by SPIRACLES. In plants, the gases oxygen and carbon dioxide are involved in both respiration and photosynthesis, and gas exchange occurs through pores in the leaves called stomata (*see* STOMA). *See also* BREATHING.

restriction endonuclease See RESTRICTION ENZYME.

restriction enzyme, *restriction endonuclease* An enzyme derived from a bacterium that cuts a chain of DNA between specific NUCLEOTIDE base sequences. Many restriction enzymes exist that are specific for different nucleotide sequences. They are all NUCLEASE enzymes. Any fragment of DNA produced in this way can be joined to other DNA by the use of another enzyme called DNA LIGASE. Hence manipulation of DNA is possible, and is used in GENETIC ENGINEERING.

restriction fragment length polymorphism (RFLP) The occurrence of specific RESTRICTION ENZYME cleavage sites in different locations in the DNA of different individuals. Thus cleavage of DNA from different individuals at specific base sequences produces a set of DNA fragments which are different lengths in different people. This polymorphism is due to random changes in the bases in non-coding regions (INTRONS) of the DNA, which causes the deletion of existing restriction sites or the creation of new sites. RFLP has provided genetic markers which have transformed GENETIC MAPPING studies and GENE TRACKING through families. *See also* GENETIC FINGERPRINTING.

restriction mapping A method employed to determine the sites in a length of DNA that are cleaved by RESTRICTION ENZYMES. By using various combinations of restriction enzymes, a number of different DNA fragments of various sizes are obtained. These can be separated by gel ELECTROPHORESIS and specific fragments can be identified using GENE PROBES in SOUTHERN BLOTTING. A restriction map can be deduced showing the order of restriction sites in the original length of DNA. Changes in this pattern can then be detected indicating rearrangements or deletions in the genes. This technique is a useful tool for detecting genetic abnormalities. Restriction mapping is used in CHROMOSOME MAPPING.

reticuloendothelial system (RES) The system of circulating tissue MACROPHAGES, including those in CONNECTIVE TISSUE (histiocytes), LIVER (Kupffer cells), SPLEEN, LUNGS, LYMPH NODES and the LYMPHATIC SYSTEM.

retina In vertebrates and cephalopods (higher molluscs), a light-sensitive layer at the back of the EYE. In humans, the retina contains sensory cells called rods and cones, which convert the light energy they receive into NERVE IMPULSES that travel along the OPTIC NERVE to the brain. At the point where the optic nerve leaves the retina, there are no rods or cones and this is called the blind spot.

There are about 6 million cones per eye, mostly concentrated in a region on the retina called the fovea centralis. The cones are sensitive to colour and are used mostly for day vision. Because each cone has its own NEURONE connection to the brain, visual acuity is high compared to rods, which share neurones. There are more rods (about 120 million), which are distributed throughout the retina. Rods are mostly used for night vision and cannot distinguish colour.

The basic structures of rods and cones are similar apart from the shape. Each rod possesses thousands of vesicles containing a photosensitive pigment called rhodopsin. Exposure to light causes rhodopsin to split into its constituent parts, which generates an ACTION POTENTIAL resulting in a nerve impulse. The pigment is resynthesized by energy provided from MITOCHONDRIA, which are also present in the rod.

In cones, a similar process occurs but the photosensitive pigment is iodopsin, which is less sensitive to light (so more is needed to initiate a nerve impulse).

Colour vision is explained by the trichromatic theory, in which it is thought that three forms of iodopsin exist occurring in three different types of cones. These respond to three different types of light, green, blue and red, and other colours are perceived by a combined stimulation of these.

retinol *See* VITAMIN A.

retrovirus An important group of RNA viruses that can use the enzyme REVERSE TRANSCRIPTASE to make DNA from their single-stranded RNA. The DNA form of the virus is a PROVIRUS. The human immunodeficiency virus (HIV) causing AIDS is a retrovirus. Some human cancers are thought to be caused by retroviruses that can carry (and mutate) host cell genes capable of inducing cancer in other cells. These genes are called ONCOGENES and most viruses known to cause cancer contain them, but some cellular oncogenes exist that can be activated by other non-viral factors. Many retroviruses are harmless and in fact the proviral DNA may be integrated permanently into the host cell DNA and passed through generations as an endogenous virus.

reverse osmosis The passage of a solvent through a SEMIPERMEABLE MEMBRANE from a region of high SOLUTE concentration to one lower – that is, in the reverse direction to OSMOSIS. Reverse osmosis occurs on the application of a pressure greater than the OSMOTIC PRESSURE.

reverse transcriptase An enzyme that synthesizes single-stranded COPY DNA (cDNA) from a single-stranded RNA template. The enzyme is found in RETROVIRUSES, but is also useful experimentally in GENETIC ENGINEERING.

rhesus disease, *haemolytic disease of the newborn* A condition of newborn babies in which the foetal RED BLOOD CELLS are broken down, causing anaemia, heart failure and possible brain damage. Rhesus disease arises during pregnancy when a mother who does not have the RHESUS FACTOR in her blood (is rhesus negative) carries a foetus that is rhesus positive. The mother will produce ANTIBODIES to the rhesus ANTIGEN if foetal blood crosses the placenta, and these can pass across the placenta causing rhesus disease. Complete blood transfusion with rhesus negative blood is then necessary.

A first child is usually unaffected because foetal blood only passes into the mother towards the end of pregnancy or at birth and the antibody response is slow to build up, but it does continue to build up after birth. To prevent the antibodies affecting subsequent rhesus-positive foetuses, the mother is given anti-rhesus GLOBULIN just after the first pregnancy to prevent the formation of antibodies.

rhesus factor, *antigen D* In humans, a protein on the surface of RED BLOOD CELLS involved in the rhesus BLOOD GROUP SYSTEM. The factor was first identified in rhesus monkeys, hence its name. Most individuals (75–85 per cent) possess the factor and are called rhesus positive (Rh+), while those who do not carry the factor are rhesus negative (Rh–). Anti-rhesus ANTIBODIES are produced by Rh– people after exposure to the rhesus ANTIGEN (the antibodies are not naturally occurring). The rhesus system can cause a problem during pregnancy as a Rh– mother carrying a Rh+ baby will produce antibodies to the rhesus antigen if foetal blood crosses the placenta. The antibodies can pass across the placenta causing RHESUS DISEASE.

rhesus system A classification of human blood types based on the presence or absence of the RHESUS FACTOR. Most individuals carry the rhesus factor and are termed rhesus positive (Rh+) whilst those who do not carry the factor are rhesus negative(Rh–). Presence of the rhesus factor is unrelated to the other main BLOOD GROUP SYSTEM, the ABO SYSTEM. Blood must be matched for rhesus compatibility in transfusions. The rhesus system can cause a problem during pregnancy if a rhesus negative mother carries a rhesus positive baby and RHESUS DISEASE occurs.

rheumatoid arthritis An autoimmune disease (*see* AUTOIMMUNITY) in which the joints, particularly of the hands, are affected. There is swelling of the tissue around the joint and the synovial membrane, a build up of fluid in the joint and destruction of the cartilage and bone. This causes extreme pain and deformity. Treatment with drugs can help to reduce pain and inflammation and slow down or halt the disease. Rheumatoid arthritis is not yet fully understood. The disease sometimes subsides spontaneously for long periods, notably during pregnancy, but flare-ups also occur.

rhizoid A simple hair-like outgrowth on a plant that serves as a ROOT because it is able to absorb water and nutrients. Rhizoids are often found on BRYOPHYTES, some FUNGI and ALGAE.

rhizome An underground horizontal, branching stem, protected by scaly leaves. An example is the iris. The stem grows horizontally year after year, and buds form between the scaly leaves and develop into new vertical shoots. It is a structure of VEGETATIVE REPRODUCTION.

Rhizopod A member of the phylum RHIZOPODA.

Rhizopoda A phylum from the kingdom PROTOCTISTA that consists of PROTOZOA. The members (rhizopods) move and trap food by means of PSEUDOPODIA. Reproduction is generally by BINARY FISSION but can be by the formation of GAMETES. Rhizopods can be free-living or PARASITES, and are an irregular shape. An example is *Amoeba*.

Rhodophyta A phylum from the kingdom PROTOCTISTA that consists of red ALGAE. Red algae are mostly marine; they contain CHLOROPHYLLS a and d and other pigments, which give them their colour. They are characterized by a complete absence of FLAGELLA at any stage.

rhodopsin A light-sensitive pigment found in the RODS of the eye. *See* RETINA.

rib A long, narrow, curved bone of vertebrates extending from the spine. In humans, there are 12 pairs, joined at the back to the vertebrae of the spine. The ribs form the ribcage, which can move to allow chest expansion by movement of INTERCOSTAL MUSCLES between the ribs. In humans, the upper seven ribs are joined by cartilage to the breast bone (sternum) at the front of the THORAX; the next three are joined to each other at the ends by cartilage; and the last two (floating ribs) are not attached at the front. In fish and reptiles, the ribs extend along most of the spine, but in mammals they are confined to the thorax where they provide protection for the lungs and heart.

ribcage The structure formed by the RIBS that encloses, supports and protects the heart and lungs. The ribcage consists of pairs of ribs attached at the back of the THORAX to the vertebrae of the spine (*see* VERTEBRAL COLUMN) and at the front to the breastbone (STERNUM). The ribcage can move up and down to allow it to expand, by movement of the INTERCOSTAL MUSCLES between the ribs.

riboflavin *See* VITAMIN B.

ribonuclease *See* RNASE.

ribonucleic acid *See* RNA.

ribose ($C_5H_{10}O_5$) A PENTOSE sugar that is a component of RNA.

ribosomal RNA (rRNA) A type of RNA with a large molecular mass, making up more than half the mass of a cell's total RNA and more than half the mass of the RIBOSOMES. Ribosomal RNA can be either a single or double helix and its base sequence is similar in all organisms.

ribosome Any one of many granules of PROTEIN and RIBOSOMAL RNA (rRNA) in a cell that are the site of PROTEIN SYNTHESIS. Ribosomes are associated with ENDOPLASMIC RETICULUM (as rough ER) in EUKARYOTIC cells, and are present in the cytoplasm of both eukaryotic and PROKARYOTIC cells. Ribosomes of prokaryotic cells are slightly smaller than those of eukaryotic cells. Ribosomes are often linked together to form chains of polyribosomes or polysomes.

ribozyme RNA molecules that can act as ENZYMES as well as PROTEIN.

ribulose bisphosphate A carbon dioxide acceptor in the CALVIN CYCLE.

rickets A bone disorder caused by lack of calcium deposits, in which the bones do not harden and therefore bend out of shape. It results from a deficiency of VITAMIN D or PHOSPHATES. Rickets is also associated with kidney disease.

RNA (ribonucleic acid) A NUCLEIC ACID associated mainly with the synthesis of proteins from DNA. RNA is found in the nucleus and cytoplasm of cells. It is usually a single-stranded chain of NUCLEOTIDES synthesized from DNA by the formation of BASE PAIRS. The organic bases in RNA are adenine, guanine, cytosine and uracil (which replaces the thymine of DNA) and a PENTOSE sugar that is always RIBOSE. There are three main forms of RNA, all concerned with PROTEIN SYNTHESIS. These are RIBOSOMAL RNA (rRNA), TRANSFER RNA (tRNA) and MESSENGER RNA (mRNA). In some viruses, for example RETROVIRUSES, RNA can make up the hereditary material, instead of DNA.

RNA polymerase A POLYMERASE enzyme that catalyses the synthesis of RNA from a DNA template or from an RNA strand. There are three types of RNA polymerase. Type I is involved with the synthesis of RIBOSOMAL RNA; type II with the synthesis of MESSENGER RNA (*see* TRANSCRIPTION) and type III with TRANSFER RNA synthesis.

RNAse (ribonuclease) One of many enzymes that hydrolyse RNA by breaking down the sugar–phosphate bonds. *See also* RESTRICTION ENDONUCLEASE.

rod A type of light-sensitive cell found in the RETINA of most vertebrates. Rods are mostly used for night vision and cannot distinguish colour.

rodent A member of the order Rodentia in the class MAMMALIA. It includes most of the species of mammals. Rodents are characterized by the presence of a single pair of incisors at the front of the upper and lower jaw that continue to grow as they are worn down. They are placental mammals and can be omnivores (plant or meat eating) or herbivores (plant eating). Examples include rats, mice, squirrels, beavers, porcupines and guinea pigs.

rodenticide A PESTICIDE that kills RODENTS.

root In botany, the part of a plant that is usually underground, that anchors the plant and absorbs water and dissolved mineral ions. Roots usually grow downwards and towards water (*see* TROPISM) but some plants produce aerial roots that absorb moisture from the atmosphere or provide further anchorage. The root has a central column of vascular tissue (XYLEM and PHLOEM) surrounded by a single-celled ring of ENDODERMIS. Inside the endodermis is a layer of PARENCHYMA, called the PERICYCLE, from which lateral roots originate. The bulk of the root is made of cortex (parenchyma) storing starch, which is surrounded by the outer exodermis and a one-cell thick epidermal layer with ROOT HAIRS. The end of the root is protected by a root cap, where the actively dividing cells push through the soil (*see* MERISTEM).

Water and dissolved mineral ions are absorbed into the root hairs, through the cortex, endodermis and pericycle into the xylem, to travel upwards through the roots, stems and to the leaves by TRANSLOCATION. The roots of certain plants can form a symbiotic (*see* SYMBIOSIS) relationship with some bacteria (*see* ROOT NODULE). *See also* ADVENTITIOUS ROOT, CONTRACTILE ROOT, FIBROUS ROOT, PNEUMATOPHORE, TAP ROOT.

root hair A delicate extension from the epidermal cells (*see* EPIDERMIS) on the surface of a plant ROOT. The layer of cells producing the root hairs is called the piliferous layer. Root hairs are found near the tip of young roots and increase the surface area for absorption of water and mineral ions. They function for only a few weeks and are continually replaced nearer the growing tip. *See also* TRANSLOCATION.

root nodule A swelling on the ROOTS of leguminous plants (LEGUMES), such as beans, peas and clover, that is caused by infection with nitrogen-fixing bacteria such as *Rhizobium* (*see* NITROGEN FIXATION). A symbiotic relationship exists: the bacteria convert atmospheric nitrogen into nitrates that the plant can use, independently of soil nitrates, and in return the bacteria get a source of nutrition. The nodule develops like a lateral root and during the early stages of development central cells fill up with the bacteria enclosed in a membrane (thus the bacteria remain extracellular). Neither the legume nor the bacteria alone can fix nitrogen, and it is thought that the site of fixation is this membrane separating the two.

Leguminous plants are very important in improving soil fertility because they can use atmospheric nitrogen. Without them the level of nitrates in the soil would not be sufficient to support vegetation cover. *See also* NITROGEN CYCLE.

root tuber In certain plants, a swollen region of an underground ROOT used for storing food during the winter or a dry season, thus ensuring survival of the plant. A root tuber gives rise to new plants. An example of a root tuber is the dahlia. *See also* TUBER. *Compare* STEM TUBER.

roughage, *dietary fibre* Plant material, mainly CELLULOSE from plant cell walls, that passes through the human digestive tract almost unchanged (it is not digested). The function of roughage is to provide bulk to the other material passing through the digestive system, stimulate PERISTALSIS, and ease movement of food throughout the digestive system. Roughage is important for healthy living and too little can lead to intestinal disorders and constipation.

round window A membrane-covered opening between the middle EAR and the inner ear. It is displaced by movement of another membrane called the OVAL WINDOW. *See also* EAR OSSICLES, COCHLEA.

RQ *See* RESPIRATORY QUOTIENT.

rRNA *See* RIBOSOMAL RNA.

rumen The first of four chambers that form the stomachs of RUMINANTS.

ruminant An even-toed mammal with a complex digestive system, including several stomach chambers, for the digestion of plant food. Cattle, goats, deer and giraffes all have a four-chambered stomach, while camels have three chambers. Food is stored and initial digestion carried out in the first chamber, called the rumen. This involves the enzyme cellulase (*see* CELLULOSE), which is provided by symbiotic bacteria. The food is then returned to the mouth for further chewing before being passed to the next stomach.

runner A vegetatively reproducing structure similar to a STOLON. Unlike in the stolon, stems develop from the parent plant and travel across the soil surface, periodically producing ADVENTITIOUS ROOTS, and new plants grow at these points. Examples include the strawberry and buttercup.

S

saccharide *See* MONOSACCHARIDE, DISACCHARIDE and POLYSACCHARIDE.

saccharin An artificial sweetener that is 500 times sweeter than sugar but with a bitter aftertaste and potentially carcinogenic (*see* CARCINOGEN). It has therefore largely been replaced by other sweetening agents. *See also* ASPARTAME.

sacculus A sac-like structure within the inner EAR concerned with balance and from which the COCHLEA arises. The sacculus lies below the UTRICULUS and together they provide a link between the cochlea and SEMI-CIRCULAR CANALS. The sacculus contains a patch of sensory hair cells by which movement in relation to gravity is detected. *See also* MACULA.

saliva A fluid produced by the SALIVARY GLANDS that helps digestion of food in the mouth. Saliva is over 99 per cent water, and contains an enzyme, called salivary AMYLASE, that converts starch to sugar. It also contains mineral ions to maintain the correct pH for the enzyme, and a sticky MUCIN that helps bind food and make it easier to swallow.

salivary amylase, *ptyalin* An AMYLASE enzyme present in the SALIVA of some mammals, including humans.

salivary gland Any EXOCRINE GLAND in the mouth that produces SALIVA. There are three pairs of salivary glands: sublingual, submaxillary and parotid. *See also* MEROCRINE GLAND.

Salmonella A genus of BACTERIA that cause typhoid and paratyphoid fevers and salmonella food poisoning. The bacteria are carried in food and water and can be passed on by human carriers in the unhygenic preparation of food. Domestic pets can also be carriers. Food poisoning can be severe but is usually not fatal. Vaccination against typhoid fever is available.

salt Any compound produced by a reaction in which some or all of the hydrogen in an acid is replaced by a metal or other positive ion.

SA node *See* PACEMAKER.

saprophyte *See* SAPROTROPH.

saprotroph, *saprophyte* An organism that feeds on dead organisms or excrement. Saprotrophs cannot make food for themselves and are a type of HETEROTROPH. Most saprotrophs are bacteria or fungi and they absorb material through their cell walls, some directly and others after breakdown by enzymes that they release onto the food material to cause extracellular digestion.

Saprotrophs play an essential role in recycling elements in dead organisms, making them available again to living organisms, but they can also cause spoilage of food or other materials at great economic loss. Some are useful to humans because of the by-products in their breakdown of organic material, for example they are used in brewing, baking and yoghurt production.

Compare DETRITIVORE. *See also* DECOMPOSER.

sarcolemma The membrane that surrounds a muscle fibre (*see* MUSCLE). Infoldings of this membrane, called transverse tubules (T tubules), project into the fibre over the Z-discs (*see* MUSCULAR CONTRACTION). When those tubules make contact with the SARCOPLASMIC RETICULUM, an ACTION POTENTIAL is transferred from the sarcolemma to the sarcoplasmic reticulum. This leads to the release of stored calcium ions and subsequent contraction of the muscle fibres.

sarcoma A malignant TUMOUR of CONNECTIVE TISSUE, for example of bone, cartilage or blood.

sarcomere The repeating unit of cross-striations in a MUSCLE fibre.

sarcoplasm The CYTOPLASM of MUSCLE fibres.

sarcoplasmic reticulum A system of internal membranes found within MUSCLE fibres that controls calcium ion concentration and is important in MUSCULAR CONTRACTION.

scanning electron microscope A type of ELECTRON MICROSCOPE developed in the 1960s, in which a fine beam of electrons passes over the surface of the specimen and some electrons are absorbed and others are reflected. Secondary electrons may also be emitted by the specimen, and these and the reflected electrons are amplified to form an image showing the three-dimensional exterior of the specimen on a screen. The RESOLUTION is not as good as in the TRANSMISSION ELECTRON MICROSCOPE (about 10 nm) and the overall magnification is 10–200,000. *See also* SCANNING TRANSMISSION ELECTRON MICROSCOPE, SCANNING TUNNELLING MICROSCOPE.

scanning transmission electron microscope (STEM) A type of ELECTRON MICROSCOPE that combines features of both the TRANSMISSION ELECTRON MICROSCOPE and the SCANNING ELECTRON MICROSCOPE to produce a magnification of 90 million. A fine beam of electrons moves over the specimen but a thin slice is used so that the electrons also pass through it. A computer interprets the electrical signal formed by reflected electrons and those penetrating the specimen, to form an image on a screen.

scanning tunnelling microscope A type of ELECTRON MICROSCOPE invented in 1981 that magnifies an image by passing a small tungsten probe over the surface of the specimen. The tip of the probe may be as fine as one atom, and it moves very close to the specimen so that electrons jump (or tunnel) between the specimen and probe. The magnitude of the electron flow depends on how close the probe is to the specimen, and therefore the contours of the surface can be determined and transmitted to a screen. Images can be magnified 100 million times which means individual atoms can be resolved.

scapula In humans, the shoulder blade. *See* PECTORAL GIRDLE.

Schwann cell A specialized GLIAL CELL responsible for the formation of the MYELIN SHEATH surrounding some nerve AXONS.

sclera *See* SCLEROTIC COAT.

sclerotic coat, *sclera* The white of the EYE; the tough outer layer of the eye.

sclereid A cell type of plant SCLERENCHYMA. Sclereid cells are round and found singly or in small clusters in the hard shells of fruit, seed coats and bark. The walls of sclereids contain LIGNIN, which makes them impermeable to water and gases, so mature sclerenchyma cells die.

sclerenchyma Plant tissue consisting of thick-walled cells providing strength and support. When mature, the sclerenchyma cells die and the cell contents are lost, leaving only the wall.

The CELL WALL contains large deposits of LIGNIN that provides extra strength. Lignin is difficult to digest and so provides protection from attack by many organisms. There are some regions of the cell wall where PLASMODESMATA are present and lignin is not deposited, so forming pits, through which water and dissolved minerals pass from one cell to another. Some sclerenchyma cells are spherical (sclereids) and are found singularly or in small clusters in hard shells of fruit, seed coats, bark and stem CORTEX. Other cells are elongated (fibres) and are found in bundles providing the main supporting tissue of mature stems. Sclerenchyma cells are associated with the vascular tissue (*see* PHLOEM, XYLEM).

SCP *See* SINGLE CELL PROTEIN.

scrapie A disease of sheep affecting their CENTRAL NERVOUS SYSTEM, causing progressive degeneration. Scrapie is characterized by the spongy appearance of the brain after death, due to the presence of numerous holes in the tissue.

The disease is thought to be caused by a 'prion': a protein fragment capable of self-replication in animal cells. It is possible that the causal agent switches on a gene that encodes the prion protein.

See also BOVINE SPONGIFORM ENCEPHALOPATHY, CREUTZFELDT–JAKOB DISEASE.

scrotum In male mammals, a skin sac that contains the TESTES and keeps the SPERM cooler than body temperature. The muscles in the wall of the scrotum can contract and relax to bring the sac closer to or further from the ABDOMEN as the temperature varies.

scurvy A deficiency of VITAMIN C in which an abnormal type of COLLAGEN is made. Scurvy leads to weakness of muscles and joints, skin sores and ulcers, bleeding of teeth, gums and other organs due to burst capillaries, and dry skin and hair.

Scyphozoa A class of the phylum CNIDARIA, including the jellyfish.

sebaceous gland In mammals, a gland in the SKIN that produces an oily secretion called sebum, which gives skin and hair its water-resistance and lubrication. Over-secretion of sebum is a cause of acne.

sebum *See* SEBACEOUS GLAND.

secondary growth, *secondary thickening* In plants, the increase in thickness of shoots and roots derived from the CAMBIUM. Secondary growth includes the development of secondary XYLEM and PHLOEM, new water – and food – conducting vessels, and accounts for most of the mature structure of the plant. It is seen in DICOTYLEDONS and GYMNOSPERMS but rarely in MONOCOTYLEDONS. *Compare* PRIMARY GROWTH.

secondary thickening *See* SECONDARY GROWTH.

second filial generation *See* F$_2$ GENERATION.

secretin A hormone secreted by the SMALL INTESTINE in response to acid CHYME from the STOMACH. Secretin inhibits production of GASTRIC JUICE and stimulates production of PANCREATIC JUICE and BILE. *See also* PANCREAS.

secretion The production and release of a substance, for example hormones and enzymes, from a cell or specialized gland. The substance produced can also be called a secretion. *Compare* EXCRETION.

secretory cell A general name for any cell that produces and releases a specific substance. *See also* SECRETION.

seed The reproductive structure containing the EMBRYO and food stores that develops from a fertilized OVULE in higher plants (ANGIOSPERMS and CONIFERS). The seed is protected by a hard, impermeable outer seed coat or TESTA. After fertilization (*see* DOUBLE FERTILIZATION) the diploid ZYGOTE divides by MITOSIS and develops into the embryo, and the food source for the growing embryo is the ENDOSPERM or the COTYLEDONS.

In flowering plants (angiosperms), the seeds are enclosed within a FRUIT that protects them during development. In conifers, the seed is unprotected. When the conditions (such as temperature and water) are favourable, the seed will germinate (*see* GERMINATION) and develop into a new plant. Until the conditions are right, the seed will remain in a state of DORMANCY.

seed leaf *See* COTYLEDON.

seed plant Any plant bearing SEEDS. There are two phyla of seed-bearing plants, the CONIFEROPHYTA and ANGIOSPERMOPHYTA. Together, these form the largest group of plants, including most of the Earth's land vegetation, and the structurally most complex plants.

selection pressure The effect that the environment has on the transmission of particular genes, or the organisms possessing them, whether it be to eliminate or to favour them. *See* NATURAL SELECTION.

self-pollination The transfer of pollen from the male part to the female part of the same plant. *See* POLLINATION.

semen The fluid in which SPERM are carried during their ejaculation from the PENIS at copulation or mating. Semen is secreted by the SEMINAL VESICLES, PROSTATE GLAND and COWPER'S GLANDS. It contains alkaline chemicals (to neutralize the acidity of the vagina which would kill sperm), sugars (which nourish the sperm and help make them mobile), MUCUS (in which the sperm swim) and PROSTAGLANDINS (which cause muscular contractions of the UTERUS and OVIDUCTS, so helping the sperm reach the OVUM).

semi-circular canal Any one of three tubes in the inner EAR concerned with balance. Semi-circular canals are linked to the COCHLEA by two sac-like parts, the UTRICULUS and the SACCULUS, both of which contain a fluid called endolymph. If there is movement of the head, such as a turning movement, the endolymph displaces a gelatinous plate, called the cupula, that lies within the ampulla, a swollen region in each canal. This movement of the cupula is detected by sensory hairs and transmitted to the brain by the auditory nerve so that the imbalance can be corrected. The utriculus and sacculus provide information on the position of the body relative to gravity, as well as movement due to acceleration or deceleration. This is achieved by deposits of calcium carbonate called OTOLITHS, which are embedded in a jelly-like substance in a small region called the MACULA. The otoliths respond to vertical and lateral movements of the head and displace sensory hairs, thereby transmitting a message to the brain.

semi-conservative replication The method by which DNA replicates under the control of the enzyme DNA POLYMERASE. During this process DNA polymerase causes the two strands of DNA to separate. The NUCLEOTIDE bases on each strand then join with new, free complementary bases and the strands rejoin. Part of the strand is rejoining while the remaining unpaired bases continue to attract their complimentary nucleotides. The result is four strands making two identical DNA molecules, each consisting of one original strand and one new strand.

Evidence for semi-conservative replication comes from the experiments of Meselsohn and

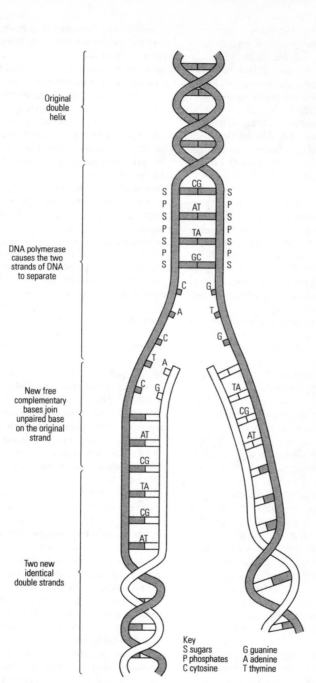

Original
double
helix

DNA polymerase
causes the two
strands of DNA
to separate

New free
complementary
bases join
unpaired base
on the original
strand

Two new
identical
double strands

Key
S sugars   G guanine
P phosphates   A adenine
C cytosine   T thymine

Semi-conservative replication of DNA.

Stahl, who followed the incorporation of the ISOTOPE nitrogen–15 (heavy nitrogen) into DNA of the bacteria *Escherichia coli.*

semi-lunar valve Any one of two crescent-shaped non-return valves, one in the AORTA and one in the PULMONARY artery, that prevent blood from re-entering the heart.

seminal vesicle A sac-like structure extending from each of the VAS DEFERENS in the reproductive system of male mammals. The seminal vesicles produce a MUCUS secretion that forms part of the volume of SEMEN. *See also* SEXUAL REPRODUCTION.

seminiferous tubule One of many coiled tubes in the male TESTIS that are lined with germinal EPITHELIUM cells from which SPERM are produced. In between the tubules lie INTERSTITIAL CELLS, which secrete the male hormone TESTOSTERONE. The seminiferous tubules merge to form the vasa efferentia, a group of small ducts through which mature sperm pass into the EPIDIDYMIS to be stored.

semipermeable membrane A material through which one type of molecule can pass but not another. Typically, such membranes are considered in processes such as OSMOSIS, where the SOLVENT molecules can pass through the membrane, but the SOLUTE cannot.

One explanation of this effect is that the solute molecules are too large too pass through the membrane, which acts as a 'molecular sieve'. However, some membranes still work with IONIC solutions where the solute ions are smaller than the solvent molecules, suggesting that this cannot be the explanation for every case. *See also* CELL MEMBRANE, DIALYSIS.

sense organ An organ containing nervous tissue that has specialized RECEPTORS for detecting specific stimuli and converting them into a NERVE IMPULSE, so that an animal can gain information about its surroundings. In humans, the main sense organs are the EYE (which detects light and colour), the EAR (which detects sound and balance), the NOSE (which detects smell), the TONGUE (which detects taste) and small sense organs in the SKIN (which detect temperature, pressure and pain).

Chemoreceptors are receptor cells that detect chemicals. They are involved in taste and smell and are found on the tongue and nose. Olfactory (smell) receptors are linked with taste and are found at the back of the nose. Photoreceptors detect light. Mechanoreceptors detect

pressure changes, gravity and vibrations (sound) and there are many in the ear. Thermoreceptors detect temperature changes, for example in the skin. Electroreceptors detect electrical fields (which is important in fish).

Many senses, for example hearing and smell (important for detecting food, danger and mates), are more acutely developed in animals other than humans. Some species see outside the human spectrum, for example insects can see in the ultraviolet range and snakes can see in the INFRARED range. Most mammals, however, cannot distinguish colours. In some animals, the PINEAL GLAND can detect light and keeps a track of daylength and seasons.

sensory neurone A NEURONE that transmits NERVE IMPULSES containing information about the internal and external environment to the CENTRAL NERVOUS SYSTEM.

sensory system The part of the NERVOUS SYSTEM that consists of RECEPTORS collecting information from the internal and external environment, and SENSORY NEURONES carrying this information to the CENTRAL NERVOUS SYSTEM where it is processed.

sepal A part of a flower that forms the outer layer of the PERIANTH. The collective term for sepals is the calyx. Sepals are derived from modified leaves and are usually green and capable of PHOTOSYNTHESIS. Their main function is to protect the other floral parts when the flower is a bud. In some flowers, the sepals are brightly coloured and assist in attracting insects.

septum (*pl. septa*) Any dividing partition in a plant or animal.

serine An AMINO ACID found in many proteins.

serotonin, 5-hydroxytryptamine A substance derived from the amino acid TRYPTOPHAN. It is secreted by the PINEAL GLAND in the brain and acts as a NEUROTRANSMITTER. Serotonin is found in the brain, blood PLATELETS and the intestines. It causes a general lack of activity (opposes NORADRENALINE), influences mood and also induces the constriction of blood vessels. The hallucinogenic drug LSD acts as an antagonist of serotonin.

serous membrane A thin membrane lining a closed body cavity, such as the PLEURAL CAVITY and PERITONEAL CAVITY. The membranes are the PLEURA and PERITONEUM respectively. A serous membrane consists of a thin layer of

MESOTHELIUM attached to a surface by a thin layer of CONNECTIVE TISSUE.

Sertoli cells The cells lining the SEMINIFEROUS TUBULES that protect and nourish immature SPERM cells. *See* TESTIS.

serum The fluid that remains after blood has been allowed to clot and the clot removed. Serum is a watery, yellowish liquid, with the same composition as blood PLASMA minus the clotting substances. It contains many dissolved proteins, including antibodies, sugars, fats, hormones and salts.

Antiserum is blood serum containing antibodies to a specific ANTIGEN, and can be made artificially by administrating the required antigen. This can be useful in protection against some diseases (*see* VACCINATION).

severe combined immune deficiency (SCID) *See* IMMUNODEFICIENCY.

sewage disposal The removal of any water-borne waste products passing through sewers, including human excrement, household waste and industrial waste. The sewage passes through sewers to a sewage works to be treated before its discharge into rivers and the sea. Untreated (or raw) sewage causes water POLLUTION and EUTROPHICATION.

In the treatment of sewage, large pieces of debris are first filtered out or broken down and heavier inorganic material (detritus and grit) is allowed to deposit out. The remaining sewage flows into large tanks for primary sedimentation, where sand, silt and organic material settles out, forming sludge that can be digested into simpler compounds and semi-dried to be used as fertilizer or dumped at sea. Methane gas is produced during the digestion of sludge, which can be used to generate electricity, so potentially making sewage treatment works self-sufficient. The remaining sewage, with most solid material removed, is called effluent and is removed for further break down of dissolved organic materials by a variety of aerobic micro-organisms, for example urea is broken down into nitrates by *Nitrosomonas* and *Nitrobacter* (*see* NITRIFICATION, AMMONIFICATION). These micro-organisms are allowed to settle out, forming HUMUS that is digested with the sludge. The remaining effluent is then safe to pass into rivers and seas.

Treatment of sewage successfully removes complex chemicals such as DDT, organic compounds and most PATHOGENS. However, some potential pollutants, such as phosphorus and copper, are only partially removed and small numbers of some pathogens may survive (including those causing paratyphoid and dysentery).

sex chromosome A CHROMOSOME that determines the sex of an organism. Most animals, including humans, have two sex chromosomes – the X-chromosome and the Y-chromosome. In humans, a pair of X-chromosomes produces a female (XX) and a non-identical pair of an X- and a Y-chromosome produces a male (XY). The sex chromosomes carry GENES which regulate development of sex organs and secondary sexual characteristics. Some other genes carried on the sex chromosomes have nothing to do with the sex of an organism, but are linked by virtue of their position (*see* SEX LINKAGE). The X-chromosome is longer than the Y-chromosome and carries more genes. *See also* SEX DETERMINATION.

sex determination The process by which the sex of an organism is determined. In humans, as in many other species, sex determination is dependent on two sex CHROMOSOMES, the X-chromosome and the Y-chromosome. The Y-chromosome is shorter than the X-chromosome and lacks many of the GENES carried on the X-chromosome. A pair of identical X-chromosomes (XX: homogametic) produces a female, whereas a non-identical pair of an X- and Y-chromosome (XY: heterogametic) produces a male. Maleness is caused by a single gene (of 14 BASE PAIRS) on the Y-chromosome that is present but inactive on the X-chromosome.

Sex determination is different in other groups of animals. In birds, fish, butterflies and reptiles, the males are mostly XX and the females XY. In other insects, the male has no Y-chromosomes and is XO. In some fish and reptiles, environmental factors such as temperature can affect sex determination.

sex linkage The association of a particular characteristic with the sex of an organism. Sex-linked characteristics are carried by GENES present on the SEX CHROMOSOMES, which may have nothing to do with the sex of an organism but are linked to it by their position. The X-CHROMOSOME carries many such genes. Two examples of X-linked genes in humans are those causing HAEMOPHILIA and red/green colour blindness. Both genes are RECESSIVE and

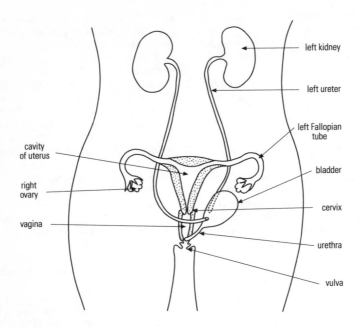

Human female reproductive system.

are expressed almost exclusively in males because the males usually have only one X- and one Y-chromosome (XY). If the recessive ALLELE is present on the X-chromosome then it is expressed because there is no possibility of it being masked by a DOMINANT allele on another X-chromosome. For haemophilia or red/green colour blindness to occur in females (usually XX), both X-chromosomes would have to carry the recessive allele, which is very rare. Usually one of the female X-chromosomes has the dominant allele that masks the recessive one. Such a female is called a carrier, because she can pass the recessive allele onto her offspring, where it may be expressed in a male.

Dominant mutant X-linked genes are unusual but can occur, for example the absence of incisor teeth, and these are more often seen in females because there is twice the chance of it occurring. Genes linked to the Y-chromosome are rare and will only occur in males, where they will be inherited whether they are dominant or recessive.

See also LINKAGE, MENDEL'S LAWS.

sexual reproduction The production of offspring, requiring the union of two GAMETES, usually from different individuals, that are genetically different from either of the parents. Most organisms (except BACTERIA and CYANOBACTERIA) show some form of sexual reproduction. The gametes are usually female ova (eggs; *see* OVUM) and male SPERM. Some primitive organisms, for example earthworms and sponges, are HERMAPHRODITES, with one individual producing both gametes, although cross-fertilization still often occurs. The genetic VARIATION provided by sexual reproduction allows the adaptation to changed environmental conditions (and therefore greater chance of survival; *compare* ASEXUAL REPRODUCTION). However, the rate of reproduction is slower than asexual since a partner has to be found before the process of mating (copulation or coitus) can begin.

In some animals, the breeding cycle is affected by seasonal changes, such as a rise in temperature. In birds, there is seasonal growth of the GONADS during the spring. These control mechanisms optimize the chances of survival

Human male reproductive system.

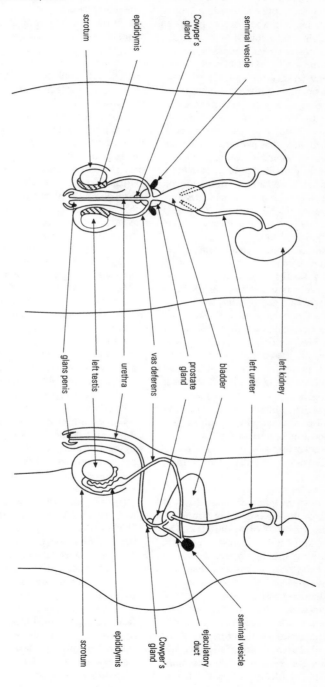

scrotum

epididymis

Cowper's gland

seminal vesicle

glans penis

left testis

urethra

vas deferens

prostate gland

bladder

left ureter

left kidney

scrotum

epididymis

Cowper's gland

ejaculatory duct

seminal vesicle

of the young. In some non-human mammals, the female is only receptive to breeding at certain times in her OESTROUS CYCLE. In these species, courtship behaviour by the males is used to determine whether a female is receptive or not, judged by her behavioural response. In some species, for example cats and rabbits, the act of mating stimulates OVULATION.

Many species recognize a sexually mature individual by their development of secondary sexual characteristics, such as bright feathers in birds, growth of antlers in deer, the mane of a lion. In humans, secondary sexual characteristics include the growth of pubic hair in both sexes, growth of facial hair and deepening of the voice in males. Human females develop breasts and wider hips.

In mammals, sexual stimulation causes an increase in the blood supply to the genitals of both male and females, in males causing the PENIS to become hard and erect. Stimulation of sensory cells at the tip of the penis cause muscular contractions that move the sperm from the VAS DEFERENS into the URETHRA, where they mix with the other constituents of SEMEN. The contractions of the penis cause ejaculation (expulsion of the semen) and sperm are propelled through the CERVIX and into the UTERUS, where they swim to the OVIDUCTS, and if successful fertilize an egg (*see* FERTILIZATION). Ejaculation in humans is accompanied by a pleasurable feeling called orgasm, which the female can also experience by the contraction of an equivalent set of muscles.

See also ACROSOME REACTION, DOUBLE FERTILIZATION, MENSTRUAL CYCLE, POLLINATION, PREGNANCY.

sheath In plants, the base of a LEAF.

short-sightedness, *myopia* A common defect of vision in which the eye lens is too strong for the size of the eyeball. Nearby objects can be seen clearly, but the sufferer cannot see distant objects clearly. It can be corrected using a diverging lens in front of the eye, effectively weakening the eye lens.

shoulder girdle *See* PECTORAL GIRDLE.

sickle-cell disease An inherited blood disorder in which the RED BLOOD CELLS are sickle-shaped and very fragile, and are therefore easily lost from the circulation, causing anaemia and possibly death. Sickle-cell disease is most common in people of African origin but is also found in people from north-east India and the east Mediterranean. The disease is caused by a point MUTATION in HAEMOGLOBIN. Substitution of a single NUCLEOTIDE causes the wrong amino acids to be incorporated into two of the POLYPEPTIDE chains making up the haemoglobin molecule. In the HETEROZYGOUS state not all cells are sickle-shaped and the anaemia is less debilitating than in the HOMOZYGOUS state.

In regions of Africa where death from MALARIA is a threat, the heterozygous sickle-cell condition is an advantage because the sickle-shaped cell is less easily invaded by the malarial parasite.

sieve element A type of plant cell found in the PHLOEM, concerned with transporting food throughout the plant. Sieve elements are long, thin-walled cells, joined end to end to form SIEVE TUBES. The cross-walls of the sieve elements partially break down, resulting in the formation of SIEVE PLATES. The sieve element is living but has no nucleus and very little cytoplasm, so it depends on adjacent COMPANION CELLS to sustain it.

sieve plate The end wall of a SIEVE ELEMENT which is perforated like a sieve. This allows the transport of food materials from one sieve element to the next along the SIEVE TUBE.

sieve tube One of a series of tubes in the PHLOEM of plants connecting the leaves, shoots and roots and responsible for the transport of food throughout the plant. Sieve tubes consist of cells called SIEVE ELEMENTS joined end to end and interconnected by SIEVE PLATES.

simple microscope *See* MICROSCOPE.

single cell protein (SCP) A food source rich in proteins and vitamins that is manufactured by the large-scale FERMENTATION of a variety of raw materials by different micro-organisms, then dried and sold as an animal or human food. As well as being a good, healthy food, SCP is theoretically economical to make because the raw materials can be agricultural or industrial waste, alcohol, sugars or petroleum chemicals. The micro-organisms used are varied and include fungi, yeasts, algae and bacteria.

One example of an SCP is Pruteen, manufactured as an animal food. The manufacture of Pruteen uses methanol, which is a waste product of other processes, making it cheap to make. The aerobic bacterium *Methylophilus methylotrophus* ferments the methanol to yield the tasteless, cream-coloured Pruteen. A human SCP is mycoprotein, manufactured by the action of the fungus *Fusarium graminearum* on flour waste.

The success of SCP has been limited because other food surpluses, for example butter and grain mountains, make it unnecessary. Greater demand would make it less economical because not enough raw materials would be available through waste.

single circulation *See* CIRCULATORY SYSTEM.

sinoatrial node *See* PACEMAKER.

sinusoid A channel or cavity in certain organs, such as the liver, where blood mixes and allows exchange of materials between the blood and the tissue it is supplying.

site of special scientific interest (SSSI) An area of land in the UK worthy of protection due to the animals, plants or geological features present there. *See also* CONSERVATION.

SI unit The unit of measurement in the internationally agreed METRIC SYSTEM (*Système International*). All units in the SI system are expressed in terms of seven base units and two supplementary units. The base units are metre (length), second (time), kilogram (mass), ampere (current), kelvin (temperature), mole (amount of substance) and candela (luminous intensity). The supplementary units are the radian (angle) and steradian (solid angle).

Any quantity that cannot be expressed directly in terms of one of these units can be expressed in terms of a derived unit, such as the metre per second for velocity.

skeletal muscle *See* VOLUNTARY MUSCLE.

skeleton A rigid structure in animals that provides support and protects internal organs. A skeleton may be internal (endoskeleton), as in vertebrates, and made of BONE and CARTILAGE that provide attachment points for MUSCLES in addition to support and protection of vital organs (such as the skull and ribs). The axial skeleton refers to the head and body, while the appendicular skeleton refers to the limbs (including the hips and shoulder bones). In some vertebrates, for example sharks, the entire skeleton is made of cartilage to provide extra flexibility.

The skeleton can also be external (exoskeleton), as in ARTHROPODS, providing the same supporting function. A third type of skeleton is a HYDROSTATIC SKELETON, as found in invertebrates such as earthworms, where the body cavity is full of fluid.

skin The external body covering of vertebrates. Skin consists of an outer EPIDERMIS and an inner DERMIS. The epidermis consists of several layers, the surface of which is a tough waterproof layer of dead cells that provides protection and prevention of water loss. This is constantly worn away and replaced by the living cells of the underlying tissue. The deepest layer of epidermis is called the MALPIGHIAN LAYER, which determines skin colour and protects the tissues beneath from ultraviolet light. The dermis contains blood vessels, hair follicles, nerves, SEBACEOUS GLANDS and SWEAT GLANDS embedded in CONNECTIVE TISSUE. Beneath the dermis is an insulating layer of subcutaneous fat that is also a long-term food reserve. The study of skin and its disorders is called dermatology. *See also* SENSE ORGAN.

sliding filament theory *See* MUSCULAR CONTRACTION.

small intestine Part of the DIGESTIVE SYSTEM, located after the STOMACH. In humans, the small intestine consists of a 6-m long, 4-cm diameter, coiled muscular tube situated in the lower abdomen. It is responsible for DIGESTION of food and absorption of the products. It consists of the DUODENUM, JEJUNUM and ILEUM. The duodenum is a short length of intestine where most digestion occurs. The ileum is concerned mostly with absorption and is specialized for this. The jejunum is not anatomically distinct from the ileum.

The inner lining of the walls of the small intestine are folded to increase their surface area, and consists of projections called VILLI, which contain smooth muscle fibres to allow repeated relaxation and contraction to help food and enzyme mixing. When food enters the duodenum, it mixes with BILE and PANCREATIC JUICES that help to neutralize the acid CHYME from the stomach and contain a variety of enzymes needed for the breakdown of proteins, fats, carbohydrates and nucleic acids. Further digestive juices are made by the intestinal wall itself. Some non-digestive enzymes are also made in the small intestine activate other enzymes entering from the PANCREAS. The contents of the small intestine are passed along by PERISTALSIS.

See also INTESTINAL JUICE, SECRETIN.

smooth muscle *See* INVOLUNTARY MUSCLE.

sodium (Na) A soft, silvery metallic element. It occurs in nature mostly as sodium chloride in sea water. Sodium is essential to all living organisms. In plants, sodium, like potassium, is a constituent of the sap vacuole and therefore helps to maintain turgidity. Deficiency is rare in plants as sodium is so abundant in the soil. In animals, sodium plays a crucial role, along with potassium, in the maintenance of electrical, osmotic and cation/anion balance across cell membranes. It is necessary for the functioning of the kidney, nerves and muscle. Deficiency may cause muscle cramps. *See also* SODIUM PUMP.

sodium pump, *cation pump* A mechanism in the CELL MEMBRANE of most animal cells in which sodium ions (Na^+) are pumped out of the cell and potassium ions (K^+) are pumped into the cell by ACTIVE TRANSPORT, an energy-requiring process. The sodium pump is important in the regulation of cell volume, by altering the osmotic potential (*see* OSMOSIS) and establishing a resting potential (*see* NERVE IMPULSE).

soil Small particles of rock, formed by weathering, mixed with organic material overlying the bedrock of the Earth's surface. The grains of soil have water and air between them, and the texture and properties of soil determine which plants can grow in it. Different types of soil develop under different climatic and physical conditions, for example deep soils develop in warm, wet climates and in valleys, and shallow soils develop in cool dry areas and on slopes. The effects of the nature of soil on the ECOSYSTEM are collectively called edaphic factors.

Soil consists of a number of layers. The top soil is the uppermost layer that contains most of the nutrients needed by plants, but it is susceptible to SOIL EROSION. Below the top soil lies the sub-soil, which does not provide many nutrients but is important in the drainage of water in the soil.

Clay soil has very small particles (mostly aluminium silicate) and water passes through slowly. Sandy soil has larger particles (mostly silicon dioxide) through which water can pass quickly. The best soils for plant growth are a mixture of sand and clay (loam soil). The HUMUS content of soil determines its fertility. The BIOTIC (living) element of soil plays an important role in soil texture and fertility (as well as in decomposition); burrowing animals, for example earthworms, improve aeration and drainage and assist humus breakdown.

soil erosion The wearing away or redistribution of SOIL caused by the action of water, wind, ice and human intervention. Erosion is a serious problem if left unchecked, as it may result in decline of the land, infertility and eventually in the formation of deserts. Destruction of forests and other vegetation and mechanization of farming causes unnecessary soil erosion.

soil profile The series of layers, or horizons, seen in a vertical section through SOIL to the rock beneath. The soil profile provides important information about the character of soils.

sol A COLLOID in which small solid particles are dispersed in a liquid. Sols can be lyophobic (solvent hating) in which the solid particles and the liquid do not have an affinity for each other. Lyophobic sols are unstable and with time the particles aggregate to form a precipitate. Lyophilic (solvent loving) sols are more stable and the solid particles have an affinity for the liquid phase; an example is starch in water.

soluble (*adj.*) Describing a material that will DISSOLVE in a particular SOLVENT (usually water) to form a SOLUTION.

solution A liquid that comprises a SOLVENT and a dissolved solid or gas.

solute A solid that is dissolved to form a solution.

solute potential (φ_s) A measure of the change in WATER POTENTIAL of a system due to the presence of solute molecules. Solutes lower the water potential of a system and therefore solute potential always has a negative value. *See also* OSMOSIS.

solvent A liquid in which a substance will dissolve to form a SOLUTION. Most solvents are organic liquids.

somatic (*adj.*) Relating to the body. A somatic cell is any cell other than a GAMETE.

somatic nervous system The part of the NERVOUS SYSTEM concerned with sensation and control of the skeletal muscles.

somatotrophin *See* GROWTH HORMONE.

soredium (*pl. soredia*) A very small structure characteristic of LICHENS, consisting of a cluster of algal cells surrounded by fungal hyphae. Soredia are the structures of non-sexual reproduction and are released like SPORES to be dispersed by air.

Southern blotting A technique devised by E.M. Southern for detecting the presence of specific DNA sequences among the DNA fragments obtained by cleavage with RESTRICTION ENZYMES. The mixture of DNA fragments is separated by ELECTROPHORESIS in an AGAROSE gel and the fragments are then denatured to form single-stranded fragments. These are transferred, or 'blotted', onto a nitrocellulose filter, where they remain immobilized. A specific RNA or DNA probe sequence, labelled with a radioisotope, is then added to the filter in a suspension and allowed to hybridize with any complementary fragments on the filter. After washing off any excess probe substances, hybridized fragments can be detected by AUTORADIOGRAPHY. This technique is widely used in GENETIC ENGINEERING and in GENETIC FINGERPRINTING.

speciation The emergence of new species during EVOLUTION. The cause of speciation is isolation of a group within a POPULATION. ADAPTIVE RADIATION is the result of geographical separation that prevents the separated groups interbreeding; the groups adapt differently to their surrounding and eventually a new species emerges. Isolation can also be due to physiological factors, such as incompatibility of genitalia, or behavioural factors, such as incompatibility of timing of breeding; the isolation of groups causes them to breed amongst themselves to survive, and therefore form new species.

species The lowest level in the CLASSIFICATION scheme. Different populations can exist within a species for example different breeds of dog. Members of the same species can all interbreed to produce fertile offspring. A native species has existed in a country since prehistoric times, whereas a naturalized species has been introduced to a country by humans and adapted to the new country. Exotic species need human intervention to survive.

About 1.4 million species have been identified. Species become extinct, for example through destruction of habitats, and new ones can form (*see* SPECIATION) by diversifying from other members of their group. There are more species in tropical regions than temperate ones. *See also* BINOMIAL NOMENCLATURE, GENUS.

sperm, spermatozoon (*pl.* **spermatozoa**) The male GAMETE of animals that is produced in large numbers in the TESTIS. A sperm is much smaller than the female gamete (OVUM). It consists of a head (containing the nucleus), a middle region with many MITOCHONDRIA for energy, and a tail (FLAGELLUM) by which it moves. At the tip of the head is a specialized structure called an ACROSOME that contains enzymes ready for release when the sperm meets the ovum. These enzymes dissolve the hard outer membrane of the ovum, allowing the sperm head and ovum to fuse. When the sperm leave the body (*see* SEXUAL REPRODUCTION) there are about 500 million sperm in about 3 cm^2 of SEMEN. Before FERTILIZATION of the ovum can occur, the sperm must go through a final maturation stage in the female tract called 'capacitation'. *See also* ACROSOME REACTION.

spermatid A non-motile cell in the TESTIS formed by SPERMATOGENESIS. Secondary SPERMATOCYTES divide by MEIOSIS to give rise to spermatids, which are HAPLOID. Spermatids subsequently differentiate into mature SPERM. *See also* TESTIS.

spermatocyte A cell in the TESTIS that results from the mitotic division of SPERMATOGONIA. Primary spermatocytes are produced first which then divide by MEIOSIS to give secondary spermatocytes, which themselves divide again by meiosis to give rise to two SPERMATIDS each. Thus one DIPLOID spermatocyte gives rise to four HAPLOID spermatids. *See also* TESTIS, SPERMATOGENESIS.

spermatogenesis In animals, the formation and maturation of SPERM. Spermatogenesis occurs within the SEMINIFEROUS TUBULES of the TESTIS, where germ cells grow and divide by MITOSIS to form SPERMATOGONIA. These subsequently divide again by mitosis to produce primary SPERMATOCYTES which then divide twice by MEIOSIS to give secondary spermatocytes and then SPERMATIDS. The spermatids have half the number of CHROMOSOMES of the original germ cells. *See also* TESTIS.

spermatogonia (*sing.* **spermatogonium**) Cells within the TESTIS, lining the walls of the SEMINIFEROUS TUBULES, which divide by MITOSIS to produce primary SPERMATOCYTES.

spermatozoon *See* SPERM.

sperm duct *See* VAS DEFERENS.

Sphenophyta A phylum of the plant kingdom consisting of the horsetails. There is only one surviving genus of this group, called *Equisetum*. Horsetails are found in damp conditions and consist of an underground RHIZOME that gives rise to jointed aerial stems with prominent nodes and small leaves whorled at the nodes. They are HOMOSPOROUS with SPORANGIA in CONES.

sphincter Any ring of smooth muscle in the wall of a tubular structure such as the ALIMENTARY CANAL. Sphincters contract and relax to control the opening of the tube, for example for the passage of food or FAECES. Examples include the pyloric and cardiac sphincters (*see* STOMACH) and the anal sphincter (*see* ANUS). Control of sphincters can sometimes be voluntary, for example of the anal sphincter.

spider A member of the class Arachnida of the phylum ARTHROPODA. A spider possesses eight legs and an unsegmented ABDOMEN connected by a thin 'waist' to the cephalothorax (head and THORAX merged). The eyes are usually simple. Many species of spider exude a liquid from the underside of the abdomen, which hardens in air to form the thread seen in webs; the spider traps prey in its web, injects substances into the prey to subdue it, and then digests and sucks up the juices.

There are about 30,000 species of spiders found all over the world, some of which produce poisonous toxins, for example the black widow and tarantula.

spinal cord A major component of the CENTRAL NERVOUS SYSTEM (CNS), linking the PERIPHERAL NERVOUS SYSTEM (outside the CNS) to the BRAIN. The spinal cord consists of a cylinder of nervous tissue protected by VERTEBRAE, with a small central canal running through the centre containing CEREBROSPINAL FLUID. Surrounding this canal is GREY MATTER, usually forming an H-shape in cross-section, and around this WHITE MATTER. Surrounding all of this are membranes called the MENINGES. Pairs of spinal nerves extend along the length of the spinal cord; the uppermost of these is called the dorsal root (which only carries sensory nerves) and the lowest is called the ventral root (which only carries effector nerves). The swelling within the dorsal root, called the dorsal root ganglion, is formed by cell bodies of sensory neurones. *See also* VERTEBRAL COLUMN.

spindle The group of protein fibres formed in a cell during MITOSIS and MEIOSIS. The spindles draw the pairs of CHROMOSOMES apart as the cell divides.

spine *See* VERTEBRAL COLUMN.

spiracle 1. In insects, a pore in their hard exterior through which oxygen enters the body

and carbon dioxide leaves. Spiracles form a complex series of tubes called tracheae that divide to form smaller tubes called tracheoles, which then enter the tissues directly. Oxygen is carried in this tracheal system and not in the blood. This system is considered to be equivalent to the TRACHEA of vertebrates.

2. In many fish, the opening of the remains of the first GILL slit.

spirillum (*pl.* **spirilla**) Any spiral-shaped bacterium. *See* BACTERIA.

spleen The largest mass of lymphoid tissue (*see* LYMPH NODE), found near the stomach in vertebrates. The spleen has blood circulating through it rather than LYMPH and is important in destroying worn-out blood cells and storing new RED BLOOD CELLS and LYMPHOCYTES to be pumped into the circulation when needed. *See also* RETICULOENDOTHELIAL SYSTEM.

sponge A member of the phylum PORIFERA.

spongy mesophyll A layer of spherical, loosely arranged cells within the MESOPHYLL of a leaf blade. The spongy mesophyll lies below the palisade mesophyll. There are fewer CHLOROPLASTS here than in the palisade mesophyll but more air spaces between the cells for rapid diffusion of gases (entering through the STOMATA).

spontaneous generation theory, *abiogenesis* A theory put forward to explain the ORIGIN OF LIFE, which suggests that living matter arose from non-living matter. For example it was believed that vermin spontaneously developed from household rubbish. This theory has now been disproven.

sporangium (*pl.* **sporangia**) In fungi and plants, the structure within which SPORES are produced.

spore The reproductive structure of many primitive plants, including FUNGI, FERNS and mosses (*see* BRYOPHYTA), and also some BACTERIA and ALGAE. Spores usually consist of a single cell and they can develop without the need for fusion with another cell. Spores can form sexually or asexually and are a means of very rapid increase in the size of a population. They are light and easily dispersed, and can remain dormant (*see* DORMANCY) until the conditions are favourable for their development. *See also* GERMINATION.

sporophyll A leaf bearing sporangia (*see* SPORANGIUM). Examples include the scale-like leaves of CONES of CONIFERS, and the STAMENS and CARPELS of flowering plants.

sporophyte One form of a plant that shows ALTERNATION OF GENERATIONS. The sporophyte is the DIPLOID generation that produces HAPLOID spores by MEIOSIS. *See also* GAMETOPHYTE.

sporozoan A member of the phylum APICOMPLEXA.

sporozoite An infective stage in the life cycle of protozoans from the phylum APICOMPLEXA, such as the *Plasmodium* parasite. *See* MALARIA.

sporulation The formation of SPORES (the reproductive body of many lower plants, such as mosses, ferns and fungi). Sporulation is a common means of ASEXUAL REPRODUCTION, whereby numerous small unicellular spores are produced (by MEIOSIS, so providing genetic VARIATION) that are easily dispersed and grow to new organisms when the conditions are suitable.

SSSI *See* SITE OF SPECIAL SCIENTIFIC INTEREST.

stamen The essential male reproductive structure in a flowering plant (ANGIOSPERM). The stamen surrounds the CARPELS and consists of a long stalk called the filament, with an ANTHER at its apex that produces the POLLEN grains that contain the male GAMETES. There are variable numbers of stamens and they can be in different positions, which is useful in classification of flowering plants. The collective name for the stamens is androecium.

standard deviation In statistics a value which gives an indication of the spread of data, above or below the AVERAGE. It is calculated according to the formula:

$$\sqrt{SD} = \Sigma \delta^2 / n$$

where Σ is the sum of; δ = difference between each value in the sample and the mean and n is the total number of values in the sample.

These terms are all associated with NORMAL DISTRIBUTION (Gaussian) curves, bell-shaped curves obtained when plotting continuous variation graphically, such as height in humans – most people are an intermediate height with some being taller and some shorter.

stapes, *stirrup* One of the three EAR OSSICLES in the mammalian middle EAR.

starch A complex POLYSACCHARIDE that is the main food reserve of green plants. Starch consists of hundreds of GLUCOSE polymers, some of which are straight chain molecules (amylose) and some branched molecules (amylopectin). The main dietary sources of starch for humans are cereals, potatoes and legumes. Starch is used in many ways in industry, for example to stiffen paper and textiles, as a thickening agent in foodstuffs, and as glucose syrups (by HYDROLYSIS of starch). *See also* CARBOHYDRATE.

start codon The CODON AUG that is always present at one end of a MESSENGER RNA (mRNA) molecule and at which PROTEIN SYNTHESIS begins. The codon codes for the amino acid methionine, and the first RIBOSOME to be attracted to the mRNA in TRANSLATION attaches to this AUG codon. *See also* GENETIC CODE.

stationary phase 1. *See* BACTERIAL GROWTH CURVE.

2. *See* CHROMATOGRAPHY

statistics The branch of mathematics that deals with the collection and interpretation of large quantities of data, and the prediction of the outcome of sampling processes, where a small number of measurements are taken on a much larger group. *See also* CHI-SQUARED TEST, NORMAL DISTRIBUTION, STANDARD DEVIATION.

stearic acid A saturated FATTY ACID that occurs in many plant and animal fats.

Stelleroidia A class of the phylum ECHINODERMATA which consists of the starfish and brittle stars. Members are flattened with arms. *Compare* ECHINOIDIA.

stem The main axis of a plant, which supports the leaves, buds and flowers. The stem contains vascular tissue (*see* VASCULAR BUNDLE), which conducts food, water and minerals between the roots and the leaves.

stem tuber In certain plants, a swollen region of an underground stem that is modified for storing food during the winter or a dry season. A stem tuber also gives rise to new plants the following season from the terminal or axillary buds. A potato is an example of a stem tuber. *Compare* ROOT TUBER. *See also* TUBER.

sterilization 1. The killing or removal of living organisms. Sterilization can be achieved by heat treatment (boiling), by chemical disinfectants, irradiation with gamma radiation and filtration. It is important in food processing, medicine and research.

2. A surgical operation to prevent reproduction. In males this is called vasectomy and the passage of sperm is blocked. In females, the FALLOPIAN TUBES are tied to prevent FERTILIZATION. Sterilization can be used as a means of CONTRACEPTION.

sternum In humans, the breast bone. *See* PEC-
TORAL GIRDLE.

steroid hormone Any one of a group of
hormones that are LIPIDS derived from a compound called cyclopentanoperhydrophenan-
threne. They have a complex structure
consisting of four carbon rings. A large subgroup of steroids is the steroid alcohols, or
STEROLS, of which CHOLESTEROL is a member.
The sex hormones OESTROGEN, PROGESTERONE
and TESTOSTERONE are another important group
of steroid hormones, as are the CORTICOSTEROIDS
produced by the cortex of the ADRENAL GLAND,
including CORTISONE and ALDOSTERONE. Many of
these hormones, or their synthetic variants,
have therapeutic uses, for example the contraceptive PILL, HORMONE- REPLACEMENT THERAPY, as
anti-inflammatory drugs and in adrenal failure.
See also ANABOLIC STEROIDS.

sterol A large subgroup of STEROID HORMONES of
which CHOLESTEROL is a member. Sterols are
steroid-based alcohols.

stigma In flowers, a structure within the CARPEL
(female reproductive structure) that is supported by the STYLE and is specialized for
receiving POLLEN. The stigma often has hairs
and produces a sticky secretion to attract
pollen grains.

stirrup *See* STAPES.

stolon A long vertical stem, for example in the
blackberry, that reproduces vegetatively by
bending over to touch the soil and developing
ADVENTITIOUS ROOTS, at which point a new
plant grows. A stolon is similar to a RUNNER.

stoma (*pl. stomata*) One of many pores in the
EPIDERMIS of plants that is the main route of
water loss in a plant and the site of carbon
dioxide and oxygen exchange. Stomata can be
opened or closed by a pair of GUARD CELLS,
which surround them. The guard cells change
shape (by changes in turgidity) to adjust water
loss. Factors affecting stomatal opening
include levels of carbon dioxide, light, temperature, humidity, air currents and water availability. For example, the pores are open during
cold weather but closed during hot weather to
reduce evaporation of water. Stomata close at
night, when PHOTOSYNTHESIS cannot take place,
to reduce water loss. Stomata are also found in
large numbers on the aerial parts of a plant.
See also LEAF, TRANSPIRATION.

stomach A muscular cavity in animals, situated
just below the DIAPHRAGM, that produces acids

and enzymes to digest food entering it from
the OESOPHAGUS, and then passes the food to
the SMALL INTESTINE for further digestion and
absorption.

The lining of the stomach is a folded layer
called the gastric mucosa, embedded in which
are gastric pits lined with secretory cells
producing GASTRIC JUICE, in response to production of GASTRIN. Gastric juice consists of
mostly water mixed with hydrochloric acid,
and provides the acid environment needed for
the digestive enzymes to function. The
enzymes PEPSIN and RENNIN are made by the
gastric pit as inactive precursors, which are
activated by the acidity of the stomach. Goblet
cells in the stomach lining secrete MUCUS,
which helps protect the lining from its own
gastric juice and also helps lubricate movement of food. Cells of the stomach lining are
replaced continually to avoid damage by the
acidity.

Food is thoroughly mixed in the stomach
until a creamy chyme forms, which is released
slowly into the small intestine in manageable
amounts. The cardiac and pyloric sphincters
are two rings of muscles at the entrance and
exit of the stomach respectively, which can
relax and contract to allow food to enter and
leave. The stomach itself contracts periodically
throughout digestion (to aid mixing and food
movement).

Some HERBIVORES have stomachs with
more than one chamber to allow CELLULOSE
breakdown to occur without affecting the
environment needed for the other micro-
organisms vital for digestion.
See also CROP, GIZZARD.

stone fruit *See* DRUPE.

stonewort Any member of the phylum CHARO-
PHYTA.

**stop codon, *termination codon, nonsense
codon*** Any one of three CODONS that do not
code for an AMINO ACID and cause PROTEIN SYN-
THESIS to come to an end. The codons are
UAA, UAG and UGA. When a stop codon is
reached on a MESSENGER RNA (mRNA) molecule during TRANSLATION, the POLYPEPTIDE
chain is released from the mRNA. *See also*
GENETIC CODE, MUTATION.

storage granule Insoluble material found in all
cells as a store of food energy. STARCH granules
are present in the CYTOPLASM of plant cells and
in CHLOROPLASTS. GLYCOGEN granules are found

in animal cells. Oil or LIPID droplets can be found in both plant and animal cells.

streptomycin An ANTIBIOTIC that affects PROTEIN SYNTHESIS.

striated muscle *See* VOLUNTARY MUSCLE.

striped muscle *See* VOLUNTARY MUSCLE.

stroma In a CHLOROPLAST, the gel-like matrix in which the grana (*see* GRANUM) are embedded. In animals, the matrix (intercellular material) of an organ.

style In flowers, a structure at the centre of the CARPEL (the female reproductive structure). The style can be a long slender stalk that supports the STIGMA at the top in a position that is optimal for receiving POLLEN. In some flowers it is short or even absent.

suberin A complex waxy substance found in the cell walls of many plants, especially corky tissues. Suberin is impermeable to water and provides a protective layer. It constitutes the CASPARIAN STRIP in the endodermis of ROOTS, where it plays an important role in the movement of water through the root.

sublingual gland Any SALIVARY GLAND situated beneath the tongue.

submaxillary gland Either of a pair of SALIVARY GLANDS situated close to the lower jaw.

sub-soil *See* SOIL.

substitution In genetics, a type of POINT MUTATION in which a single base (*see* NUCLEOTIDE) in the DNA is replaced by another. If a PURINE base is substituted by another purine or a PYRIMIDINE base is replaced by another pyrimidine then this is termed a transition mutation. If a purine base is substituted by a pyrimidine or vice versa, this is a transversion mutation. Such mutations can have serious effects since they usually alter an amino acid in a protein, which can affect its functioning. For example, SICKLE-CELL DISEASE is caused by a substitution of thymine by adenine in the CODON coding for an amino acid in HAEMOGLOBIN. *See also* MUTATION.

substrate 1. A substance that is acted upon by an ENZYME.

2. The material on which micro-organisms grow, for example agar, or the surface to which cells in TISSUE CULTURE attach.

succession The series of changes occurring within a COMMUNITY, from its origin to its stable climax. The first colonization of an area is called primary succession, and recolonization of an established area destroyed, for example

by fire or flood, is called secondary succession. A climax community, such as a deciduous oak woodland, eventually forms in which there is a balance between the species (both plant and animal) sustained in the area, and new varieties only replace established species. Usually one or two dominant plants and animals form the greatest biomass.

    A series of plant successions developing in a particular area is called a sere. For example, a lithosere develops on bare rock surfaces; a hydrosere begins in fresh water; a xerosere grows under dry conditions; and a halosere develops in a salt-marsh.

sucrase An enzyme that breaks down SUCROSE into GLUCOSE and FRUCTOSE.

sucrose ($C_{12}H_{22}O_{10}$) A DISACCHARIDE sugar made up of the MONOSACCHARIDE units GLUCOSE and FRUCTOSE. Sucrose is found in the pith of sugar cane and in sugar beet. Sucrose is what is commonly referred to as 'sugar'.

Sudan III test *See* LIPID.

sugar, *saccharide* Any one of a group of CARBOHYDRATES with relatively low RELATIVE MOLECULAR MASSES and a typically sweet taste. The term sugar commonly refers to SUCROSE. *See also* MONOSACCHARIDE, DISACCHARIDE, POLYSACCHARIDE.

sulphate A salt containing the sulphate ion, SO_4^{2-}, together with either a metal or the ammonium ion, NH_4^+. Plants obtain sulphur by absorbing sulphates from the soil. Without these there is poor root development and chlorosis (yellowing of leaves). *See also* SULPHUR CYCLE.

sulphite, *sulphate(IV)* A compound containing the sulphite ion, SO_3^{2-}, together with a cation, either a metal or the ammonium ion, NH_4^+. Sulphites are easily oxidized to SULPHATES.

sulphur (S) A yellow, non-metallic element with a distinctive odour. It burns in air and is insoluble in water but is otherwise fairly unreactive. Sulphur is abundant in nature in volcanic regions and is a constituent of some proteins and biochemical compounds such as ACETYL COENZYME A. Sulphur is often found combined with oxygen as SULPHATE, in which form it is absorbed from the soil by plants. It is an essential element for both plants and animals. *See also* SULPHUR CYCLE.

sulphur cycle The natural cycling of sulphur between the biological (BIOTIC, living) and the geological (ABIOTIC, non-living) components of the environment. Most of the sulphur in the non-living environment occurs in rocks with some in the atmosphere as sulphur dioxide (SO_2). Sulphates (SO_4^{2-}) are formed from the weathering of rocks and oxidation of sulphur or sulphur dioxide.

CHEMOAUTOTROPHIC sulphur-oxidizing bacteria such as *Thiobacillus* are able to convert sulphur, or its compounds, to sulphates. Sulphates are then taken up by plants and used to make certain sulphur-containing proteins. In this way, sulphur can pass along the FOOD CHAIN to animals. Dead animals and their faeces are decomposed by the action of anaerobic HETEROTROPHIC bacteria such as *Desulfovibrio*, which are able to reduce sulphur to generate hydrogen sulphide. This can then be converted back to sulphates by the sulphur-oxidizing bacteria.

Another group of anaerobic PHOTOAUTOTROPHIC bacteria, the green sulphur and the purple sulphur bacteria can convert hydrogen sulphide to sulphur. Sulphur can then again become incorporated into rocks.

suppressor T cell (Ts) A subset of T CELLS that suppress the IMMUNE RESPONSE and are therefore important in maintaining tolerance to self tissues.

suprarenal gland *See* ADRENAL GLAND.

surface tension The force that appears at the surface of a liquid and tends to pull the liquid into spherical droplets. It is caused by unequal cohesive forces (*see* COHESION) between molecules at the surface, resulting in an inward pull of molecules towards one another and creating a skin-like layer at the surface. The effect of surface tension allows small objects, such as some insects, to rest on a water surface.

suture joint An immovable JOINT, such as those between the bones of the skull.

sweat The secretion from the SWEAT GLANDS.

sweat gland In mammals, a gland located in the SKIN that is responsible for perspiration and therefore heat loss when necessary. Sweat glands are found all over the body of primates, but are more localized in other mammals, for example the feet and face of cats and dogs. There are more sweat glands in male humans than females.

Sweat glands consist of coiled tubes leading, via a duct, to a pore in the skin surface. They absorb water fluid from the capillaries and transfer it to the skin as sweat (water, salt

and urea) to evaporate, which aids heat loss when needed. Sweat is also thought to contain PHEROMONES, which are important in the communication of social messages. Production of sweat is controlled by the AUTONOMIC NERVOUS SYSTEM.

swim bladder An air-filled sac lying between the gut and the spine of a fish that regulates the buoyancy of the fish.

symbiont Any organism living in a state of SYMBIOSIS.

symbiosis A permanent or prolonged close association between two organisms of different species where both benefit from the association. Some definitions also include associations with deleterious effects (*see* PARASITISM). The organisms in association are called symbionts.

Examples of symbiosis occur in the digestive system of animals where PROTOZOANS, bacteria and fungi live. These micro-organisms synthesize their own vitamins (K and B), which the host animal can use and provide enzymes essential for digestion of the host's dietary material, and in return receive food and protection. In HERBIVORES, these symbionts are essential for the break down of CELLULOSE. They live in the CAECUM, APPENDIX or rumen of RUMINANTS. Ruminants also benefit by using the symbionts as a source of protein.

Another important example occurs in the NITROGEN CYCLE, where nitrogen-fixing bacteria live in nodules on stems, roots or leaves of flowering plants (ANGIOSPERMS) or CONIFERS. LEGUMES, such as beans, peas and clover, have ROOT NODULES where the bacterium *Rhizobium* lives; the bacteria get a source of carbon and the plant is provided with nitrate (independently of soil nitrates). Other examples are LICHENS and MYCORRHIZAS. *See also* COMMENSALISM, MUTUALISM.

symmetry The property of a system or object remaining unchanged when certain changes are made. In particular, it is the property of certain shapes that remain unchanged under specified transformations. For example, a square has a four-fold symmetry of rotation.

sympathetic nervous system Part of the AUTONOMIC NERVOUS SYSTEM which responds to stress. It speeds up the heart rate, increases blood pressure and prepares the body for action. The sympathetic nervous system usually opposes the PARASYMPATHETIC NERVOUS SYSTEM.

symplast pathway One of three pathways by which water and minerals move upwards through a plant. The symplast pathway carries less water than the APOPLAST PATHWAY but is more important than the VACUOLAR PATHWAY. The symplast pathway is the movement of substances through the cytoplasm of cells. Water is lost from the cells surrounding the stomata (*see* STOMA) by evaporation, which makes the WATER POTENTIAL of these cells lower than adjacent cells. Water moves from a cell with a higher water potential to one with a lower water potential, via tiny strands of cytoplasm called PLASMODESMATA, which link adjacent cells. Water thus moves up through the plant from the XYLEM along a water potential gradient. The symplast pathway also occurs in roots where a water potential gradient exists between the root hair cells and the xylem. This pathway is the only means by which water moves across the ENDODERMIS in a root (*see* CASPARIAN STRIP). *See also* TRANSLOCATION.

synapse The point at which two NEURONES or a neurone and a muscle meet to transmit a NERVE IMPULSE. The gap between the two cells is called the synaptic cleft and is at least 15 nm in width. The AXON of the transmitting neurone expands at the synapse to form a bulbous synaptic knob, within which are MITOCHONDRIA and synaptic vesicles (released from the GOLGI APPARATUS of nerve cells) containing NEUROTRANSMITTERS. When the nerve impulse reaches the synaptic knob, the neurotransmitter is released and crosses the synaptic cleft to bind to RECEPTOR molecules on the receiving cell's membrane. The receiving cell (after the synapse) is called the post-synaptic neurone and the transmitting cell (before the synapse) is the pre-synaptic neurone.

If the receiving cell is another neurone, then the impulse will be carried from the synaptic cleft of the axon to DENDRITES on the receiving cell. If the receiving cell is a muscle, then the release of neurotransmitters will cause it to contract. Most synapses are between axons and dendrites, but axon–axon or dendrite–dendrite synapses can occur. In addition, those axons and dendrites can have synapses with several dendrites and axons, respectively. Less often, the synapses are electrical rather than chemical.

There are a number of drugs that act on the synaptic transmission, either by mimicking the neurotransmitter or affecting its release or receptor binding.

synaptic (*adj.*) Relating to the SYNAPSES.

syncytium (*pl. syncytia*) Animal tissue formed by the fusion of cells to form a multinucleate mass of PROTOPLASM, for example VOLUNTARY MUSCLE and TROPHOBLAST.

synergistic (*adj.*) Describing the interaction of two or more substances, organs or organisms to produce an effect greater than the sum of their individual effects. *Compare* ANTAGONISTIC.

synovial fluid A fluid that is found in movable JOINTS in vertebrates. The synovial fluid provides lubrication and nutrients for the cartilage at the end of each bone. It is secreted by the synovial membrane.

synovial joint A freely movable JOINT such as the hip and shoulder (ball and socket joints) or the elbow and knee (hinge joints).

synovial membrane A connective tissue membrane that surrounds a SYNOVIAL JOINT. The membrane secretes SYNOVIAL FLUID, which lubricates the cartilage at the ends of the bones forming the JOINT.

T

tapetum A layer behind the RETINA of the eye in some nocturnal animals, such as cats. It reflects light and improves vision in dim light by providing more opportunities for the light-sensitive cells of the retina to absorb light (hence cat's eyes light up when the light is shone in to them).

tap root The single main ROOT of plants which grows vertically downwards. The tap root is often used for food storage, for example in the carrot.

tartrazine, E102 A dye used for giving yellow colour to food. It is associated with hyperactivity in children and skin and respiratory problems in those with an allergy to it.

tastebud A chemoreceptor (a RECEPTOR cell that detects chemicals) specialized for taste. In vertebrates, tastebuds are concentrated on the upper surface of the TONGUE.

taxis (*pl. taxes*) The directional movement of a freely motile organism, or part of one, in response to an external stimulus. Movement towards the stimulus is positive taxis and movement away from a stimulus is negative taxis. Phototaxis is the movement of an organism in response to light, for example green ALGAE swim towards light to increase the rate of photosynthesis. Chemotaxis (chemical stimulus) is common in many bacteria, which move towards higher concentrations of nutrients. Thermotaxis is movement in response to temperature. *See also* NASTIC MOVEMENT, TROPISM.

taxonomy The study of the CLASSIFICATION of living organisms.

TCA cycle *See* KREBS CYCLE.

T cell, *T lymphocyte* A type of LYMPHOCYTE that is formed in BONE MARROW and passes through the THYMUS (hence *T* cell) before settling in the SPLEEN or a LYMPH NODE.

T cells are involved in CELL-MEDIATED IMMUNITY. They only recognize foreign ANTIGENS in association with cell antigens of the MAJOR HISTOCOMPATIBILITY COMPLEX (MHC). This is called MHC restriction. There are several subsets of T cells. CYTOTOXIC T CELLS (Tc) recognize foreign antigen on the surface of tumour or virus-infected cells and destroy the cell by releasing proteins that cause CYTOLYSIS. HELPER T CELLS (T_H) help B CELLS in their ANTIBODY production. SUPPRESSOR T CELLS (Ts) that specifically suppress the IMMUNE RESPONSE; and macrophage-activating cells that produce LYMPHOKINES to activate MACROPHAGES.

Compare B CELLS.

T-cell growth factor (**TCGF**, *interleukin 2*) A GROWTH FACTOR required for the proliferation of T CELLS. *See* INTERLEUKIN.

tear gland *See* LACHRYMAL GLAND.

telophase The final stage of MITOSIS or MEIOSIS.

tendon A band of CONNECTIVE TISSUE consisting of parallel COLLAGEN fibres with FIBROBLASTS in between, attaching muscle to bone. Tendons have great strength but less elasticity than some LIGAMENTS.

tentacle A long, flexible protrusion used by some organisms for feeding, for example in *Hydra* and sea anemone.

tepal In plants, a subdivision of a PERIANTH that is not clearly differentiated into a CALYX and COROLLA.

termination codon *See* STOP CODON.

testa The protective outer coat of a SEED that is usually hard and impermeable. It is formed after fertilization of the OVULE.

testis (*pl. testes*) The male GONAD that produces SPERM and sex hormones. Most male mammals, including humans, have a pair of testes (or testicles) that descend from the body cavity shortly before birth and hang below the abdomen in a sac called the SCROTUM. Each testis is divided into lobules containing SEMINIFEROUS TUBULES that are lined with germinal EPITHELIUM from which sperm develop by a process called SPERMATOGENESIS.

Spermatogenesis consists of a series of cell divisions in which SPERMATOGONIA divide by MITOSIS to produce primary SPERMATOCYTES. These then divide by MEIOSIS to give secondary spermatocytes, and again to give SPERMATIDS (which are HAPLOID). Spermatids are protected by the surrounding Sertoli cells and they are modified into spermatozoa. Between the tubules are INTERSTITIAL CELLS, which produce the male hormone TESTOSTERONE.

The seminiferous tubules merge to form the vasa efferentia, a group of small ducts through which the sperm pass into the EPIDIDYMIS. Sperm are stored in the epididymis until they gain their motility ready for use in SEXUAL REPRODUCTION.

testosterone In male vertebrates, a STEROID HORMONE produced by the INTERSTITIAL CELLS, of the TESTIS. Testosterone is responsible for the development of secondary sexual characteristics, such as hair growth, deepening of the voice, sexual behaviour and muscle development, and also for stimulating SPERM production. The production of testosterone is controlled by another hormone, the LUTEINIZING HORMONE made by the PITUITARY GLAND. Synthetic testosterone has been used to help muscular development in athletes, although its use has now been banned.

tetrapod Any vertebrate that has evolved from four-legged ancestors. Tetrapods include mammals, birds, reptiles and amphibians. In some cases one pair of limbs has been lost or modified, for example into wings or flippers.

thalamus Part of the vertebrate FOREBRAIN that is a relay centre for other regions of the brain. The thalamus interprets sensory information and compares it to previously stored information, and sends it to the appropriate area of the CEREBRUM. The thalamus consists of mainly GREY MATTER and is associated with pain and pleasure.

thalassaemia An inherited blood disorder in which there is a gene defect involving the production of HAEMOGLOBIN. In the HOMOZYGOUS state thalassaemia causes severe anaemia and is usually fatal. It is more often presented in the HETEROZYGOUS state. There are several types of thalassaemia.

thallus A very simple, undivided plant body with no stem, leaves or roots and often thin and flattened. It is typical of some liverworts (*see* BRYOPHYTA).

thermonasty The NASTIC MOVEMENT of plants in response to temperature. For example the crocus flower opens at 16°C and closes at temperatures below 16°C.

thermoreceptor A RECEPTOR cell that detects temperature changes. *See* SENSE ORGAN.

thermotaxis The directional movement of an organism in response to temperature. *See* TAXIS.

thiamine *See* VITAMIN B.

thigmonasty, *haptonasty* The NASTIC MOVEMENT of plants in response to localized contact, for example, the leaf movements of the Venus flytrap in response to contact with an object.

thigmotropism, *haptotropism* The directional growth of a plant (or part of it) in response to physical contact, for example, tendrils of climbing plants winding around a support. *See* TROPISM.

thin-layer chromatography (TLC) A CHROMATOGRAPHY technique widely used for analysing the components in liquid mixtures. The stationary phase is a thin layer of an absorbent solid, such as aluminium oxide, supported on a vertical glass plate.

thoracic duct The main vessel of the LYMPHATIC SYSTEM, running longitudinally in front of the VERTEBRAL COLUMN. The thoracic duct carries LYMPH from the trunk and hindlimbs up through the THORAX and drains into the superior VENA CAVA in humans (or another main vein in other animals). *See also* LYMPHATIC DUCT, LYMPH VESSEL.

thorax In vertebrates, the chest cavity containing the heart and lungs and protected by the ribcage. In mammals, the thorax is separated from the ABDOMEN by a muscular DIAPHRAGM. In ARTHROPODS the separation is less clear and the thorax represents the body region between the head and the abdomen. In insects, the thorax carries the legs and wings.

The upper area of the thorax is called the PECTORAL GIRDLE and is made up of the muscles and bones needed to move the arms and forelimbs.

threonine An ESSENTIAL AMINO ACID found in many proteins and an essential dietary requirement.

threshold The minimum level of intensity of a stimulus that is required to produce a response.

thrombin An enzyme involved in the BLOOD CLOTTING CASCADE that converts FIBRINOGEN to FIBRIN.

thrombocyte *See* PLATELET.

thrombokinase An enzyme that converts the inactive PROTHROMBIN into THROMBIN, and therefore plays an important role in the clotting of blood. *See* BLOOD CLOTTING CASCADE.

thylakoid One of a number of closed flattened sacs containing photosynthetic pigments such as CHLOROPHYLL. *See also* PHOTOSYNTHESIS, CHLOROPLAST.

thymidine A PYRIMIDINE NUCLEOSIDE, consisting of the organic base THYMINE and the sugar RIBOSE.

thymine An organic base called a PYRIMIDINE that occurs in DNA but not in RNA.

thymosine A HORMONE that stimulates the activity of T CELLS. It is produced in the THYMUS.

thymus An organ found in the upper chest cavity in humans and consisting of primary lymphoid tissue (*see* LYMPHATIC SYSTEM). The thymus is responsible for the maturation of T CELLS, which pass through the thymus before they settle in secondary lymphoid tissue (e.g. SPLEEN and LYMPH NODES).

The thymus reaches full size at puberty and then shrinks, since its role in T cell maturation is completed early in life. Thereafter it acts purely as an endocrine organ (*see* ENDOCRINE GLAND) producing the hormone thymosine, which stimulates the activity of T cells.

thyroid gland An ENDOCRINE GLAND of vertebrates, located in the neck region, the main role of which is in the regulation of the body's metabolic rate (*see* METABOLISM). In mammals it is a single gland but in amphibians and birds it is paired.

The thyroid gland is under the control of the PITUITARY GLAND, which produces THYROID-STIMULATING HORMONE. In response to this hormone, the thyroid produces thyroxine (T_4) and triiodothyronine (T_3), both of which are hormones derived from the amino acid tyrosine and contain four and three molecules of iodine respectively. Iodine is therefore needed in the diet for the normal functioning of the thyroid. If iodine is limited, T_3 is preferentially formed, otherwise more T_4 is produced. Some of the circulating T_4 is converted to T_3 in the lungs and liver. These hormones stimulate growth and the metabolic rate of cells by increasing the rate at which glucose is oxidized by cells. A third hormone called CALCITONIN is also produced by the thyroid, which is concerned with regulation of calcium ions in the blood. This is done in conjunction with parathormone produced by the PARATHYROID GLANDS.

An underactive thyroid causes a condition termed HYPOTHYROIDISM, a reduction in an individual's metabolic rate leading to other symptoms. An overactive thyroid causes HYPERTHYROIDISM, which leads to an increase in metabolic rate.

thyroid-stimulating hormone (TSH), *thyrotrophin* A protein produced by the anterior PITUITARY GLAND that stimulates growth of the THYROID GLAND and its production of hormones, such as thyroxine. TSH is itself stimulated by the thyrotrophin-releasing factor (TRF) from the HYPOTHALAMUS.

thyrotrophin *See* THYROID-STIMULATING HORMONE.

thyrotrophin-releasing factor A substance, secreted by the HYPOTHALAMUS, that stimulates the THYROID-STIMULATING HORMONE.

thyroxine A hormone of the THYROID GLAND.

tissue A group of CELLS, the same or different, that together perform a specific function. Tissue cells are bound together in animals by the EXTRACELLULAR MATRIX, and in plants by the CELL WALLS. Examples of tissues include squamous EPITHELIUM (consisting of a single cell type) and MUSCLE (consisting of more than one cell type) in animals, and PARENCHYMA (single cell type) or XYLEM (more than one cell type) in plants. There are four main types of animal tissues – epithelial, connective, muscular and nervous – and two main types of plant tissues – simple and compound. Simple tissues consist of one cell type (such as plant parenchyma), whilst compound tissues are of mixed cell types (such as xylem, PHLOEM). The study of tissues is called histology. *See also* CONNECTIVE TISSUE, NERVOUS SYSTEM.

tissue culture A technique for maintaining living cells taken from a plant or animal under controlled, sterile conditions in the laboratory. *In vitro* refers to tissue culture outside the plant or organism.

Cells are usually grown in culture dishes in a liquid media that provides nutrients and a balanced pH, and placed in an incubator to maintain a constant temperature, oxygen and carbon dioxide balance and humidity close to that within the organism. Cells can be grown as a monolayer, where they attach to the base of the culture vessel, or as a cell suspension, where no attachment is made and the culture vessel is rotated during the culture period.

A primary culture refers to the original culture established from the tissue of the organism. So that the cells can continue to survive and grow, once they have filled the culture vessel they have to be divided, or subcultured, between more vessels. These are then called secondary cultures. Normal cells grow for a limited number of cell divisions (50–100) but if a culture is treated with a chemical or virus to induce the formation of cancer cells, these are said to be transformed and will divide indefinitely.

Animal tissue culture is used as a research tool for understanding cell functions and interactions better. Tissue culture also has many practical uses, such as in vaccine preparation, production of drugs and in the production of MONOCLONAL ANTIBODIES.

Plant cultures can be used to manufacture useful products such as codeine from poppies for pain relief, but many of these processes are uneconomical. Plant tissue culture is also used to generate plants for agricultural or horticultural use.

See also CONTACT INHIBITION.

tissue fluid, interstitial fluid The fluid that bathes cells at the correct PH and salt concentration. Tissue fluid is derived from blood PLASMA by filtration through the capillaries in tissue; it therefore does not have any cells and has a lower protein level than plasma. It acts as a route for diffusion of substances between cells and blood.

T lymphocyte See T CELL.

tocopherol See VITAMIN E.

tomography The use of X-RAYS to take a photograph through a selected plane of any object. Crystal detectors and amplifiers are used in place of less sensitive X-ray film. Tomography is used in medical imaging, where there are several types, for example the CAT SCAN.

tongue In vertebrates, a muscular organ attached to the floor of the mouth and is the SENSE ORGAN for taste. There is a MUCOUS MEMBRANE covering the tongue that contains nerves and tastebuds. These are chemoreceptors (RECEPTOR cells that detect chemicals) involved in detecting taste. Chemoreceptors that detect sweet chemicals are found at the tip of the tongue, those that detect sour chemicals at the sides, those that detect bitter chemicals at the back, and those that detect salt are found all over the tongue. The tastebuds ensure rejection of unsuitable food.

The tongue is also important in assisting chewing and swallowing of food, by directing food to the teeth and then pushing chewed food to the back of the mouth and into the PHARYNX. In humans, the tongue is important for speech; in other animals, it is important for lapping water and grooming.

tonoplast The membrane surrounding the VACUOLE in plant cells.

tonsils In humans, masses of LYMPHOID TISSUE at the back of the mouth and throat (palatine tonsils) and at the back of the tongue (lingual tonsils). ADENOIDS at the back of the nose are also called pharyngeal tonsils. Tonsils are sites of LYMPHOCYTE production and form part of the body's defence against infection. If they repeatedly become infected, surgical removal may become necessary; this is tonsillectomy. See also LYMPH NODE.

tooth A hard structure in the mouth of vertebrates that is embedded in the bones of the jaws and is used for biting, chewing food and in defence. Dentition is the type and number of teeth in a species.

In humans, there are two sets of teeth. The first milk teeth or deciduous teeth appear from the age of 6 months and consist of 20 teeth. The milk teeth are replaced by the permanent teeth, including a further 12 teeth, from the age of 5 years. There are 32 permanent teeth in total.

The tooth is composed of a crown, the part that is seen, and a root within the jawbone. The outer layer of the crown is called enamel and is extremely hard and covers the dentine, which although softer is still harder than bone. The pulp cavity is a hollow region in the centre of the tooth within which are nerves and blood vessels that provide nutrients and oxygen to the living cells of the dentine and remove waste products. The roots are surrounded by cement to anchor them firmly in their sockets and the gum (soft tissue) that surrounds the base of the tooth.

In some vertebrates, the teeth have a similar shape and differ only in size, but in mammals there are four distinct types of teeth. These are incisors, canines, premolars and molars (see separate entries). The development of these teeth varies between mammals, for example CARNIVORES have well developed canines but in HERBIVORES they may be absent.

Tooth decay is a modern-day problem due mainly to the consumption of sugary foods, which stick to teeth and gums and provide an area for bacteria to thrive.

topsoil The uppermost layer of SOIL. It contains most of the nutrients of the soil but is susceptible to SOIL EROSION.

toxin A poison released by plants, animals and bacteria that can cause a disease. *See also* ANTITOXIN.

trace element An element that is required in small quantities by plants and animals. *See* MICRONUTRIENT. *Compare* MACRONUTRIENT.

trachea, *windpipe* In air-breathing vertebrates, the main airway leading into the lungs, extending from the LARYNX. In humans the trachea is about 10 cm long and is strong but flexible due to its reinforcing rings of CARTILAGE along its length. The trachea branches into two tubes called bronchi (*see* BRONCHUS) that carry the air into each of the lungs. The bronchi and trachea have glands secreting a sticky liquid called MUCUS that collects dust and other unwanted particles which are then pushed out towards the mouth for swallowing, aided by CILIA on the walls of both.

In insects, the equivalent system is the tracheal system that consists of small tubes called tracheae opening to the outside at SPIRACLES.

See also BREATHING, RESPIRATORY SYSTEM.

tracheid A non-living cell type found in the XYLEM of non-flowering VASCULAR PLANTS. Tracheids are spindle-shaped, overlapping cells with heavily lignified walls (*see* LIGNIN) which provide strength and support to the plant. They also conduct water in plants with no conducting vessels, such as ferns, mosses and conifers. In these, water flows from one tracheid to another through unthickened regions in the cell wall called PITS.

tracheole A small branch of the tracheal system in insects (*see* SPIRACLE).

transamination The conversion of one AMINO ACID to another to replace deficient non-essential amino acids. Transamination occurs in the vertebrate liver.

transcription Part of the process of PROTEIN SYNTHESIS in living cells, involving the formation of a strand of MESSENGER RNA (mRNA) from a DNA template. In EUKARYOTIC cells, the mRNA may be modified by GENE SPLICING to produce a functional mRNA. The functional mRNA then carries the information necessary for the actual synthesis of proteins. Transcription involves the action of the enzyme RNA polymerase, which breaks the HYDROGEN BONDS holding the two DNA strands together

so that a portion of the DNA strand unwinds to expose the NUCLEOTIDE bases. Complementary RNA nucleotides are then attracted and form BASE PAIRS with the DNA nucleotides. The mRNA moves from the cell nucleus, where it was made, to the cytoplasm through the nuclear pores, and towards the RIBOSOMES, where it is then translated into proteins (*see* TRANSLATION).

transect A systematic method of sampling the numbers and types of animals and plants in a HABITAT. A line transect consists of a piece of string placed along the ground. Any plant or animal touching or covering the line is recorded. This type of sampling is useful where there is a transition of organisms across an area, for example a sea shore. The height of the line can be varied where this is a major factor in determining the distribution of species. A belt transect is similar to a line transect except that a second line is placed parallel to the first. The species between the lines are systematically recorded a metre at a time, sometimes using a QUADRAT alongside the line transect. *See also* MARK, RELEASE, RECAPTURE.

transfection 1. In PROKARYOTES, the uptakes of BACTERIOPHAGE DNA by TRANSFORMATION to produce a bacteriophage infection.

2. In EUKARYOTES, a term often used to refer to the uptake of foreign DNA by cultured mammalian or other animal cells. *See also* TRANSFORMATION.

transferase Any one of a group of enzymes that transfer a chemical group from one substance to another, for example phosphorylases transfer PHOSPHATE groups.

transfer RNA (tRNA) A type of RNA, making up 10–15 per cent of the total RNA of a cell, concerned with PROTEIN SYNTHESIS. Transfer RNA is a small single-stranded molecule of about 80 NUCLEOTIDES that forms a clover leaf shape, with one end being the point where a specific amino acid attaches and the other end carrying a specific three base ANTICODON sequence. There are about 20 different types of tRNA, each one combining with a different amino acid and carrying it to MESSENGER RNA (mRNA) where the anticodon sequence matches a three-base codon sequence carried on mRNA. *See also* TRANSLATION.

transformation 1. A change in certain cells (often bacteria) due to their uptake of foreign DNA. This can be as a result of culturing the cells with other killed cells or culture filtrates. The cells acquire characteristics encoded by the foreign DNA and are able to pass these onto their offspring. *See also* TRANSFECTION.

2. An event resulting in the production of a CANCER cell from a normal animal cell. This may be due to infection of the cell or to other factors, such as a chemical stimulus. *See* CARCINOGEN.

transgenic (*adj.*) Describing an organism whose GENOME has been altered experimentally (*see* GENETIC ENGINEERING) to incorporate and express GENES from another species. Using a suitable VECTOR, the foreign DNA is inserted into a fertilized egg or early embryo of the host and integrates with its genome. This technique

has great commercial potential, for example farm animals can be engineered to produce more growth hormone to improve meat production and plants can be altered to be disease resistant. The technique is also valuable in genetic studies.

translation The stage of PROTEIN SYNTHESIS in which the information carried in the MESSENGER RNA (mRNA) strand formed during TRANSCRIPTION is used to form chains of AMINO ACIDS that are ultimately assembled into PROTEINS.

The order in which amino acids are linked to form POLYPEPTIDES is dictated by the CODONS (a triplet of nucleotide bases) on the mRNA. There is always a START CODON, AUG, at one

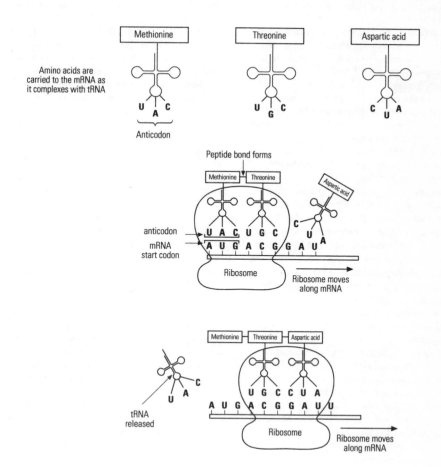

Protein synthesis – translation.

end of the mRNA, which codes for the amino acid methionine. RIBOSOMES attach to the mRNA at this start codon and attract the appropriate complementary anticodon on the TRANSFER RNA (tRNA), which is carrying the amino acid specific to that codon. This continues along the mRNA, with the ribosome moving along holding the mRNA/tRNA complex together until the amino acids have been released and PEPTIDE bonds formed between them. When a STOP CODON is reached on the mRNA, the polypeptide chain is released from the mRNA. The tRNA is free to go and get another amino acid. Many polypeptide chains maybe synthesized at the same time by several ribosomes moving along the mRNA one after another. The polypeptides are then assembled into proteins.

translocation 1. In botany, the long-distance transport of water and minerals in a plant. There are three pathways responsible for this movement throughout the plant. (i) The apoplast pathway is the movement of substances in the cell walls of plants. Water enters air spaces between the CELLULOSE fibres of the cell wall of one cell, and as it evaporates through the stomata (*see* STOMA) a tension is created that pulls the water to the next cell wall. This pathway carries the most water. (ii) The symplast pathway is the movement of substances through the cytoplasm of cells, which are connected by tiny strands of cytoplasm called PLASMODESMATA. Water passes from cell to cell along a water potential gradient that results from loss of water by TRANSPIRATION. It is of less importance than the apoplast pathway. (iii) The vacuolar pathway is the movement of water by OSMOSIS along a water potential gradient through the VACUOLE of adjacent cells and through the cell walls and cytoplasm. This contributes least to water movement.

Water and minerals move through the roots and leaves of a plant by these pathways. Water is drawn from the ROOT HAIRS to the XYLEM and upwards to the leaves in the TRANSPIRATION STREAM. Minerals move through the plant dissolved in water; they can be absorbed by root hairs either passively (DIFFUSION) or by ACTIVE TRANSPORT, and can travel through the xylem to the cells where they are needed. The organic products of photosynthesis are also translocated throughout the plant from the leaves where they are made, through the PHLOEM, to where they are needed for growth or storage. The mechanism by which materials are translocated in the phloem is not fully understood. The rate of flow is too fast to be explained by diffusion. The widely accepted theory is the MASS FLOW hypothesis, which suggests that the soluble organic products move from their source (the leaves) to a sink (other tissues where they are needed) along pressure potential gradient.

2. In genetics, a type of CHROMOSOME MUTATION in which a section of one chromosome is broken off and becomes attached to another chromosome. Thus some genetic information is lost from the original chromosome. This differs from CROSSING-OVER since the chromosomes are non-homologous.

transmission electron microscope A type of ELECTRON MICROSCOPE where the material to be examined is preserved in a suitable fixative and then embedded in a plastic resin, such as Araldite, so that ultrathin sections can be cut, using an ULTRAMICROTOME. This is necessary because electrons cannot penetrate materials very well. Sections are then stained by various methods to improve their electron scattering ability and supported on a metal grid that allows electrons to penetrate the section. Electrons are absorbed by some regions of the material (electron dense regions) but penetrate other electron transparent regions to hit the viewing screen and fluoresce. *See also* SCANNING ELECTRON MICROSCOPE, SCANNING TRANSMISSION ELECTRON MICROSCOPE.

transpiration The loss of water from a plant by evaporation. Most water (up to 90 per cent) is lost through pores in the leaves or some stems called STOMATA and the rest through the CUTICLE on the surface of leaves. In woody stems a very small amount of water is lost through LENTICELS. Loss of water from the leaves causes water to be drawn upwards from the roots in a continuous TRANSPIRATION STREAM, although this is not an essential means of obtaining water.

Transpiration seems to be a side-effect of the need to have holes in the leaves for gaseous exchange. The opening and closing of the stomata (and thus the control of transpiration) is determined by the GUARD CELLS that surround them. Internal differences (or adaptations) between plants can affect the rate of transpiration, for example, leaf area, thickness of cuticle and density of stomata (*see* XERO-PHYTE, HYDROPHYTE).

See also TRANSLOCATION.

transpiration stream The movement of water in plants by TRANSLOCATION from the ROOTS upwards to the leaves via the XYLEM. Movement is due to the cohesive forces between water molecules (causing them to stick together) and the adhesive forces between the molecules and the wall of the xylem. This creates a tension in the xylem, pulling the water up as some is drawn out of the xylem and across the leaves by TRANSPIRATION. This is called the 'cohesion tension theory'. The movement of water by the transpiration stream is not an essential means of obtaining water, as OSMOSIS would serve this purpose.

transplantation The artificial transfer of a tissue or organ from one animal to another or to a different part of the same animal. A transplantation is usually life-saving.

Most body cells have ANTIGENS on their surface that are unique to the individual (self-antigens). These are coded for by a gene complex called the MAJOR HISTOCOMPATIBILITY COMPLEX (MHC). If skin, for example, is transplanted from one part of the body to another, or between identical twins, these antigens are the same and therefore the transplanted tissue can establish and grow. However, if an organism or tissue is transplanted to another individual, the antigens would be recognized as foreign by the recipient's IMMUNE SYSTEM and rejection of the transplant would occur and it would die. To avoid this, the MHC antigen on the donor's and recipient's cells are matched as much as possible and immunosuppressive drugs are given to reduce the IMMUNE RESPONSE mounted. Despite these precautions, transplants are not always successful.

Kidneys are the organs most successfully transplanted (since 1950), but hearts, lungs, livers, pancreas, bone and bone marrow are also transplanted. The term 'grafting' is used when a small part of an organism is transplanted to another, and usually involves a closer union of tissue than with organ transplants, for example corneal grafting to restore sight to a damaged or diseased eye, and skin grafting. Most transplant material is taken from cadavers (dead bodies).

transverse tubules, *T tubules* Infoldings of the outer membrane (SARCOLEMMA) of a muscle fibre. T tubules make contact with the SARCOPLASMIC RETICULUM, causing release of calcium ions, which allows muscle contraction. *See also* MUSCULAR CONTRACTION.

Trematoda A class of the phylum PLATYHELMINTHES (flatworms). The Trematoda are all PARASITES and are of great economic importance. Examples are the liverfluke (*Fasciola hepatica*) of sheep and cattle, with the snail as an intermediate host, and the blood-fluke (*Schistosoma*) of humans, which causes bilharzia or schistosomiasis.

tricarboxylic acid cycle (TCA cycle) *See* KREBS CYCLE.

tricuspid valve Three membranous flaps between the right ATRIUM and VENTRICLE in the mammalian heart. The tricuspid valve closes when the ventricle contracts (*see* HEART), forcing blood into the pulmonary artery and preventing the backflow of blood into the atrium. *See also* BICUSPID VALVE.

triglyceride A GLYCERIDE in which all three of the hydroxyl groups on GLYCEROL have combined with a FATTY ACID. The fatty acids can be the same or mixed, and the nature of the triglyceride depends on the constituent fatty acids. Triglycerides occur naturally as the main constituents of fats and oils, providing an energy store in living animals. They also provide cooking oils, fats and margarine. The term triglyceride is often used synonymously with FAT. *See also* LIPID, ESTER.

triose A MONOSACCHARIDE containing three carbon atoms in the molecule.

triploblastic (*adj.*) Of an animal, having a body that develops from three GERM LAYERS – the ECTODERM, ENDODERM and MESODERM.

triploid A nucleus, cell or organism that has three sets of CHROMOSOMES.

tRNA *See* TRANSFER RNA.

trophic (*adj.*) Relating to NUTRITION.

trophic level The position occupied by a species or group of species in a FOOD CHAIN, for example producers and consumers. *See also* FOOD WEB.

trophoblast In mammals, the outer layer of cells of the BLASTOCYST that develops into the embryonic membrane (CHORION), the VILLI of which invade the wall of the UTERUS and develop into the PLACENTA. In humans, the trophoblast secretes HUMAN CHORIONIC

GONADOTROPHIN, which may be detected in the urine of pregnant women and forms the basis of pregnancy tests.

tropism The directional growth of a plant (or part of it) in response to an external stimulus. The tropism can be positive (growing towards the stimulant) or negative (growing away) and is caused by greater growth on one side of the plant than the other. Light is a major stimulus, the response being PHOTOTROPISM. GEOTROPISM is the response of a plant to gravity; HYDRO-TROPISM is the response to water; CHEMO-TROPISM is the response to a chemical stimulus; and THIGMOTROPISM (haptotropism) is the response to physical contact. *Compare* NASTIC MOVEMENT.

tropomyosin A protein attached to ACTIN filaments in muscles that has a role in MUSCULAR CONTRACTION. Tropomyosin consists of two strands that run along the length of the actin filament and covers the binding sites for MYOSIN. During muscular contraction, tropomyosin is displaced by another protein called TROPONIN. This reveals the myosin binding sites and enables actin and myosin to interact.

troponin A complex protein attached to ACTIN filaments in muscles that has a role in MUSCULAR CONTRACTION. During muscular contraction, troponin binds to calcium ions, displacing another protein called TROPOMYOSIN and revealing binding sites for MYOSIN. This enables actin and myosin to interact.

trypsin A PROTEASE enzyme in the vertebrate digestive system that breaks down proteins during digestion. Trypsin is secreted by the PANCREAS in its inactive form as trypsinogen. This is activated in the SMALL INTESTINE by enterokinase, an enzyme secreted by the DUO-DENUM. Trypsin does not need an acid environment to function.

trypsinogen The inactive precursor of TRYPSIN.

tryptophan An ESSENTIAL AMINO ACID found in certain pulses.

TSH *See* THYROID-STIMULATING HORMONE.

tuber A swollen region of an underground stem (a stem tuber, e.g. the potato) or root (a root tuber, e.g. the dahlia) that is modified for storing food and gives rise to new plants, and so is a structure for VEGETATIVE REPRODUCTION. A tuber lasts for one season only, with new tubers forming the following year in a different place.

tumour, *neoplasm* A swelling or lump caused by an overgrowth of cells in a specific area of the body. Tumours can be malignant or benign. Malignant tumours show unlimited, rapid growth and invade surrounding tissues. They shed cells that can be transported through the blood or LYMPHATIC SYSTEMS to form a secondary tumour (or metastasis) in another part of the body. Benign tumours are not cancerous, are slower growing and non-evasive. They can therefore be surgically removed more easily and do not usually re-occur. Any tumour, even benign, in a difficult site (e.g. the brain) or causing a physical blockage, can be life-threatening. *See also* CARCI-NOMA, SARCOMA.

tundra A terrestrial BIOME of high latitudes, characterized by treeless expanses and a permanently frozen subsoil.

Turbellaria A class of the phylum PLATY-HELMINTHES (flatworms). Unlike the other members of this phylum, which are PARASITES, the Turbellaria are free-living, for example *Planaria*.

turgor The rigid condition of a plant cell when it is full of water. Water enters by OSMOSIS, causing the CYTOPLASM and CELL WALL to be pressed together, and the cell is said to be turgid. The pressure exerted by the fluid against the cell wall is called the turgor pressure. Turgor is important in providing support for some plants. *See also* PLASMOLYSIS.

Turner's syndrome A genetic disorder affecting females in which an X-chromosome is missing and individuals are genetically XO instead of the normal XX. They are phenotypically small, sexually immature females, lacking ovaries and menstrual cycle and are infertile.

tympanic membrane *See* EARDRUM.

tyrosine An AMINO ACID found in many proteins. It is precursor of ADRENALINE, THYROXINE and MELANIN.

U

ultracentrifuge A CENTRIFUGE that operates at very high speeds. It can be used in the laboratory to separate COLLOIDS, submicroscopic particles or particles as small as a NUCLEIC ACID, or protein.

ultramicrotome A machine for cutting very thin sections of tissue (embedded in a resin such as Araldite) for use with an ELECTRON MICROSCOPE. The sections are 20–100 nm thick. A glass or diamond knife is used.

ultrasound High frequency pressure waves used in medicine in ULTRASOUND IMAGING. Other uses include detection of flaws in metals or to measure the depth of the sea.

ultrasound imaging A technique for studying the interior of opaque structures, widely used in medicine. ULTRASOUND is used to investigate various body organs and is especially useful in the routine examination of human foetuses. The technique relies on the reflection of ultrasound from boundaries between materials of differing densities. The DOPPLER EFFECT also enables motion, particularly blood flow rates, to be measured by ultrasonic techniques – the shift in frequency of the reflected wave gives a measure of the motion of the reflecting particles. Ultrasound is non-ionizing, unlike X-RAYS and CAT SCANS, so cannot produce mutations. However, high levels can produce tissue changes, so intense levels are used in physiotherapy and to break up kidney stones.

ultraviolet radiation ELECTROMAGNETIC RADIATION with wavelengths less than those of visible light and greater than those of X-RAYS, that is between 4×10^{-7} m and 10^{-9} m. They are produced by the more energetic changes in energy in atomic electrons. Ultraviolet radiation from the Sun is mostly absorbed in the upper layers of the atmosphere (the OZONE LAYER), so relatively little reaches the Earth. That which does reach ground level is responsible for the tanning and burning effect of exposure to sunlight and, with prolonged exposure, is believed to be responsible for skin cancer. In human skin it causes the formation of VITAMIN D. Ultraviolet radiation can be detected by photographic film and can be made visible by FLUORESCENCE. Some fish and birds can detect ultraviolet radiation and use it to locate food or territories. Ultraviolet radiation is used in the laboratory to sterilize equipment as it is strongly germicidal. Ultraviolet light of wavelength 260 nm is absorbed by DNA and alters the structure of the PYRIMIDINE BASES, causing MUTATIONS.

umbilical cord The connection between the FOETUS and PLACENTA of placental mammals throughout pregnancy. The umbilical cord has one vein and two arteries, which transport oxygen and nutrients to the young and transport waste products away. After birth, the cord falls off, leaving a mark called the naval.

unicellular (*adj.*) Describing organisms or their parts that consist of only one cell. *Compare* ACELLULAR, MULTICELLULAR.

unstriated muscle *See* INVOLUNTARY MUSCLE.

uracil An organic base called a PYRIMIDINE that occurs in RNA but not in DNA.

urea ($CO(NH_2)_2$) A waste product formed from the break down of ammonia, NH_3, in the mammalian LIVER, which is then excreted in the URINE. Ammonia is itself a waste product derived from the breakdown of proteins and NUCLEIC ACID, but is very toxic and therefore converted in many vertebrates to urea, which is harmless. Many aquatic animals excrete ammonia as their main nitrogenous waste product and are said to be ammoniotelic, in contrast to ureotelic animals, which excrete urea.

Urea is made by liver cells in a cyclic process called the urea or ORNITHINE CYCLE, which is closely linked to the KREBS CYCLE. Urea is a white solid when purified and has some industrial uses, for example in fertilizers and pharmaceuticals.

See also URINARY SYSTEM.

urea cycle, *ornithine cycle* The series of biochemical reactions that convert ammonia to UREA as part of the excretion of metabolic waste products in UREOTELIC animals. Ammonia is a waste product from the breakdown of proteins and NUCLEIC ACID but is very toxic and therefore needs to be converted to the less toxic urea, which can then be excreted in solution as URINE. The urea cycle occurs in the mammalian LIVER and is closely linked to the KREBS CYCLE. Ammonia enters the urea cycle in two places, firstly in combination with carbon dioxide as carbamyl phosphate, which combines with ornithine to form citrulline. The second molecule of ammonia enters as aspartic acid, which combines with the citrulline to form arginosuccinate. Fumaric acid is removed from the arginosuccinate and can enter the Krebs cycle or be used to regenerate aspartic acid. The removal of fumaric acid produces the amino acid arginine, which is split into urea and ornithine by the enzyme arginase.

ureotelic (*adj.*) Describing an animal that excretes UREA as its main nitrogenous waste product.

ureter A fine tube through which URINE passes from the KIDNEY to the BLADDER. The ureter contains smooth muscle fibres in its wall that contract to assist the movement of urine to the bladder.

urethra A tube in mammals (males and females) that carries URINE from the BLADDER to the outside, the opening of which is a SPHINCTER (constricting muscle) under voluntary control. The urethra in males joins the VAS DEFERENS and also carries SEMEN. *See also* URINARY SYSTEM.

uric acid ($C_5H_4N_4O_3$) A semi-solid nitrogenous waste produced by most land animals that develop in a shell, including reptiles, insects and birds. Uric acid is produced instead of URINE where water is scarce. Humans also produce some uric acid, which if in excess can build up as crystals in joints and tissues, causing gout, or it can form kidney or bladder stones.

uricotelic (*adj.*) Describing an animal that excretes URIC ACID as its main nitrogenous waste product.

uridine A PYRIMIDINE NUCLEOSIDE, consisting of the organic base URACIL and the sugar RIBOSE.

urinary system The system of organs and tubes that removes nitrogenous waste and excess water from the bodies of animals. In vertebrates, the urinary system consists of a pair of KIDNEYS, two URETERS, a BLADDER and the URETHRA. The kidneys produce URINE, which passes through the ureter, a fine tube extending from each kidney, to the bladder where it is stored (up to 0.7 litres in humans) before it is discharged. In mammals, urine is then carried

to the outside through the urethra, a tube with a SPHINCTER (constricting muscle) at its opening which is under voluntary control.

In other vertebrates (most reptiles, birds, amphibians and many fish), urine drains from the bladder into the CLOACA, a chamber containing all excretory products (digestive and urinary) and into which the reproductive tracts also enter.

See also BOWMAN'S CAPSULE.

urine A fluid made by the KIDNEYS that contains excess water, salt, proteins, some acid and UREA. Reptiles, insects and birds and most land mammals developing in a shell excrete nitrogenous waste as the semi-solid URIC ACID.

urino-genital system The URINARY SYSTEM and REPRODUCTIVE SYSTEM considered together.

urticaria An ALLERGIC REACTION of the skin characterized by a red rash which itches intensely. It is caused by the release of HISTAMINE into the skin as a result of the allergic reaction. It is also called 'hives'.

uterus, *womb* In female mammals, a muscular organ within which the EMBRYO implants and develops during PREGNANCY. It is located between the BLADDER and RECTUM. The uterus is held in place by ligaments joined to the PELVIS. At the base of the uterus is a ring of muscle, called the CERVIX, that opens into the VAGINA, which connects the uterus to the outside. The uterus is connected above to the FALLOPIAN TUBES. In humans there is a single uterus of about 5×8 cm, but in many mammals there are two uteri joined at the cervix.

The outer wall of the uterus consists of smooth muscle that enables it to expand, to accommodate a growing embryo and contract under hormone stimulation during childbirth. The inner lining of the uterus is called the ENDOMETRIUM, a glandular tissue the structure of which changes in response to hormone stimulation. It is this that is shed at menstruation or thickens during pregnancy to eventually form part of the PLACENTA.

See also MENSTRUAL CYCLE.

utriculus A sac-like structure within the inner EAR concerned with balance and from which the SEMI-CIRCULAR CANALS arise. The utriculus lies above the SACCULUS, together linking the COCHLEA and semi-circular canals. The utriculus contains a patch of sensory hair cells, which detect changes in the direction and speed of movement. *See also* MACULA.

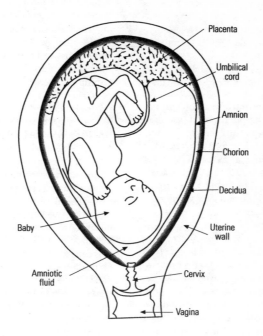

Placenta

Umbilical cord

Amnion

Chorion

Decidua

Uterine wall

Baby

Amniotic fluid

Cervix

Vagina

The pregnant human uterus.

V

vaccination, *immunization* The administration of ANTIBODY or a preparation of modified ANTIGEN (the vaccine) to induce specific antibody production and therefore provide artificial IMMUNITY against a particular disease. Edward Jenner (1749–1823) in 1796 was the first English physician to introduce a vaccine, against smallpox. As a result of a world-wide programme of vaccination, smallpox is now eradicated.

A vaccine is usually given orally or by injection and provides immunity lasting from 6 months to 6 years, depending on the disease. Booster injections are often given to maintain the level of protection. Antibodies are given by injection (passive immunity) to fight an established disease or to provide immediate protection to a person at high risk from a particular disease. If antigen is given to induce antibody production (active immunity) then the antigen is modified in one of several ways so that antibodies are made without onset of the disease.

Living micro-organisms can be administered in a vaccination. These have been treated, for example by heating, so that they multiply without causing the disease symptoms. Examples of this include vaccination against poliomyelitis, rubella (German measles), tuberculosis and measles. Dead micro-organisms can also be used, for example for typhoid, cholera, influenza and whooping cough. Protection against some diseases, for example diphtheria and tetanus, is by injection with a detoxified toxin produced by the PATHOGEN. More recently, artificial antigens have been made by GENETIC ENGINEERING so that they are harmless but immunogenic.

Protection from vaccination is limited when the micro-organism (usually a virus) shows ANTIGENIC VARIATION, for example influenza.

vaccine *See* VACCINATION.

vacuolar pathway One of three pathways by which water and minerals move upwards through a plant. It is the least important of the three pathways (*see also* APOPLAST PATHWAY and SYMPLAST PATHWAY). In the vacuolar pathway, water passes by OSMOSIS through the vacuole of one cell to the vacuole of the adjacent cell via the cell wall, membrane and cytoplasm. A water potential gradient is established (as in the symplast pathway) along which water moves up through the plant. *See also* TRANSLOCATION.

vacuole A fluid-filled cavity within the cell CYTOPLASM that is bounded by a membrane. In many plant cells, a single vacuole takes up most of the volume of the cell and contains cell sap, which can be water and other components, such as amino acids and sugars as food stores, organic waste materials for later release to the outside, or colour pigments. In animal cells, the vacuoles are smaller but there may be more than one, and they can store food or be PHAGOCYTES.

Contractile vacuoles of single-cell freshwater organisms (e.g. *Amoebae*) are important for OSMOREGULATION because they are able to slowly fill with water and suddenly contract to expel their contents from the cell, so preventing excess water building up following its absorption into the cell by OSMOSIS.

vagina The front passage of female mammals that connects the UTERUS to the outside, into which the PENIS releases its sperm during intercourse and out of which a FOETUS is born. The wall of the vagina contains muscle (and can therefore expand) and secretes a lubricating fluid to neutralize the acidity of the vagina (which would kill the sperm) and allow easy penetration of the erect penis. The vagina opens to the outside through the VULVA.

valine An ESSENTIAL AMINO ACID that must be provided in the diet for normal growth and health.

variation The differences between individuals of the same species, arising as a result of SEXUAL REPRODUCTION. GENOTYPES and PHENOTYPES can show variation, and the variation may be minor or more noticeable. Variations can exist, for example, in colouring, size and behaviour.

There are two types of variation: continuous and discontinuous. Continuous variation is where individuals show a gradation from one extreme to another, such as height, and these characteristics are often controlled by POLYGENES. Discontinuous variation is where there is a number of distinct forms within a population such as blood groups. These characteristics are usually controlled by only one gene and there are no intermediate forms.

The most important cause of variation is genetic as a result of the mixing of two parental genotypes during sexual reproduction and the random distribution of chromosomes during MEIOSIS. Further variation comes from CROSSING-OVER and RECOMBINATION of chromosomes during meiosis, and some variation occurs as a result of MUTATION. Environmental factors can cause variation but these are not passed on. Genetic variation is the basis upon which NATURAL SELECTION can work, and provides species with an opportunity to adapt. It is therefore an evolutionary advantage.

vasa recta The blood capillaries branching from the arterioles leaving the GLOMERULUS of the KIDNEY and which run parallel with the LOOP OF HENLE, eventually draining into the renal vein. The vasa recta contribute to the COUNTERCURRENT SYSTEM that operates in the loop of Henle.

vascular bundle The main conducting tissue of VASCULAR PLANTS (for example flowering plants, ferns, mosses and conifers). The vascular bundle extends from the roots of the plant to the leaves and stems. The XYLEM is towards the centre of the bundle and PHLOEM is nearest the epidermis. *See also* CAMBIUM, PROCAMBIUM, TRANSLOCATION.

vascular cambium A type of CAMBIUM (a lateral MERISTEM) in VASCULAR PLANTS that results in the growth of secondary XYLEM and PHLOEM, new conducting vessels for water and food. *See also* CORK CAMBIUM.

vascular plant A plant with a VASCULAR BUNDLE for conducting water and sugars. Vascular plants include CONIFERS and ANGIOSPERMS.

vas deferens, *sperm duct* Either one of a pair of muscular tubes in male vertebrates that carry SPERM (during sexual intercourse) from each testis to the URETHRA and then to the outside. The sperm are carried in a fluid called SEMEN.

See also EPIDIDYMIS, SEMINAL VESICLE, SEXUAL REPRODUCTION.

vasectomy *See* STERILIZATION.

vas efferens (*pl. vasa efferentia*) One of many small tubes in male vertebrates which carry sperm from the seminiferous tubules of the TESTIS to the EPIDIDYMIS.

vasoconstriction A narrowing of the diameter of a BLOOD VESSEL, usually of the ARTERIOLES. Vasoconstriction is under the control of the VASOMOTOR CENTRE.

vasodilation An increase in the diameter of a BLOOD VESSEL, usually of the ARTERIOLES. Vasodilation is under the control of the VASOMOTOR CENTRE.

vasomotor centre A group of NEURONES in the vertebrate brain that maintain a constant blood pressure by controlling VASOCONSTRICTION (narrowing of blood vessels) and VASODILATION (widening of blood vessels).

vector 1. An organism, for example an insect, that acts as a host to a PARASITE and transmits it to another host. *See also* MALARIA.

2. A PLASMID or BACTERIOPHAGE that can carry foreign DNA inserted into its own genetic material and which therefore replicates with it. Plasmid vectors are valuable tools in GENETIC ENGINEERING.

vegetative reproduction A form of ASEXUAL REPRODUCTION in plants in which a new organism develops from structures formed by the parent plant, without the production of spores. Vegetative reproduction is important in the propagation of plants (vegetative propagation). The structures formed are numerous and varied and include BULBS, CORMS, RHIZOMES, TUBERS, STOLONS and RUNNERS. Structures that act as food stores (not stolons and runners) are called perennating organs and allow the plant to survive through the winter.

Examples of the use of vegetative reproduction by humans are grafting and cuttings. Cuttings are sections of the stems, roots or leaves put into soil that produce roots from a node and develop from this into new plants. Grafting involves placing a shoot or bud of one plant into another; they combine to form a plant with some of the advantages of both plants. Grafting is often used to propagate or modify woody plants, for example roses and fruit trees.

vein A vessel in animals with a CIRCULATORY SYSTEM that carries blood from the rest of the body towards the heart. Veins have thin muscular walls with few elastic fibres (a vein cannot alter its diameter as an ARTERY can) and a large LUMEN. All veins (except the PULMONARY vein leaving the lungs and entering the heart) carry deoxygenated blood from the main organs of the body to the heart. All contain valves to ensure the blood flows in one direction only. The blood in veins is not under high pressure and moves slowly and steadily and not in pulses, in contrast to the blood in arteries. The main vein entering the heart is the VENA CAVA. Further from the heart, small veins are called venules.

The term vein can also refer to other vessels, for example in leaves of plants, that do not contain blood.

vena cava Either of the two main VEINS of vertebrates that carry blood to the heart. The anterior vena cava carries deoxygenated blood from the rest of the body to the right side of the heart. The posterior vena cava carries oxygenated blood from the lungs to the left side of the heart.

venom A poison released by animals under attack.

ventral (*adj.*) 1. Of an animal, relating to the front or ABDOMEN.

2. Of a plant, relating to the ANTERIOR or lower surface. *Compare* DORSAL.

ventral root The lowest pair of spinal nerves along the length of the SPINAL CORD which carry only effector nerves (*see* EFFECTOR SYSTEM). *Compare* DORSAL ROOT.

ventricle 1. Either one of two chambers in the HEART, with thick muscular walls. Ventricles contract to force blood into the ARTERIES. *See also* ATRIUM, BICUSPID VALVE.

2. Any one of the four main cavities of the vertebrate BRAIN, containing CEREBROSPINAL FLUID.

venule A small branch of a VEIN.

vernalization In plants, the stimulation of flowering by exposure to low temperatures. *See* PHOTOPERIODISM.

vertebra (*pl. vertebrae*) Any one of a series of bones forming the VERTEBRAL COLUMN of vertebrates. Vertebrae consist of a large central mass (the centrum) and a hollow (neural canal or arch) through which the SPINAL CORD passes. In humans, there are seven cervical vertebrae (in the neck, with very short ribs), 12 thoracic vertebrae (in the THORAX, bearing the main ribs),

five lumbar vertebrae (in the lower back, with no ribs), the sacrum or sacral vertebrae (five fused bones joined to the PELVIS) and the coccyx or caudal vertebrae (four fused vertebrae forming the tailbone). The vertebrae in humans have different shapes according to their position in the body, but in fish they are more similar along the whole length of the vertebral column.

vertebral column, *spine, backbone* The main support of vertebrates connecting the skull, ribs, back muscles and PELVIS and enclosing and protecting the nerve fibres of the SPINAL CORD. The vertebral column is made up of a series of small bones (26 in most mammals) called vertebrae that are linked by LIGAMENTS and have CARTILAGE (intervertebral discs) inbetween the bones. The spine is curved to accommodate the larger chest and pelvic regions, and muscles attached to finger-like outgrowths on the vertebrae control the limited movement of the spine.

Vertebrata A major subgroup of the phylum CHORDATA. *See* VERTEBRATE.

vertebrate Any animal with a backbone, skull and well-developed brain from the phylum CHORDATA, including mammals, birds, fish, amphibians and reptiles. There are about 41,000 species of vertebrates. Vertebrates form the dominant species of land, sea and air, not in numbers but in ecological importance and BIOMASS (because they include most of the larger animals).

vesicle Any small sac or cavity, especially one filled with fluid, within the CYTOPLASM of a living cell.

vestigial organ An ORGAN whose size and structure has diminished during evolution because of reduced selection pressure (*see* NATURAL SELECTION). An example is the human APPENDIX.

vibrio Any comma-shaped bacterium. *See* BACTERIA.

villi (*sing. villus*) One of the many finger-like projections from the inner lining of the walls of the SMALL INTESTINE that serve to increase the surface area over which absorption of food can occur. Each villus can be up to 1 mm long in humans.

The villi are covered by EPITHELIUM that also has minute projections called MICRO-VILLI, together forming a BRUSH BORDER, which further increases the area for absorption. Each villus has blood vessels, a small vessel called a lacteal and smooth muscle fibres to allow repeated relaxation and contraction to help food and enzyme mixing.

viroid A small VIRUS consisting of a single strand of NUCLEIC ACID and no protein coat. Viroids cause some important plant diseases and some rare diseases of animals.

virology The study of VIRUSES.

virus A small (20–300 nm) infectious particle containing NUCLEIC ACID (DNA or RNA) within a protein shell. Viruses are obligate PARASITES that are unable to multiply except within the living cell of a host. They are smaller than bacteria and cannot be seen through a light microscope.

A mature virus is called a virion and consists of a DNA or RNA core surrounded by a protein coat or capsid, which is sometimes surrounded by host CELL MEMBRANES gained (and maybe modified) on exit from the host cell. The capsid is made of a variable number of capsomere subunits, the symmetry of which is also variable, for example some are spherical (e.g. poliomyelitis), and others are rod-shaped (e.g. tobacco mosaic virus).

Viruses can be classified according to their nucleic acid, which is usually DNA either single-stranded, for example parvovirus, or double-stranded, for example poxvirus, herpesvirus, adenovirus and papovavirus. Some viruses have RNA as their nucleic acid, again either single-stranded, for example the picornavirus (which causes poliomyelitis) and the common cold, or double-stranded, for example reovirus (which causes diarrhoea).

RETROVIRUSES are an important group of RNA viruses that cause AIDS and some human cancers (*see* ONCOGENE). The tobacco mosaic virus (TMV) contains RNA and has been widely studied because of its economic importance. It infects tobacco, potato, tomato, blackcurrant and orchid plants and is highly infectious. Other viruses cause an array of diseases in humans, other animals and plants, for example chicken pox, mumps, measles, rabies, smallpox (although now eradicated), influenza, common cold, Lassa fever, herpes, yellow fever and poliomyelitis. Cancer can also be caused by some DNA viruses (e.g. adenovirus and papovavirus).

VACCINATION has helped to prevent the spread of some viral infections (and has eradicated smallpox), but as viruses mutate continuously it is difficult for the body to develop resistance and vaccination may only be effective against one form. ANTIBIOTICS are ineffective. Viruses can infect different species, causing different symptoms. Some antiviral drugs have been developed but these often affect the host cell too. Interferon is the human body's natural antiviral protein; it has some commercial use but is expensive to produce. *See also* BACTERIOPHAGE, PROVIRUS.

visible light ELECTROMAGNETIC RADIATION that can be detected by the human eye.

vital capacity The total amount of air that can be forcibly exhaled. *See* BREATHING.

vitamin Any one of a group of unrelated organic compounds essential in small amounts for normal body growth and metabolism. Vitamins are classified as water-soluble (B, C, H) or fat-soluble (A, D, E, K). Excess water-soluble vitamins are excreted in the urine; fat soluble vitamins can be stored (in the liver in humans), but can build up to lethal concentrations if taken in excess. A normal balanced diet usually provides the vitamin requirements but if there are inadequate levels then deficiency disease results (*see separate entries*).

vitamin A, *retinol* A VITAMIN that is important in skin structure and to form visual pigments. It is found in dairy foods, liver, fruits and vegetables. Lack of vitamin A causes night-blindness (or more severe blindness, xerophthalmia) and dry skin.

vitamin B A complex of B_1 (thiamine), B_2 (riboflavin), B_3 (niacin), B_5 (pantothenic acid), B_6 (pyridoxine) and B_{12} (cyanocobalamin) and biotin and folic acid. They are mostly COENZYMES in cellular respiration.

Thiamine is found in seeds and grain, and deficiency causes BERIBERI. Riboflavin is a precursor of FAD; niacin forms NAD and NADP; and pantothenic acid ($C_9H_{17}NO_5$) forms part of acetyl coenzyme A (*see* KREBS CYCLE). Lack of niacin causes pellagra, which results in skin lesions, diarrhoea and mental disorders. Pyridoxine ($C_8H_{11}NO_3$) is a coenzyme in amino acid metabolism and lack of it causes nervous disorders. Cyanocobalamin is important in synthesis of RNA and is needed for red blood cell formation. Lack of vitamin B_{12} (usually made by micro-organisms in the digestive system or found in meat and dairy products) causes PERNICIOUS ANAEMIA.

Biotin (also called vitamin H) is found in yeast and liver and is a coenzyme for certain enzymes that incorporate carbon dioxide into various compounds. Lack of it causes dermatitis (inflammation of the skin). Folic acid is found in liver and green leafy vegetables and is made by intestinal bacteria. It is concerned with nucleoprotein synthesis and red blood cell formation, so lack of it causes anaemia. It is often given to pregnant women to prevent anaemia and neural tube defects (spina bifida).

vitamin C, *ascorbic acid* ($C_6H_8O_6$) A VITAMIN found in fresh fruit and vegetables (but is destroyed by soaking or overcooking). It is needed for the synthesis of COLLAGEN. SCURVY results from a deficiency of vitamin C.

vitamin D, *cholecalciserol* A VITAMIN that is found in fatty fish and margarine and made in the skin if exposed to enough sunlight. It is needed for the absorption of calcium and phosphorous and therefore is important for the formation of bones and teeth. Lack of vitamin D causes RICKETS.

vitamin E, *tocopherol* A VITAMIN found in vegetable oil. It has an unclear function in humans, but causes sterility in rats.

vitamin K, *phytomenadione* A VITAMIN found in leafy vegetables and liver and synthesized by intestinal bacteria. It is involved with blood clotting and lack of it causes haemorrhaging. It is often given to newborn babies to prevent brain haemorrhage, although its routine use has been questioned. *See also* BLOOD-CLOTTING CASCADE.

vitreous humour A transparent, jelly-like substance found in the vertebrate EYE in a chamber behind the lens. It helps to maintain the shape of the eyeball.

vivipary Reproduction in animals where the FOETUS develops and is nourished inside the female. Vivipary involves a PLACENTA in mammals such as humans. Other examples include some reptiles and amphibians. *See also* OVIPARY, OVOVIVIPARY.

voicebox *See* LARYNX.

voluntary muscle, *skeletal muscle, striated muscle, striped muscle* Muscle activated by MOTOR NERVES under voluntary control. *See* MUSCLE.

vulva The female external genital organs that comprises pairs of outer and inner folds of skin called the labia majora, labia minora and the clitoris (erectile tissue analogous to the male PENIS).

WXYZ

water cycle The chain of events by which water is re-used in the atmosphere and on the Earth's surface. The condensation of water vapour to produce rain and the evaporation of water from oceans are the key stages in this process. The absorption of water in soil and its transport to the oceans by rivers are also important, as is the role of respiration in plants and animals, which absorb water from their surroundings and return it to the atmosphere by evaporation.

water potential (φ) A measure of the tendency of water in a biological system to go to its surroundings. Water potential can be expressed as the sum of the SOLUTE POTENTIAL (φ_s) and the PRESSURE POTENTIAL (φ_p). The water potential of pure water is zero and solutions of increasing concentration have increasing negative values. Water tends to move from areas of high (less negative) water potential to areas of low water potential. OSMOSIS in plants is described in terms of water potential.

watt The SI UNIT of power. One watt represents the energy converted from one form to another at a rate of one joule per second.

wax A natural solid fatty substance made from ESTERS, FATTY ACIDS or ALCOHOLS. In nature, waxes provide a protective, waterproof covering to many animals and plants. Animal waxes such as beeswax, lanolin and wax from sperm-whale oil are used in cosmetics, ointments and polishes. Mineral waxes are obtained from petroleum.

Western blotting A technique similar to SOUTHERN BLOTTING except that it is used to detect proteins. The protein mixture is separated by POLYACRYLAMIDE GEL ELECTROPHORESIS. After blotting onto a filter, radiolabelled ANTIBODY is added which is specific for the protein of interest. The target protein can then be detected using AUTORADIOGRAPHY.

white blood cell, *leucocyte* A type of blood cell made in the white BONE MARROW, the main function of which is defence. White blood cells are larger than RED BLOOD CELLS, are colourless, contain a nucleus and are capable of independent amoeboid (*see* AMOEBA) movement.

In humans, there are about 11,000 white blood cells per millilitre of blood. Their numbers increase (leucocytosis) in response to

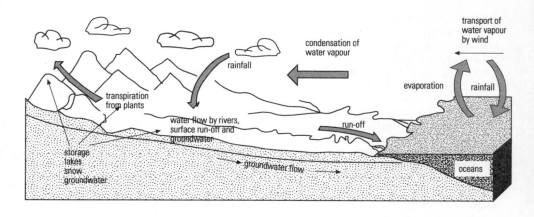

Summary of the water cycle.

blood loss, cancer or most infections, or decrease (leucopenia) during starvation, PERNICIOUS ANAEMIA and certain infections. There are two main types of white blood cell: GRANULOCYTES and AGRANULOCYTES. *See also* LYMPHOCYTE, MACROPHAGE, NATURAL KILLER CELL, PHAGOCYTE.

white matter The tissue of the vertebrate BRAIN and SPINAL CORD that consists of AXONS (protected by a MYELIN SHEATH), GLIAL CELLS and blood vessels. It is lighter in appearance than GREY MATTER, due to the myelin sheath. It usually forms an outer layer around the grey matter, although in the cerebral CORTEX of the brain of higher primates it forms an inner layer (*see* CEREBRUM).

windpipe *See* TRACHEA.

womb *See* UTERUS.

woody perennial *See* PERENNIAL PLANT.

woody plant A general term for those plants possessing secondary XYLEM. Woody plants are trees and shrubs, the distinction between these being mostly size. The secondary xylem forms a hard tissue, wood, under the BARK of these plants, which is strengthened by deposits of LIGNIN. The xylem in some trees, for example conifers, does not contain conducting vessels, and TRACHEIDS are used instead to conduct water. These trees, such as pine, provide what is commercially called softwood, in contrast to the hardwood of trees such as oak. Non-woody plants do not undergo secondary growth. *See also* CAMBIUM.

wool A natural protein (KERATIN) fibre mainly obtained from the fleece of sheep. It consists of coiled protein chains joined by HYDROGEN BONDS, which gives it a springy texture.

worm A general term for a long, limbless invertebrate. Worms can be from several phyla. *See* ANNELID, FLATWORM, NEMATODE.

xanthophyll A yellow pigment in plants, of the CAROTENOID group. It functions like CHLOROPHYLL in PHOTOSYNTHESIS.

Xanthophyta A phylum of the kingdom PROTOCTISTA that consists of the yellow-green ALGAE.

X-chromosome *See* SEX DETERMINATION.

xerophyte A plant that has adapted to living in dry conditions. Xerophytic plants may reduce water loss by TRANSPIRATION by having small leaves (such as in pine trees) or no leaves (for example in many cacti), by the orientation of leaves to avoid direct sunlight, or by having a covering of small hairs on the leaf's surface to trap moist air. The STOMATA of xerophytes may be sunken. Many are also able to store water in their leaves or stems (for example in succulents and cacti) and have shallow but extensive root systems to capture surface water. Some of these xerophytic features may be exhibited by other plants where the water supply is limited, for example due to freezing or to high salt concentrations in salt marshes. *See also* HYDROPHYTE.

xerosere A series of plant SUCCESSIONS growing under dry conditions.

X-ray Electromagnetic radiation with a wavelength in the range of 10^{-11} to 10^{-9} m, which is between GAMMA RAYS and ULTRAVIOLET RADIATION. X-rays are a form of IONIZING RADIATION and the shorter wavelengths (hard X-rays) are highly penetrating. They can be detected by photographic film or with a fluorescent screen.

The penetrating quality of X-rays has led to their use for examining the internal structure of various objects, including the human body. X-rays pass through most body tissues (depending on their nature, thickness and density) but bone, being dense, prevents their passage and shows up white on X-ray photographs. Thus X-rays are an invaluable tool in examination of fractures of bones. X-rays can destroy tissues and can cause cancer.

xylem A compound plant tissue (made of a number of different cell types), the main function of which is to carry water from the roots to other parts of the plant. Xylem consists of PARENCHYMA cells, SCLERENCHYMA fibres, TRACHEIDS and CONDUCTING VESSELS. Tracheids are dead, spindle-shaped, overlapping cells with heavily lignified walls (*see* LIGNIN) and provide strength and support to the plant. They also conduct water in plants with no vessels, such as ferns, mosses and conifers.

The conducting vessels are made up of cells joined end to end whose cross-walls have be broken down to leave long tubes used for carrying water. The vessels are thickened to different extents. Protoxylem has rings or spirals of lignin so can still expand, but metaxylem has more lignin arranged in a reticulate pattern. Protoxylem and metaxylem form the primary xylem found in non-woody plants. In trees and shrubs further growth occurs and the secondary xylem forms, providing extra support (*see* CAMBIUM). Most flowering plants use

vessels for conducting water. *See also* TRANSPI-
RATION, TRANSPIRATION STREAM.

Y-chromosome *See* SEX DETERMINATION.

yeast A group of FUNGI of the phylum ASCOMY-
COTA that is used by humans as a fermenting
agent in baking, brewing and making wines
and spirits. Yeasts reproduce asexually by
BUDDING or sexually by the formation of a
ASCOSPORES developing within a structure
called an ASCUS. When growth is rapid, the new
daughter cells remain attached to the parent
cells forming long chains.

Yeasts produce enzymes that convert
starch or sugars to alcohol and carbon dioxide;
different yeasts act upon different substrates
(for example, cereal grain in the case of beer
and grapes in the case of wine). *Saccharomyces
cerevisiae* (brewer's yeast) is a yeast commonly
used in baking and brewing. Some yeasts are
human PATHOGENS, for example *Candida albi-
cans* causes thrush, and some are useful in
genetic research.

yellow body *See* CORPUS LUTEUM.

yolk A store of food, mostly protein and fat, in
the EGGS of many animals. The yolk provides
nourishment for the developing EMBRYO. *See
also* YOLK SAC.

yolk sac An EXTRAEMBRYONIC MEMBRANE that is
important in birds and reptiles, as it contains
YOLK to nourish the EMBRYO. In humans, the
yolk is less important and combines with the
CHORION.

zidovudine *See* AZT.

zinc (Zn) A blue-white metallic element. It is a
MICRONUTRIENT for living organisms. In animals, it is a PROSTHETIC GROUP for several
enzymes such as CARBONIC ANHYDRASE, which
is important in the transport of carbon dioxide in the blood. It is required in plants for leaf
formation, the synthesis of AUXIN and ANAERO-
BIC RESPIRATION. Deficiency in plants leads to
malformed and mottled leaves.

zoology The branch of biology that deals with
the study of animals. Zoology includes areas

such as anatomy, physiology, behaviour and
evolution.

Zoomastigina A phylum from the kingdom PRO-
TOCTISTA that consists of PROTOZOA. The members (flagellates) move by one or more FLAGELLA
and reproduce by BINARY FISSION. Flagellates can
be free-living or parasitic (*see* PARASITE). An
important example is *Trypanosoma*, the parasite
that causes sleeping sickness.

zooplankton Animal PLANKTON (small life
forms living on the surface of fresh or salt
water) that feed on PHYTOPLANKTON (plant
plankton) and themselves provide food for
larger fishes. Zooplankton are mostly able to
move by FLAGELLA.

zoospore A motile SPORE produced in a SPO-
RANGIUM, possessing one or more FLAGELLA,
that is present in some ALGAE and FUNGI.

zwitterion An ion with both a positive and a
negative charge. For example amino acids in
water form zwitterions by the loss of a proton
from COOH, making it negative, which then
goes to the NH_2 group, making it positive.

zygomorphic (*adj.*) A term describing FLOWERS
that exhibit BILATERAL SYMMETRY. An example is
the white dead nettle, which has petals of different sizes and unequal sepals and thus only one
plane of symmetry. *Compare* ACTINOMORPHIC.

Zygomycete A member of the ZYGOMYCOTA phylum of FUNGI.

Zygomycota A phylum of the kingdom FUNGI
characterized by their absence of septa (parti-
tions) in the HYPHAE and by the production of
ZYGOSPORES during SEXUAL REPRODUCTION.
Zygomycetes produce a large, branched
MYCELIUM. Examples include bread mould
(*Rhizopus*) and pin mould (*Mucor*). ASEXUAL
REPRODUCTION also occurs by CONIDIA or by
SPORES.

zygospore A SPORE produced during SEXUAL
REPRODUCTION by members of the phylum
ZYGOMYCOTA of fungi.

zygote A fertilized OVUM (egg) before it begins
cleavage. *See* EMBRYONIC DEVELOPMENT.

Appendix I

Appendix I: Amino acids

Amino acid	*Abbreviation*	*Formula*

arginine — Arg

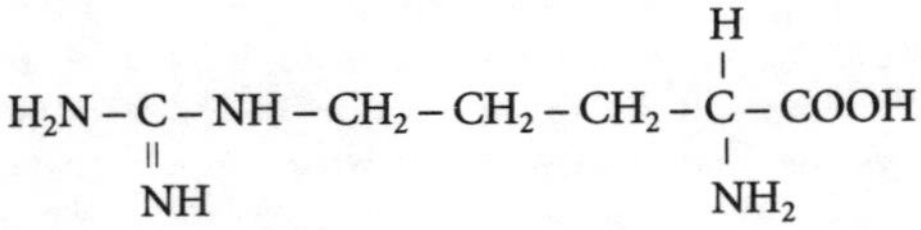

$$H_2N - C - NH - CH_2 - CH_2 - CH_2 - \underset{\underset{NH_2}{|}}{\overset{\overset{H}{|}}{C}} - COOH$$
$$\underset{NH}{\overset{\|}{}}$$

aspartic acid — Asp

$$HOOC - CH_2 - \underset{\underset{NH_2}{|}}{\overset{\overset{H}{|}}{C}} - COOH$$

histidine — His

$$HC = \underset{\underset{N}{|}}{C} - CH_2 - \underset{\underset{NH_2}{|}}{\overset{\overset{H}{|}}{C}} - COOH$$
$$\underset{\overset{N}{\underset{H}{}}}{\overset{\|}{C}} \overset{NH}{/}$$

isoleucine — Ile

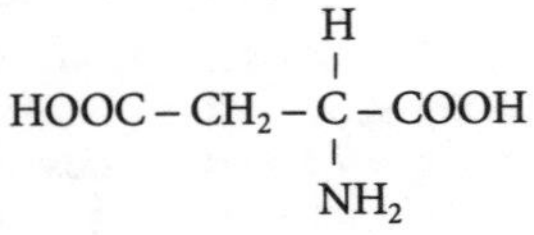

$$CH_3 - CH_2 - \underset{\underset{CH_3}{|}}{CH} - \underset{\underset{NH_2}{|}}{\overset{\overset{H}{|}}{C}} - COOH$$

leucine — Leu

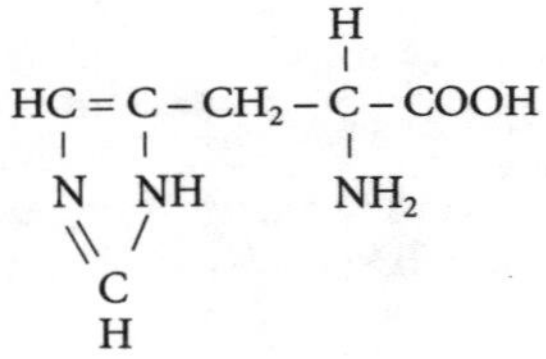

$$\begin{matrix} H_3C \\ \\ H_3C \end{matrix} CH - CH_2 - \underset{\underset{NH_2}{|}}{\overset{\overset{H}{|}}{C}} - COOH$$

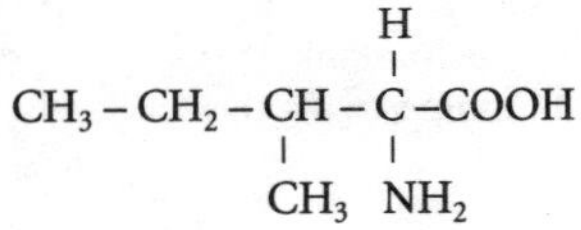

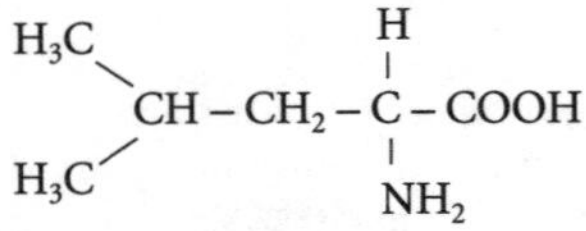

lysine — Lys

$$H_2N - CH_2 - CH_2 - CH_2 - CH_2 - \underset{\underset{NH_2}{|}}{\overset{\overset{H}{|}}{C}} - COOH$$

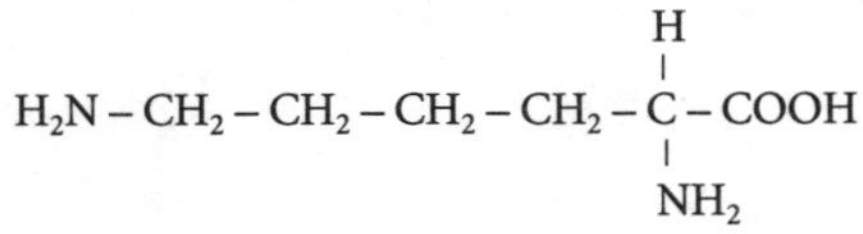

Amino acids (continued)

Amino acid	*Abbreviation*	*Formula*

methionine — Met

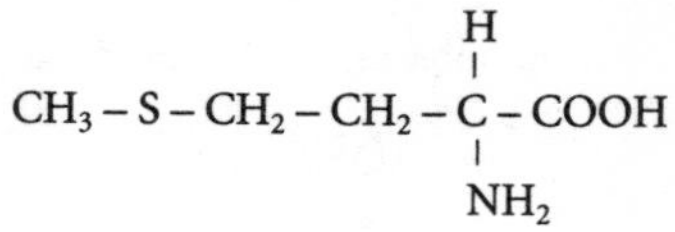

$$CH_3 - S - CH_2 - CH_2 - \underset{\underset{NH_2}{|}}{\overset{\overset{H}{|}}{C}} - COOH$$

phenylalanine — Phe

$$\text{(Phenyl)} - CH_2 - \underset{\underset{NH_2}{|}}{\overset{\overset{H}{|}}{C}} - COOH$$

threonine — Thr

$$CH_3 - \underset{\underset{OH}{|}}{CH} - \underset{\underset{NH_2}{|}}{\overset{\overset{H}{|}}{C}} - COOH$$

tryptophan — Trp

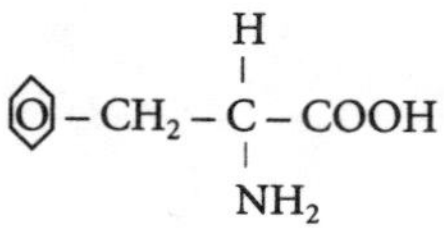

$$\underset{\underset{H}{C}}{\overset{C}{\underset{CH}{|}}} C - CH_2 - \underset{\underset{NH_2}{|}}{\overset{\overset{H}{|}}{C}} - COOH$$

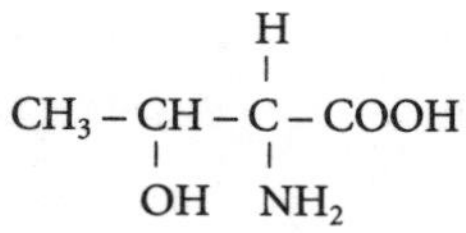

tyrosine — Tyr

$$HO - \text{(Phenyl)} - CH_2 - \underset{\underset{NH_2}{|}}{\overset{\overset{H}{|}}{C}} - COOH$$

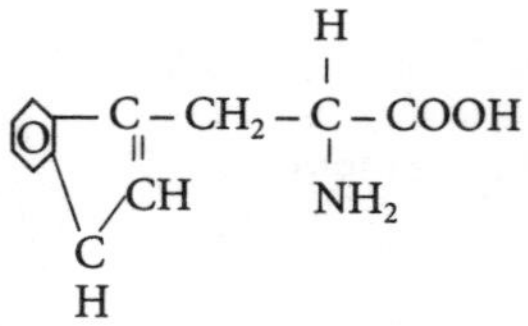

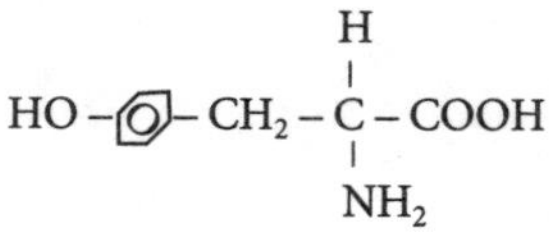

valine — Val

$$\underset{H_3C}{\overset{H_3C}{>}} CH - \underset{\underset{NH_2}{|}}{\overset{\overset{H}{|}}{C}} - COOH$$

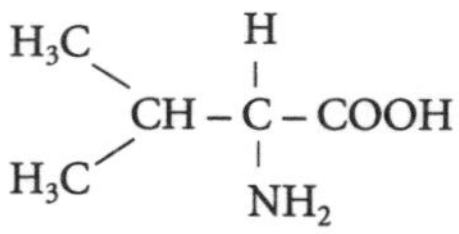

Appendix II: Classification of living organisms

Prokaryotae
(*prokaryotes*)

bacteria
e.g. *Escherichia coli*

cyanobacteria (blue-green bacteria)
e.g. *Nostoc*

Protoctista
(*protoctists*)

Phylum	Common name	Example
*Rhizopoda	rhizopods	*Amoeba*
*Zoomastigina	flagellates	*Trypanosoma*
*Apicomplexa	sporazoans	*Plasmodium*
*Ciliophora	ciliates	*Paramecium*
Euglenophyta	euglenoid flagellates	*Euglena*
Oomycota	oomycetes	*Phytophthora*
Chlorophyta	green algae	*Chlamydomonas*
Rhodophyta	red algae	*Chondrus*
Phaeophyta	brown algae	*Fucus*

* These 4 phyla together constitute the protozoans

Fungi
(*fungi*)

Phylum	Common name	Example
Zygomycota	zygomycetes	*Mucor*
Ascomycota	ascomycetes	*Neurospora*
Basidiomycota	basidiomycetes	*Agaricus* (mushroom)

Classification (continued)

Plantae
(*plants*)

Phylum	*Common name*	*Class*	*Common name*	*Example*
Bryophyta	bryophytes	Hepaticae	liverworts	*Pellia*
		Musci	mosses	*Bryum*
Lycopodophyta	club mosses			*Selaginella*
Sphenophyta	horsetails			*Equisetum*
Filicinophyta	ferns			*Pteridium* (bracken)
Coniferophyta	conifers			*Pinus* (Scots pine)
Angiospermophyta	angiosperms (flowering plants)	Monocotyledoneae	monocotyledons	*Triticum* (wheat)
		Dictotyledoneae	dicotyledons	*Ranunculus* (buttercup)

Animalia
(*animals*)

Phylum	*Common name*	*Class*	*Common name*	*Example*
Cnidaria	cnidarians			*Aurelia* (jellyfish)
Platyhelminthes	flatworms	Tubellaria	turbellarians	*Polycelis*
		Trematoda	trematodes or flukes	*Fasciola* (liverfluke)
		Cestoda	cestodes or tapeworms	*Taenia* (tapeworm)
Nematoda	nematodes or roundworms			
Annelida	annelids	Polychatae	polychaetes or marine worms or segmented worms	*Nereis* (ragworm)

Classification: Animalia (continued)

Phylum	Common name	Class	Common name	Example
		Oligochaeta	oligochaetes or earthworms	*Lumbricus* (earthworm)
		Hirudinea	leeches	*Hirudu* (medicinal leech)
Mollusca	molluscs	Gastropoda	gastropods	*Helix* (garden snail)
		Pelecypoda	bivalves	*Ostrea* (oyster)
		Cephalopoda	cephalopods	*Sepia* (cuttlefish)
Arthropoda	arthropods	Crustacea (Superclass)	crustaceans	
		Branchiopoda		*Daphnia* (water flea)
		Malacostraca		*Carcinus* (crab)
		Chilopoda	centipedes	*Lithobius* (centipede)
		Diplopoda	millipedes	*Iulus* (millipede)
		Insecta	insects	*Locusta* (locust)
		Arachnida	arachnids	*Scorpio* (scorpion)
Echinodermata	echinoderms	Stelleroidea	star fish and brittle star	*Asterias* (star fish)
		Echinodea	sea urchins	*Echinocardium* (heart urchin)
Chordata	chordates	Chondrichthyes	cartilaginous fish	*Scyliohinus* (dogfish)
		Osteichthyes	bony fish	*Clupea* (herring)
		Amphibia	amphibians	*Rana* (frog)
		Reptilia	reptiles	*Lacerta* (lizard)
		Aves	birds	*Columba* (pigeon)
		Mammalia	mammals	*Homo* (human)